Artificial Intelligence for Sustainable Supply Chain Management

The integration of artificial intelligence (AI) into supply chain management (SCM) is transforming traditional operations into intelligent, resilient, and sustainable ecosystems. This book explores the application of AI-driven models to enhance decision-making, optimize logistics, and minimize environmental impact across various stages of the supply chain. Emphasis is placed on how machine learning, deep learning, and predictive analytics enable proactive demand forecasting, waste reduction, and resource optimization. The research proposes a sustainability-centered framework that combines AI-based optimization with environmental, social, and governance principles to achieve transparency, traceability, and ethical sourcing. The findings demonstrate that AI adoption not only strengthens supply chain agility and resilience but also contributes significantly to achieving Sustainable Development Goals by promoting circular economy practices and minimizing carbon footprints.

Features:

- Includes a step-by-step procedure for developing the Blockchain Internet of Things (IoT) for industries using D-APP and hyperledgers.
- Covers different blockchain and IoT-based sustainable SCM merits, demerits, challenges, and risks.
- Delves into the integration of AI into smart contracts to foster responsible and efficient operations across industries.
- Showcases the use of AI algorithms within smart contract frameworks to optimize decision-making processes and ensure ethical considerations are prioritized in automated contract executions.
- Explores how AI-driven solutions are being tailored to meet the unique demands of different sectors, including manufacturing, healthcare, retail, agriculture, and logistics.

It is primarily written for senior undergraduates, graduate students, and academic researchers in electrical engineering, electronics and communications engineering, computer science and engineering, SCM, manufacturing engineering, and industrial engineering.

Artificial Intelligence for Sustainability

S. Velliangiri and P. Karthikeyan

Artificial Intelligence is used to process and analyze massive amounts of data and is used in a variety of industries, including finance, renewable energy, environmental health, and transportation. Its applications can process high-dimensional data in extensive datasets and analyze them with goals for predictions, recognition, and interaction challenges for more sustainable environmental protection in climate change, energy transformation, and pollution. This series discusses intelligent computing design and implements methodological and algorithmic long-term solutions by addressing sustainability problems. It showcases sustainable technologies including the IoT, cloud computing, blockchain, and machine learning. Aimed at graduate students, academic researchers, and professionals, the proposed series will focus on key topics including drone data analytics in aerial computing, deep learning techniques for Industry 4.0, autonomous robots in UAV, and applications of computer vision and natural language processing.

Applications of Blockchain and Artificial Intelligence in Finance and Governance
A M Viswa Bharathy, Dac-Nhuong Le, and P. Karthikeyan

Sustainable Farming through Machine Learning: Enhancing Productivity and Efficiency
Suneeta Satpathy, Bijay Paikray, Ming Yang, and Arun Balakrishnan

Artificial Intelligence and Sustainable Networks: Innovations Beyond 5G
Edited by R. Krishnamoorthy, Kazuaki Tanaka, and Mohamed S. Soliman

Artificial Intelligence for Sustainable Supply Chain Management
Edited by N. Nasurudeen Ahamed, P. Karthikeyan, and Polinpapilinho F. Katina

Artificial Intelligence for Sustainable Supply Chain Management

Edited by
N. Nasurudeen Ahamed,
P. Karthikeyan, and Polinpapilinho F. Katina

CRC Press
Taylor & Francis Group
Boca Raton London New York

CRC Press is an imprint of the
Taylor & Francis Group, an **informa** business

First edition published 2026
by CRC Press
2385 NW Executive Center Drive, Suite 320, Boca Raton FL 33431

and by CRC Press
4 Park Square, Milton Park, Abingdon, Oxon, OX14 4RN

CRC Press is an imprint of Taylor & Francis Group, LLC

Library of Congress Cataloging-in-Publication Data
Names: Ahamed, Nasurudeen N., 1989- editor | Karthikeyan, P., 1981- editor |
Katina, Polinpapilinho F. editor
Title: Artificial intelligence for sustainable supply chain management /
edited by N. Nasurudeen Ahamed, P. Karthikeyan and Polinpapilinho F. Katina.
Description: First edition. | Boca Raton: CRC Press, 2026. |
Series: Artificial intelligence for sustainability |
Includes bibliographical references and index.
Identifiers: LCCN 2026000193 (print) | LCCN 2026000194 (ebook) |
ISBN 9781041114772 hbk | ISBN 9781041114789 pbk | ISBN 9781003660163 ebk
Subjects: LCSH: Supply chain management—Data processing | Supply chain
management—Technological innovations | Artificial intelligence—Industrial applications |
Blockchains (Databases)—Industrial applications |
Internet of things—Industrial applications | Sustainability
Classification: LCC HD38.6 .A78 2026 (print) | LCC HD38.6 (ebook)
LC record available at https://lccn.loc.gov/2026000193
LC ebook record available at https://lccn.loc.gov/2026000194

ISBN: 9781041114772 (hbk)
ISBN: 9781041114789 (pbk)
ISBN: 9781003660163 (ebk)

DOI: 10.1201/9781003660163

Typeset in Sabon
by codeMantra

Contents

About the Editors

N. Nasurudeen Ahamed holds a Ph.D. in Computer Science and Engineering from Presidency University, Bengaluru, India (2019–2023), where he specialized in blockchain technology. He is currently a Postdoctoral Fellow at the United Arab Emirates University in Abu Dhabi, focusing on Information Systems and Security. With over a decade of academic experience, He has served as an Assistant Professor at various institutions, including Presidency University, where he taught courses on Artificial Intelligence and Blockchain. His research contributions encompass numerous peer-reviewed journal articles and conference presentations, addressing critical issues in blockchain applications within supply chain management and smart city initiatives. Additionally, He is an active reviewer for several prestigious journals and is a member of professional organizations, reinforcing his commitment to advancing technology and education in his field. His work continues to influence both academic and practical aspects of computer science.

P. Karthikeyan obtained his bachelor of engineering in computer science and engineering from Anna University, Chennai, and Tamil Nadu, India, in 2005, and received his master of engineering in computer science and engineering from Anna University Coimbatore, India, in 2009. He completed his Ph.D. degree at Anna University, Chennai, in 2018. He is currently working as an associate professor in SoCSE at RV University. He has worked as a postdoctoral research fellow at the Department of Computer Science and Information Engineering, National Chung Cheng University, Taiwan. He has contributed to national projects on using technology to promote human rights and sustainable development, supported by the National Science and Technology Council, Taiwan. He possesses a high level of proficiency in project development and research, specializing in the domains of cloud computing and the practical application of deep learning. He is well versed in programming languages, including Java, Python, R, and C. His research endeavors have led to the publication of more than 30 papers in esteemed international journals, many of which have garnered commendable impact factors. Furthermore, he has shared his research insights through presentations at over 20 international conferences, solidifying his reputation within the academic community.

Polinpapilinho F. Katina is an associate professor in the Department of Informatics and Engineering Systems at the University of South Carolina Upstate (Spartanburg, South Carolina, USA). He previously served as a postdoctoral researcher for the National Centers for System of Systems Engineering (Norfolk, Virginia, USA) and adjunct professor in the Department of Engineering Management and Systems Engineering at Old Dominion University (Norfolk, Virginia, USA).

He holds a BSc in engineering technology with a minor in engineering management, an Master of Engineering in Systems Engineering, and a PhD in engineering management/systems engineering, all from Old Dominion University (Norfolk, Virginia, USA). He has received additional training from, among others, the Environmental Systems Research Institute (Redlands, California) and Politecnico di Milano (Milan, Italy). His areas of research/teaching interests include, among others, complex system governance, critical infrastructure systems, decision-making and analysis (under uncertainty), emerging technologies, energy systems (smart grids), engineering management, infranomics, manufacturing systems, system of systems, systems engineering, systems pathology, and systems thinking.

Contributors

Afsha Anjum
Department of Master of Computer Applications
RV College of Engineering
Bangalore, India

Anushree
Department of Computer Applications, Nitte Institute of Professional Education
Nitte (Deemed to be University)
Mangaluru, India

Sanjay Mallenahalli Basavaraj
Department of MCA
RV College of Engineering
Bengaluru, Karnataka, India

Chandrani Chakravorty
Department of MCA
RV College of Engineering
Bengaluru, India

Saravanan Chinnappan
Department of MCA
RV College of Engineering
Bengaluru, Karnataka, India

Richa Choudhary
Department of Operations
Amity Global Business School
Bangalore, Karnataka, India

Sukruth Gowda
Department of Artificial Intelligence
Sai Vidya Institute of Technology
Bengaluru, Karnataka, India

Dhanalakshmi Hari
School of Allied and Healthcare Professions
RV University
Bengaluru, Karnataka, India

Rama Moorthy Hejamadi
Department of Computer Applications, Nitte Institute of Professional Education
Nitte (Deemed to be University)
Mangaluru, India

Kavana
Department of Computer Applications, Nitte Institute of Professional Education
Nitte (Deemed to be University)
Mangaluru, India

Rajkumar Kotteswaran
Department of Data Science and Business Systems
SRM Institute of Science and Technology
Chennai, Tamil Nadu, India

Gajalakshmi Kumar
Independent Researcher
Tamil Nadu, India

Divya Tarikere Lokeshwara
Department of MCA
RV College of Engineering
Bengaluru, India

Yuvaraj Mohanselvam
Department of Operations
Amity Global Business School
Bangalore, Karnataka, India

Umaselvi Muthusamy
Department of Computer Science and Engineering
P. A. College of Engineering and Technology
Pollachi, Coimbatore, India

Manju Bala Perumal
Department of Artificial Intelligence and Data Science
IFET College of Engineering
Tamil Nadu, India

Galiveeti Poornima
Department of Engineering & Technology at School for Advanced Studies
S-VYASA
Bengaluru, India

Ravi Prakash
Department of Computer Applications, Nitte Institute of Professional Education
Nitte (Deemed to be University)
Mangaluru, India

Rajesh Rangan
Department of Computer Science and Engineering
Vel Tech Rangarajan Dr. Sagunthala R&D Institute of Science and Technology
Chennai, Tamil Nadu, India

Usharani Selvaraj
Department of Artificial Intelligence and Machine Learning
IFET College of Engineering
Tamil Nadu, India

Mohammad Ayyan A. Shaikh
Department of MCA
RV College of Engineering
Bengaluru, Karnataka, India

Mounika S. Shetty
Department of Computer Applications
Nitte Institute of Professional Education, Nitte (Deemed to be University)
Mangaluru, India

Shreya
Department of Computer Applications, Nitte Institute of Professional Education
Nitte (Deemed to be University), Mangaluru, India

Saravanan Somasundaram
Department of Mechanical Engineering
Annamalai University
Annamalainagar, Tamilnadu, India

Thiruselvan Subramanian
School of Computer Science and Engineering
RV University
Bengaluru, Karnataka, India

Huay Ling Tay
School of Business, Logistics & Supply Chain Management
Singapore University of Social Sciences
Singapore, Singapore

Balaji Venkatesan
RV University
Bengaluru, India

Chapter 1

AI and Blockchain-Enabled IoT (BIoT) for Supply Chain Sustainability

Umaselvi Muthusamy

1.1 INTRODUCTION

In the 21st century, there is increasing pressure on global supply chains to increase their efficiency, sustainability, and resilience. Strict environmental regulations, growing consumer awareness, and geopolitical unpredictability have businesses searching for innovative ways to transform their supply chain models. Among the most exciting technologies that will help bring about this change are blockchain, artificial intelligence (AI), and the Internet of Things (IoT). When these technologies are integrated into a single ecosystem called Blockchain-enabled IoT (BIoT), they have the potential to drastically alter traditional operations. The blockchain's decentralized and unbreakable ledger ensures that people can trust it, see where it came from, and see what it does. These technologies work together to make a powerful trio that solves many of the common problems in supply chain management (SCM), like inefficiency, lack of transparency, and practices that aren't good for the environment.

Social justice, environmental responsibility, and economic viability are all included in the notion of supply chain sustainability. The idea of supply chain sustainability includes social fairness, environmental accountability, and economic viability. This means reducing carbon emissions, waste, and unethical sourcing, as well as boosting labor standards. AI and IoT technology help businesses reach these goals by using smart analytics, automated compliance monitoring, predictive maintenance, and circular economy techniques.

The purpose of this chapter is to present a thorough examination of the possible uses of AI and BIoT technologies to achieve sustainable supply chain results. We'll discuss their respective functions, their synergy, and real-world applications. Real-world use cases, architectural models, implementation techniques, difficulties, and future directions will all be included in the document. By doing this, it seeks to give researchers, practitioners, and policymakers working on supply chain ecosystem sustainability and digital transformation initiatives useful information.

DOI: 10.1201/9781003660163-1

1.1.1 Supply chain sustainability overview

Businesses are switching from profit-only approaches to sustainable SCM (SSCM) in today's socially and environmentally sensitive society. To minimize detrimental effects on society and the environment while generating long-term value, SSCM combines ethical, environmental, and economic objectives. Stakeholder cooperation, ethical labor, waste minimization, and environmentally sustainable procurement are all part of it. Typically, a supply chain consists of manufacturers, distributors, retailers, suppliers, and customers, all of whom present opportunities to include sustainable practices. The essential elements of a conventional supply chain are depicted in Figure 1.1.

The triple bottom line idea, which emphasizes a balance between People, Planet, and Profit, is the fundamental tenet of SSCM (Seuring & Müller, 2008).SC is aptly described as "the strategic and transparent integration of ethical, environmental, and economic goals within the traditional supply chain functions" (United Nations Global Compact, 2010)by the United Nations Global Compact. This concept emphasizes the value of taking a comprehensive approach, recognizing that a truly sustainable supply chain goes beyond simple compliance to actively improve the planet. To preserve

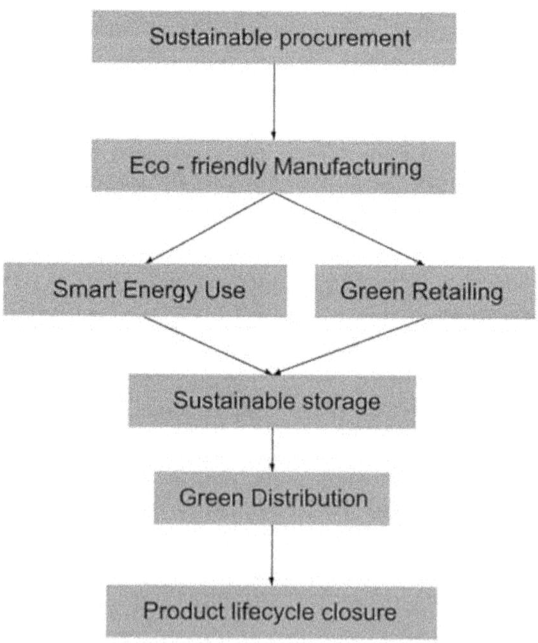

Figure 1.1 Sustainable supply chain elements.

Table 1.1 Measurements and areas of interest

Dimension	*Focus areas*
Environmental	Carbon emissions, waste reduction, renewable energy, circular economy practices
Social	Labor rights, workplace safety, community development, fair trade
Economic	Cost-efficiency, innovation, risk management, long-term profitability

profitability and competitiveness, it focuses on lowering carbon footprints, saving resources, guaranteeing moral labor practices, and boosting resilience. The many dimensions and areas of attention are displayed in Table 1.1.

1.1.2 Supply chain sustainability factors

Three interrelated dimensions form the foundation of sustainable supply chains.

1.1.2.1 Environmental sustainability

Some of the most important actions to take are to minimize greenhouse gas emissions, use more renewable energy sources, produce less waste, save water and energy, and apply ideas from the circular economy, such as recycling and reusing. For instance, studies have shown that a company's supply chain emissions are often the most significant part of its carbon footprint. The World Economic Forum released a report that identified eight significant supply chains that contribute to more than half of the world's greenhouse gas emissions. This shows that environmental measures can have a significant effect by dealing with these sources of pollution that are often ignored (World Economic Forum, 2023).

1.1.2.2 Social sustainability

Human interaction in supply chains is addressed by social sustainability. It places a strong emphasis on supporting fair trade practices across the supply chain, ensuring safe working conditions, encouraging community development, ethical sourcing, and maintaining fair labor standards. Businesses can improve their brand image, attract and retain talent, and lower the risk of human rights abuses and reputational harm by putting societal responsibility first. Recent industry reports highlight that IoT-powered smart supply chains significantly enhance real-time visibility and operational resilience (Accenture, 2022). For example, Patagonia sets a high bar for the fashion industry with its dedication to ethical labor practices and supply chain transparency.

1.1.2.3 Economic sustainability

The goal of economic sustainability is to guarantee resilience and long-term profitability. It entails maximizing resource utilization, reducing expenses, fostering innovation, and skillfully handling risks related to interruptions, laws, and harm to one's reputation. A more decisive competitive advantage, improved operational efficiency, and long-term cost savings are frequently the results of initial investments in sustainable practices, despite their seemingly high cost (Sarkis, 2020).

Multiple strong factors are driving the overall adoption of Sustainable SSCM. First, there is much regulatory pressure since governments all over the globe are passing more stringent environmental and social laws, such as the EU's Green Deal and Extended Producer Responsibility requirements, which call for accountability and compliance throughout supply chains. Second, increased consumer awareness is important; as demonstrated by Patagonia's success, knowledgeable customers are actively looking for brands that are dedicated to sustainability, influencing their decisions to buy, and demanding transparency. Third, a key tactic used by firms like IKEA is Corporate Social Responsibility, which encourages businesses to incorporate sustainability to boost brand value, foster trust, and improve corporate image. Fourth, risk management is a significant motivator since sustainable supply chains are more resilient to shocks like geopolitical unrest, which improves operational continuity (Brandenburg et al., 2014). Lastly, businesses have a strong financial incentive to prioritize SSCM due to investor demand for Environmental, Social, and Governance (ESG) criteria in investment decisions, as well as the growth of impact investing and green financing.

Organizations are using a variety of best practices to create more sustainable supply chains. In green procurement, suppliers' environmental performance is assessed, while eco-certified, sustainable, and recycled goods are sourced. Supply chain traceability, frequently facilitated by blockchain technology, enables transparency and accountability by tracing materials and emissions from source to shelf. Ultimately, circular economy practices attempt to move away from the linear "take–make–dispose" model by designing out waste and keeping resources in use, as exemplified by Interface's closed-loop recycling system.

SSCM is a good idea, but putting it into action is not easy. The high initial costs of green technologies, sustainable materials, and ethical labor practices can be a significant problem for smaller businesses. It's also hard to set consistent standards for sustainability because global supply chains are so complicated, with many different people, rules, and cultures involved. When there isn't real-time, reliable data for sustainability metrics, data transparency problems arise. This makes it harder to keep track of things and write about them. Many people don't want to change because traditional supply networks are often set up to save money, which makes people who are used to the way things are now resistant to change.

1.1.3 Benefits of supply chain sustainability

Businesses can gain significant long-term benefits by overcoming the initial problems and investing in sustainable supply chain practices. One of the main benefits is that it lowers a company's risk, making it less likely that its supply chain will be disrupted by things like resource shortages, bad weather, or negative publicity from unethical behavior. This also increases the value of the brand by making customers and employees more loyal, attracting top talent, and changing how people perceive the company, especially among the growing number of consumers who care about social and environmental issues.

Sustainable practices also make businesses run more smoothly by encouraging less waste and better use of resources. Over time, this can lead to substantial savings on materials and energy costs. Companies can avoid fines and legal problems by proactively following eco-friendly and ethical standards. This helps them stay in line with current and future laws. Lastly, efforts to make things more sustainable can help the market grow by creating new opportunities and attracting investors who care about ESG performance.

1.1.4 Real-world illustration

Unilever's Sustainable Living Plan, which came out in 2010, is a great example of how to successfully implement SSCM. The plan had very ambitious goals in a few important areas. For example, it aimed to achieve 100% Debate on Roundtable on Sustainable Palm Oil (RSPO) accreditation for sourcing palm oil in an environmentally friendly way, significantly reduce operational CO_2 emissions, and recycle a large amount of nonhazardous waste. This project not only improved social and environmental well-being, but it also had significant financial benefits, such as building trust with stakeholders, saving money, and growing the brand by more than €1 billion. Unilever's experience shows that putting sustainability at the center of the business plan has a significant impact.

SSCM is not just a trend; it's a significant change in how businesses work. Even though it has numerous potential benefits, it is hard to put into practice because of the high costs of green technologies, sustainable materials, and fair labor practices. Because supply chains are so complicated around the world, with many different stakeholders, rules, and cultural contexts, it is also hard to set and follow consistent sustainability standards. As businesses face more pressure to act ethically and responsibly, SSCM is becoming more and more important to modern business strategy. As technology gets better and customer expectations rise, it will be important for long-term profits and a good impact on the world to make the supply chain more sustainable.

1.2 SUPPLY CHAIN AI

AI is changing SCM by making it easier to make decisions based on data, using intelligent automation, predictive analytics, and operational optimization. AI is changing old supply chains, which are often slow and reactive, into networks that are strong, flexible, and able to adapt for companies that want to stay competitive in the face of global threats. This change is necessary because of the COVID-19 pandemic, political unrest, and climate change. Machine learning (ML) algorithms analyze past data, consumer behavior, patterns that occur over and over, and outside factors such as the weather or economic indicators to make specific demand forecasts. AI also improves inventory management. AI models maximize stock levels to satisfy service objectives while lowering holding costs by utilizing real-time consumption data, supply volatility, and lead-time unpredictability. Continuous learning and policy improvement for inventory control are made possible by reinforcement learning algorithms. Additionally, dynamic pricing, an additional AI-powered feature, recommends the best prices based on real-time information on market dynamics, rival pricing, and demand elasticity, increasing sales and customer satisfaction.

AI helps with supplier selection, route optimization, fraud detection, and customer service automation through chatbots and natural language processing (NLP), as well as logistics and inventory. Together, these developments improve supply chain operations from start to finish, reduce delays, and boost transparency. Supply chains will become smarter, faster, and more robust as AI technologies advance, making their incorporation into SCM even more crucial in an increasingly complicated global marketplace.

AI makes it possible for fleet management, autonomous delivery, and route optimization in logistics and transportation. Delivery times, poisons, and fuel consumption can all be decreased with AI-powered solutions such as intelligent transportation systems and route optimization algorithms. For example, Dalsey, Hillblom and Lynn (DHL) may reduce delivery times by up to 20% by using AI for route planning and real-time tracking.

AI greatly improves supplier relationship management (SRM) by making it easier to evaluate supplier performance, risk, and loyalty. AI systems analyze a large amount of unstructured data from emails, contracts, news reports, and other sources to find early signs of possible supplier risk or disruption. They do this by using advanced methods such as NLP. This lets businesses take steps to lower their risks and strategically grow their supplier base. IBM Watson is a good example because it analyzes a large amount of different data sources to figure out when things will go wrong. It has also been used to manage suppliers' risks. This ability to predict helps companies make better choices and build a supply chain that is more flexible and able to respond to changes.

Virtual assistants and chatbots powered by AI are also being utilized more and more in customer service and procurement to answer questions, process orders, and make recommendations. These AI agents increase customer satisfaction and responsiveness by operating around the clock. By automating repetitive tasks such as order entry, invoice processing, and compliance checking, robotic process automation and AI increase operational efficiency and lower error rates.

AI makes it possible to track carbon footprints, reduce waste, and plan for a circular economy in sustainability initiatives. AI, for instance, can simulate how various supply chain routes will affect the environment and recommend more environmentally friendly options. The adoption of AI, in sustainable supply chains, may reduce waste by 25% and emissions by up to 15%, according to a McKinsey (2022) report. Table 1.2 lists the primary uses of AI in supply chain operations.

Despite its benefits, integrating AI in supply chains poses several challenges. Data quality and integration remain significant hurdles. AI systems require large volumes of high-quality, real-time data from disparate sources, which may be fragmented across legacy systems. Ensuring data privacy, especially with vendors and partners, adds complexity. Moreover, a lack of technical expertise and high initial investments deter small and medium-sized enterprises (SMEs) from adopting AI technologies. Ethical concerns, including bias in algorithms and explainability of AI decisions, also pose significant challenges, especially in supplier evaluations or pricing decisions.

Moreover, AI implementation requires organizational change management. Employees need to be upskilled to work alongside AI tools, and companies must cultivate a culture of data-driven decision-making. AI systems

Table 1.2 AI in supply chain operations

Supply chain function	*AI application*	*Benefits*
Demand forecasting	Predictive analytics, ML regression models	Reduced stockouts, improved accuracy
Inventory management	Reinforcement learning, optimization algorithms	Lower holding costs, better service levels
Logistics & transportation	Route optimization, ITS, autonomous vehicles	Cost savings, faster delivery
Warehouse management	Robotics, computer vision, AI-based scheduling	Increased efficiency and accuracy
Supplier management	NLP for risk detection, predictive scoring	Proactive risk mitigation
Customer service	AI chatbots, virtual assistants, recommendation engines	24/7 support, improved user experience
Sustainability	Emissions monitoring, green optimization models	Lower environmental impact

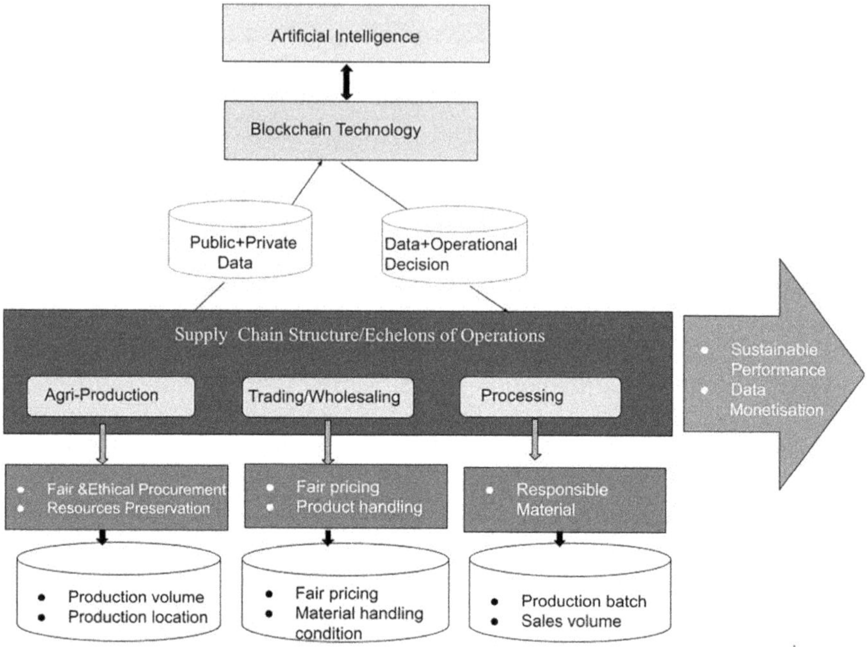

Figure 1.2 AI in the supply chain.

should not fully replace human judgment, especially in strategic areas such as supplier negotiations or crisis response. A hybrid human–AI collaboration approach, where AI augments human capabilities, is regarded as the best. A conceptual diagram of AI in the supply chain is provided in Figure 1.2.

AI is also being combined with the IoT and blockchain to enhance transparency and traceability. Smart sensors give real-time data on product location and condition, which AI algorithms use for predictive maintenance and anomaly identification. Blockchain provides tamper-proof data that feeds AI algorithms with dependable knowledge. This AI–IoT–blockchain combination is building the backbone of next-generation digital supply chains. Looking forward, AI is projected to be increasingly autonomous and prescriptive. Future supply chains will incorporate cognitive systems capable of end-to-end self-management, autonomously resolving disruptions, negotiating with suppliers, and responding to market changes without human intervention. This vision corresponds with the development of Industry 5.0, where AI enables human-centric, sustainable, and resilient production and distribution systems.

1.3 SUPPLY CHAIN AI AND NLP

1.3.1 Revolutionizing AI integration

Through the facilitation of data-driven decision-making, intelligent automation, predictive analytics, and operational optimization, AI is revolutionizing SCM. Because of AI integration, traditional supply chains, which are frequently reactive and inefficient, are changing into networks that are resilient, agile, and adaptive. For businesses hoping to stay competitive in the face of global risks such as the COVID-19 pandemic, geopolitical unrest, and climate change, this transition is essential.

1.3.2 Forecasting with predictive analytics

Demand forecasting using predictive analytics is a fundamental use of AI in SCM. To provide precise demand forecasts, ML algorithms examine past data, consumer behavior, seasonal patterns, and outside factors such as the weather or economic indicators. This enhances planning, minimizes stockouts and overstocking, and lessens the bullwhip effect. AI-enhanced forecasting models have shown 20%–50% higher accuracy than conventional techniques, according to Choi et al. (2021). AI-driven projections are used by market leaders such as Amazon and Walmart to dynamically modify their pricing, promotions, and inventory replenishment plans.

1.3.3 Inventory management and dynamic pricing

AI also significantly improves inventory management. AI models maximize stock levels to satisfy service objectives while lowering holding costs by utilizing real-time consumption data, supply volatility, and lead-time unpredictability. Continuous learning and policy improvement for inventory control are made possible by reinforcement learning algorithms. Additionally, dynamic pricing, an additional AI-powered feature, recommends the best prices based on real-time information on market dynamics, rival pricing, and demand elasticity, increasing sales and customer satisfaction.

1.3.4 SRM using AI and NLP

There are problems with implementing SSCM, such as the high costs of ethical labor, sustainable technologies, and materials, as well as the difficulty of standardizing practices across different global supply chains. However, the benefits are significant and are becoming more apparent with the help of AI. NLP is used to improve SRM, evaluate supplier performance, and predict and reduce risks by analyzing large datasets. AI also makes logistics and inventory management more efficient, finds the best transportation routes to save money and protect the environment, and improves customer service with chatbots and virtual assistants. Adding AI to the supply chain also helps efforts to be more environmentally friendly by making it easier to track carbon footprints and reduce waste. Even though there are problems with data quality, integration, privacy, and technical know-how, combining AI and SSCM gives businesses a clear way to lower risks, raise brand value, improve operational efficiency, and ultimately make supply chains that are more resilient, profitable, and ethical in a complicated global market.

1.3.5 Human–AI decision-making

AI systems should not replace human decision-making fully, especially in strategic areas such as supplier conferences or crisis response. A hybrid human–AI collaboration approach, where AI augments human capabilities, is regarded as the best.

AI has also been combined with the IoT and blockchain to enhance transparency and traceability. Smart sensors give real-time data on product location and condition, which AI algorithms use for predictive maintenance and anomaly identification. Blockchain provides tamper-proof data that feed AI algorithms with dependable knowledge. This AI–IoT–blockchain combination is building the backbone of next-generation digital supply chains.

Looking forward, AI is projected to be increasingly autonomous and prescriptive. Future supply chains will incorporate cognitive systems capable of end-to-end self-management, autonomously resolving disruptions, negotiating with suppliers, and responding to market changes without human intervention. This vision corresponds with the development of Industry 5.0, where AI enables human-centric, sustainable, and resilient production and distribution systems.

1.4 SUPPLY CHAINS USING BLOCKCHAIN TECHNOLOGY

1.4.1 Blockchain and automation

SCM is changing quickly thanks to blockchain technology, which makes it possible for all stakeholders to share data in a transparent, trustworthy, traceable, and impenetrable manner. Initially, developers created blockchain's primary architecture, a distributed ledger system, to support cryptocurrencies such as Bitcoin (Zheng et al., 2018). Its versatility makes it perfect for handling the complexity and dispersion of global supply chains. Blockchain offers a dependable digital foundation to improve the security, effectiveness, and sustainability of contemporary supply chains in light of growing worries about product authenticity, fraud, unethical sourcing, and regulatory compliance (Saberi et al., 2019).

Data are frequently fragmented across several stakeholders (suppliers, manufacturers, shipping companies, and retailers) in a typical supply chain, making it susceptible to mistakes, manipulation, or delays. With each transaction being recorded in a block, timestamped, cryptographically encrypted, and connected to the block before it, blockchain ensures that everyone involved has access to the same version of the truth (Francisco & Swanson, 2018). Once recorded, data remain unchangeable without accord, significantly reducing the risk of scams and reconciliation errors. Blockchain and smart contracts greatly improve SCM by automating agreements and making it easier to follow the flow of goods, which makes the process more open and accountable (Saberi et al., 2019). Smart contracts, which are self-executing agreements on the blockchain, can automatically initiate payments, fines, or compliance checks based on pre-defined criteria. This eliminates manual bottlenecks (Saberi et al., 2019). One important use is end-to-end product traceability, which uses blockchain to keep track of every step a product takes, from its source to the customer (Saberi et al., 2019). This is very important for valuable or sensitive items such as organic foods and medicines. For example, IBM and Walmart worked together to show how powerful this technology is by reducing the time it takes to trace contaminated food from 7 days to just 2.2 s. This changed the way food safety issues are handled (IBM, 2021). Customers can develop trust and brand loyalty by scanning QR codes to view a complete source record.

Blockchain also makes it easier to audit and verify sellers. Companies can keep unchangeable records of their labor compliance, ethical sourcing credentials, certifications, and quality inspections. This is especially crucial for sectors that are under investigation for environmental damage or human rights abuses. By connecting blockchain with IoT sensors, companies can record and permanently store real-time data such as temperature, shipment status, and location on the blockchain. This ensures data integrity in cold chains and the transportation of fragile commodities. Table 1.3 presents the main advantages and uses of blockchain technology in supply chains.

Table 1.3 Advantages and uses of blockchain technology in supply chains

Supply chain function	*Blockchain application*	*Key benefits*
Traceability	Immutable product history, provenance records	Prevents fraud, recalls, improves consumer trust
Contract enforcement	Smart contracts for payments, delivery compliance	Reduces disputes, automates settlements
Supplier verification	Tamper-proof audit trails, ethical sourcing records	Supports compliance, ESG standards
Logistics and shipping	Shipment records, cross-border tracking	Eliminates paperwork, speeds up customs
Inventory transparency	Shared inventory data across partners	Reduces stockouts, improves planning
Risk management	End-to-end visibility to detect disruptions early	Enhances agility, lowers costs

1.4.2 Function of the supply chain: key benefits of blockchain applications

- The ability to trace unchangeable provenance data and product histories enhances consumer trust while preventing fraud and product recalls.
- Enforcement of Contracts: Smart contracts for delivery compliance and payments minimize disagreements and streamline resolutions.
- The verification of suppliers, along with unchangeable audit trails and documents of ethical sourcing, encourages adherence to ESG principles.
- Shipping & Logistics: Records of shipments and cross-border monitoring reduce paperwork and expedite customs.
- Transparency of Inventory: Inventory data shared among partners enhance planning and decrease stockouts.
- Controlling Risk: End-to-end visibility to identify problems before they become serious reduces expenses and increases agility.

1.4.3 Supply chain challenges with blockchain

Every transaction creates a new block, which is then permanently recorded on the blockchain ledger. Additionally, blockchain improves carbon tracking and sustainability reporting. Businesses can promote ESG objectives by tracking energy use, carbon emissions, and material sources on-chain. Businesses are using blockchain-based platforms, such as Provenance and Everledger, to certify their supply chains against labor exploitation, conflict minerals, and deforestation (Saberi et al., 2019).

Blockchain adoption in SCM is not without its drawbacks, despite its benefits. Scalability and energy intake are two significant issues, mainly with conventional proof-of-work consensus systems such as Bitcoin (Zheng et al., 2018). However, enterprise-grade installations are made possible by more energy-efficient techniques such as Proof of Stake or Proof of Authority seen in more recent blockchain systems such as Hyperledger Fabric or Ethereum 2.0 (Mettler, 2016). Interoperability, or ensuring that blockchain solutions can connect with current cloud, IoT, or ERP platforms, is another difficulty. Cooperation among global supply chain partners is stimulating since governance and standardization are still developing (Francisco & Swanson, 2018).

High initial investment expenditures and a lack of technical competence might also prevent small and medium-sized businesses (SMEs) from taking part. Furthermore, privacy issues need to be resolved; particularly in reasonable B2B settings, not all data should be accessible to every node. Fine-grained access control is possible in IoT by permissioned blockchains or hybrid models, which maintain security and transparency (Zheng et al., 2018). When combined with AI and the IoT, blockchain's impact in supply chains is further enhanced, creating a trio of intelligent digital infrastructure. IoT devices produce reliable real-time data (such as location and temperature), which AI analyzes to predict breaks or improve operations. In turn, blockchain promises that the data are securely exchanged among parties and are auditable and unchangeable. The new BIoT paradigm for supply chain automation and sustainability is based on this convergence (Saberi et al., 2019).

Examples from the real world highlight blockchain's revolutionary potential even more: to eliminate conflict diamonds, the diamond company De Beers tracks each diamond's provenance using blockchain technology. Using blockchain technology, Nestlé and Carrefour enable customers to track the origin of milk, meat, and vegetables. Over 12 million papers were digitized by TradeLens, a blockchain-based shipping platform developed by Maersk and IBM that has reduced fraud and delays in international maritime logistics. For anti-counterfeit solutions in wine and luxury goods supply chains, VeChain combines blockchain technology with the IoT.

Blockchain technology solves a number of the innate inefficiencies in conventional supply chains, ranging from fraud and regulatory violations to a lack of visibility and trust. Blockchain improves supply chains' efficiency, sustainability, and accountability by guaranteeing unchangeable records, enabling smart contracts, and encouraging open cooperation. Blockchain will act as the digital trust layer in the future of autonomous, robust, and sustainable supply chain ecosystems as adoption picks up speed, particularly when combined with AI and IoT.

1.5 BIoT EMERGENCE

1.5.1 Blockchain-enabled IoT

A breakthrough paradigm known as BIoT has emerged as a result of the confluence of blockchain and the IoT. This meld tackles fundamental issues with data integrity, security, transparency, and decentralization in distributed systems (Reyna et al., 2018). Real-time data are key for smart operations in industries such as healthcare, manufacturing, shipping, and agriculture, and the IoT facilitates smooth communication between interconnected physical devices, including sensors, actuators, cars, and smart appliances (Panarello et al., 2018).

Because they use centralized architectures, traditional IoT systems are vulnerable to scalability problems, cyberattacks, and single points of failure (Christidis & Devetsikiotis, 2016). Lee and Lee (2015) outline key enterprise applications and investment considerations that shaped early IoT adoption frameworks. The use of blockchain technology can help. Blockchain is a decentralized ledger system that builds trust and openness without the need for middlemen by putting data in blocks that can't be changed and adding time stamps to them (Christidis & Devetsikiotis, 2016). The new field of BIoT solves these problems by combining the secure, distributed ledger of blockchain with the ability of IoT devices to create and share data on their own (Reyna et al., 2018; Conoscenti et al., 2017).

1.5.2 Accountability in smart systems

In a typical BIoT system, IoT devices send data to the blockchain network through gateways or edge devices. The data can include temperature, location, and energy use (Christidis & Devetsikiotis, 2016). Based on logic that has already been written down, they initiate payments or logistical operations (Christidis & Devetsikiotis, 2016). For example, if the temperature of a perishable item goes above a certain level, a smart contract can automatically let everyone know and start taking steps to fix the problem (Christidis & Devetsikiotis, 2016). This automation, along with blockchain's unchangeable traceability, reduces the need for people to get involved and ensures that accountability can be proven. This makes it perfect for uses that need quality control, fraud prevention, and compliance (Conoscenti et al., 2017).

The growth of BIoT is clear in many fields. For example, IBM and Maersk's TradeLens platform uses BIoT to keep track of global freight, making it easier for customs to clear and reducing paperwork (Reyna et al., 2018). AgriDigital is an example of a system that uses a combination of IoT sensors and blockchain technology to make food safer and trace its origins (Reyna et al., 2018). Blockchain keeps a record of transactions that can't be changed, which builds trust and openness throughout the supply chain (Reyna et al., 2018). This integration makes it easy to trace things, which builds trust in consumers and helps prove that farming methods are safe for the environment and fair. The healthcare industry also uses BIoT to securely share unchangeable health records with authorized providers and monitor patients' vital signs in real time (Reyna et al., 2018). These applications show that BIoT can make digital trust ecosystems in systems that are usually unclear or broken up (Reyna et al., 2018).

However, combining blockchain with IoT devices that don't have a large amount of resources is hard from both a technical and an organizational point of view (Conoscenti et al., 2017). To overcome the constraints of devices with limited memory and processing capabilities, researchers are investigating lightweight consensus algorithms such as Delegated Proof of Stake and Practical Byzantine Fault Tolerance, in addition to off-chain solutions such as sidechains (Conoscenti et al., 2017). Scalability is another issue because the huge amounts of data from millions of IoT devices can slow down blockchain networks (Conoscenti et al., 2017). People are working on solutions like only storing data hashes on-chain and using distributed storage systems such as the InterPlanetary File System (IPFS) (Conoscenti et al., 2017). Moreover, issues pertaining to data privacy, system interoperability,

Table 1.4 IoT, Blockchain, and BIoT capabilities

Feature	*Traditional IoT*	*Blockchain*	*BIoT integration*
Architecture	Centralized	Decentralized	Decentralized edge-enabled
Data integrity	Low (prone to tampering)	High (immutable)	High with verified sensor data
Trust mechanism	Third-party authentication	Consensus algorithms	Smart contracts and consensus
Scalability	High (but limited by cloud)	Low to medium	Medium (using off-chain enhancements)
Security	Vulnerable to cyberattacks	Strong cryptography	Strong with end-to-end security
Interoperability	Limited	Evolving	Depends on standards and middleware
Automation	Manual/ cloud-triggered	Rule-based (smart contracts)	Real-time autonomous execution

and adherence to regulations must be resolved to facilitate the extensive adoption of BIoT (Panarello et al., 2018). Table 1.4 presents the IoT, blockchain, and BIoT capabilities.

1.5.3 Empowering distributed intelligence through emerging technologies

Advanced technologies such as edge computing, digital twins, and AI are shaping the future of BIoT. AI greatly improves BIoT systems by analyzing sensor data to find patterns and smartly starting blockchain transactions, which makes it easier to make decisions (Reyna et al., 2018). For instance, AI-powered predictive maintenance in manufacturing uses IoT sensor data to predict problems with equipment and logs strange events on-chain (Reyna et al., 2018). Edge computing helps with this by processing data on the spot, which reduces latency and network load. Moreover, entities such as the Industrial Internet Consortium, IEEE, and ISO are creating secure and interoperable BIoT frameworks through continuous standardization initiatives (Panarello et al., 2018).

The adoption of IoT technologies presents both operational benefits and organizational risks, including security vulnerabilities, governance challenges, and scalability limitations (Brous et al., 2019). The creation of BIoT is a significant step forward for distributed computing because it lets smart devices talk to each other safely, on their own, and on a large scale in a trustless environment. This integration makes it easier to see how things work across ecosystems and gives digital economies a programmable, unchangeable base. As businesses move toward Industry 4.0 and Society 5.0, BIoT is likely to become a key part of smart, decentralized, and long-lasting infrastructure. This will lead to new ideas in fields as diverse as aerospace and agriculture. To fully realize the potential of BIoT, it is necessary to overcome major obstacles, such as technical limitations related to scalability and power constraints, the establishment of international standards, and the resolution of privacy and regulatory concerns (Panarello et al., 2018). To create a safe, open, and strong future for the BIoT, researchers, technologists, policymakers, and people in the industry need to work together (Panarello et al., 2018).

1.6 SYNERGIES BETWEEN AI AND BIoT

1.6.1 BIoT for decentralized autonomy

Businesses can make safe, self-directed, and smart digital ecosystems by combining AI and BIoT platforms (Zhang et al., 2020). AI gives IoT devices the ability to analyze, learn from, and improve the huge amounts of data they collect in real time. This makes powerful predictive analytics and autonomous decision-making possible (Zhang et al., 2020). The decentralized nature of BIoT, which combines pervasive IoT networks with secure blockchain architecture, solves important problems of security, trust, and transparency (Dorri et al., 2017, Zhang et al., 2020). This combination makes it possible to greatly improve end-to-end automation and data monetization in the industrial, environmental, and consumer sectors (Zhang et al., 2020).

This method gets around the problems that come with traditional IoT, which depends on centralized servers that are prone to latency, security risks, and single points of failure (Christidis & Devetsikiotis, 2016). BIoT makes it possible to make decisions in real time by bringing AI closer to the edge and processing data where they are generated (Zhang et al., 2020). For instance, in smart agriculture, AI-powered cameras can find problems with crops, and blockchain can then safely and permanently record these events, making it possible to respond more quickly and clearly (Zhang et al., 2020). Smart grids also use AI to predict how much energy will be needed, and blockchain keeps track of small transactions between energy producers and consumers safely (Zhang et al., 2020). This combination of AI and blockchain gives systems a solid and trustworthy foundation for making decisions on their own and improving operations (Dorri et al., 2017). For industries with a lot on the line, such as healthcare, banking, and defense, blockchain's unchangeable nature is very important because it ensures that data provenance, or the data's origin and history, is completely verifiable and can't be changed (Dorri et al., 2017, Zhang et al., 2020). These verifiable data are a strong base for AI-driven insights and decisions in fields where honesty and safety are very important (Dorri et al., 2017, Zhang et al., 2020).This integrated model turns reactive networks into strong and flexible systems that can change on their own with little help from people (Dorri et al., 2017).

1.6.2 Trust and security

We can make BIoT networks safer and more reliable by adding AI to them. This will help keep them safe from cyberattacks and mistakes that could cause harm. For instance, anomaly detection models can be used to start smart contracts that put limits on things like system lockdowns or notifications. AI can also improve the consensus mechanism in blockchains to make them more scalable and faster. AI models can also check the trustworthiness of devices in BIoT networks, which protects against rogue nodes, Sybil attacks, and data spoofing (Nguyen et al., 2021).

Wearable IoT devices track patient vitals in the medical field, and AI identifies possible threats such as cardiac arrest and immediately logs the data to blockchain for clinical access. While BIoT ensures that judgments (such as route rerouting or environmental alarms) are verified, timestamped, and unaltered, AI handles traffic, air quality, and surveillance data in smart cities. Predictive maintenance is carried out in production using AI-enabled BIoT systems, which automatically start blockchain-logged maintenance tickets and anticipate the likelihood of machinery failure. Additionally supported by this fusion are energy trading, decentralized finance, precision farming, and driverless cars.

Notwithstanding these advantages, there are a number of obstacles to integration. First, the deployment of large AI models is limited by resource constraints at edge devices, which calls for lightweight architectures (e.g., MobileNet and TinyML). Second, it is crucial to guarantee the trustworthiness of AI models through explainability (explainable AI) and fairness, particularly when judgments are permanently recorded on blockchain. Third, the necessity for massive datasets for AI training raises data privacy issues, and the transparency of blockchain technology may reveal private information. Federated learning, in which AI models are trained across decentralized nodes without sending raw data, is used to address the issue, and blockchain guarantees the auditability of training procedures. Additionally, the integration between AI ecosystems (TensorFlow, PyTorch) and blockchain platforms (Ethereum, Hyperledger, etc.) is still developing, necessitating the use of middleware and standard APIs.

1.6.3 Real-world applications

Recent studies demonstrate the growing interest in this interaction. For instance, Zhang et al. (2020) describe how AI is used to optimize smart contract triggers in "AI-Enabled BIoT for Smart Industries" (IEEE Access).

A lightweight blockchain architecture for IoT combined with AI for safe home automation was also presented by Dorri et al. (2017). Additionally, Mohanta et al. (2019) examine how AI models can be certified and validated on-chain in their paper "Blockchain Technology and Its Applications to AI" (*Procedia Computer Science*). These studies highlight how digital infrastructures can be transformed into self-optimizing, self-auditing systems through fusion.

The pharmaceutical industry, which has strict rules for handling sensitive products such as vaccines and biologics, is greatly improved by the use of AI and BIoT systems. IoT temperature sensors monitor the conditions of shipments at all times in this ecosystem. AI models then analyze the data to find any problems that might occur, such as equipment failures or any discrepancies. If a problem is found, smart contracts are automatically triggered to send notifications or even stop payment to logistics providers. All sensor readings and AI conclusions are stored on a blockchain in a way that can't be changed. This creates a record that suppliers, partners, and regulators can use to prove compliance and ensure quality. In a fragile cold chain, this closed-loop system is very important for ensuring that everyone is accountable, that operations can change quickly, and that products remain safe.

The move toward Industry 5.0, which stresses smooth cooperation between people and machines, makes AI–BIoT systems even more important. In this human-centered world, AI–BIoT networks can help apps that use decentralized data governance and algorithms that are aware of ethics. These networks will be even better when they work with Web 3.0, 5G, and quantum computing in the future. This will let devices negotiate resources on their own, adapt to new situations, and create complex system behaviors. AI agents will manage and improve these BIoT ecosystems, while blockchain technology will ensure that traceability and compliance are always possible. This will lead to a new kind of smart, safe, and strong infrastructure.

1.7 COLLABORATIONS BETWEEN BIoT AND AI

1.7.1 AI–BIoT integration

Combining blockchain technology, AI, and IoT platforms makes it feasible to create digital ecosystems that are self-sufficient, smart, and safe. AI gives people the mental tools they need to learn, optimize, and perform predictive analytics using real-time data. At the same time, BIoT uses decentralized blockchain infrastructure and widespread IoT networks to ensure that security, transparency, and trust are all present (Wang et al., 2020). This improves automation from start to finish, data monetization, system intelligence, and the ability to make decisions independently. Blockchain ensures that these insights can be trusted across different ecosystems, cannot be changed, and can be verified. AI, on the other hand, can turn a large amount of data from the IoT into meaningful information (Saberi et al., 2019).

This collaborative design changes the way current systems evolve from networks that only react to changes to frameworks that are self-aware, adaptable, and resilient. In traditional IoT systems, sensor nodes gather data and deliver it to centralized cloud servers, where AI models are used. This centralized technique, on the other hand, often has drawbacks including latency, single-point failures, and security holes. The incorporation of AI into BIoT decentralizes this knowledge by putting ML models inside smart contracts or at the edge (AIoT). It makes it possible to make safe, real-time decisions closer to where the data are stored. For instance, edge-AI models in smart agriculture use data from IoT cameras to spot problems with plants, and blockchain keeps track of these results for clear auditing and payment processing for farmers (Liu et al., 2021). Blockchain manages the small payments between energy providers and users, and smart grids use AI to predict how much energy is needed. Blockchain guarantees that data come from the right source, which is very important for fields such as healthcare, banking, and military. AI, on the other hand, makes it easier to understand the context.

1.7.2 Anomaly detection to consensus enhancement

We can make BIoT networks safer and more reliable by adding AI to them. This will help keep them safe from cyberattacks and mistakes that could cause harm. For instance, anomaly detection models can be used to start smart contracts that put limits on things like system lockdowns or notifications. AI can potentially improve the consensus mechanism in blockchain to make it more scalable and faster. AI models may also check the trustworthiness of devices in BIoT networks, which protects against rogue nodes, Sybil attacks, and data spoofing (Nguyen et al., 2021).

There are several ways that AI and BIoT can work together in industry. Blockchain keeps track of things in supply chains and logistics, while AI uses IoT-based inventory sensors to predict what people will want. In the medical field, wearable IoT devices keep track of patients' vital signs, and AI finds potential dangers such as cardiac arrest and logs the data to a blockchain for clinical use right away (Francisco & Swanson, 2018; Wang et al., 2020). BIoT ensures that decisions (such as changing routes or setting off alarms for environmental reasons) are checked, timestamped, and not changed. AI, on the other hand, deals with traffic, air quality, and surveillance data in smart cities. Predictive maintenance is carried out in production using AI-enabled BIoT systems, which automatically start blockchain-logged maintenance tickets and anticipate the likelihood of machinery failure. Additionally supported by this fusion are energy trading, decentralized finance, precision farming, and driverless cars (Liu et al., 2021).

Notwithstanding these advantages, there are several obstacles to the integration. First, the deployment of large AI models is limited by resource constraints at edge devices, which calls for lightweight architectures (e.g., MobileNet and TinyML). Second, it is crucial to guarantee the trustworthiness of AI models through explainable AI and fairness, particularly when judgments are permanently recorded on blockchain. Third, the necessity for massive datasets for AI training raises data privacy issues, and the transparency of blockchain technology may reveal private information. Federated learning, in which AI models are trained across decentralized nodes without sending raw data, is used to address the issue, and blockchain guarantees the auditability of training procedures. Additionally, the integration between AI ecosystems (TensorFlow, PyTorch) and blockchain platforms (Ethereum, Hyperledger, etc.) is still developing, necessitating the use of middleware and standard APIs.

1.7.3 Transforming digital infrastructures

Recent studies demonstrate the growing interest in this interaction. For instance, Zhang et al. (2020) describe how AI is used to optimize smart contract triggers in "AI-Enabled BIoT for Smart Industries" (IEEE Access). A lightweight blockchain architecture for IoT combined with AI for safe home automation was also presented by Dorri et al. (2017). Additionally, Mohanta et al. (2019) examine how AI models can be certified and validated on-chain in their paper "Blockchain Technology and Its Applications to AI" (*Procedia Computer Science*). These studies highlight how digital infrastructures can be transformed into self-optimizing, self-auditing systems through fusion.

1.7.4 Pharma cold chain monitoring

Biologics and vaccines in pharmaceutical supply chains need exact temperature control. IoT temperature sensors monitor shipment conditions all the time. AI algorithms can predict when equipment may break down and find problems. If AI finds something that could go bad, a smart contract sends out alerts and freezes payment to the logistics provider until more information is available. Blockchain saves all sensor data and AI inferences in a way that cannot be changed. This helps suppliers, logistics partners, and regulators check that they are obeying the regulations. This closed-loop approach ensures that individuals are accountable, that operations can change, and that quality is always guaranteed. All of these things maintain the integrity of pharmaceutical products.

As we move toward Industry 5.0, where people and machines work together without any problems, AI–BIoT solutions become even more important. Decentralized data governance and algorithms that are cognizant of ethics can help apps that are focused on people. Combining Web 3.0, 5G, and quantum computing technologies will make AI–BIoT networks even better. Devices will be able to negotiate resources, change their behavior on their own, and adapt to new situations. Blockchain will keep track of everything and ensure it follows the rules, while AI agents will manage and improve BIoT ecosystems.

1.8 APPLICATIONS OF AI AND BIoT IN SUPPLY CHAIN SUSTAINABILITY

1.8.1 IBM Food Trust

The combination of BIoT and AI technologies has moved from being a theoretical idea to being used in real life through a number of case studies across global supply chains. These deployments show how producers have used the AI–BIoT partnership to improve operational flexibility, sustainability, and transparency while also ensuring that their supply chain models are in line with the UN's Sustainable Development Goals (SDGs) and new regulatory frameworks.

The IBM Food Trust is a well-known example. It was formed with the help of Walmart, Nestlé, and Unilever. The software lets you track food goods all the way from farms to customers. This is done by using blockchain technology, IoT devices, and AI analytics. When storing and moving, IoT devices monitor things such as temperature, humidity, and location. A permissioned blockchain is used to record these readings so that they can never be changed. AI systems that analyze real-time sensor data can help reduce food waste and carbon emissions by identifying faults, predicting waste, and improving route scheduling. Using this technology, Walmart was able to locate the source of the infected mangoes in just 2.2 s instead of 7 days. This helped them waste less food and deal with safety problems right away.

1.8.2 Maritime logistics

Another strong point is the TradeLens platform, which IBM and Maersk worked together to create. It uses blockchain and the IoT to digitize international shipping. IoT-enabled containers that check temperature, vibration, location, and handling situations in real time are combined with TradeLens. AI models process the data to identify hazards, forecast delays, and enhance fleet efficiency. For end-to-end auditability and transparency, every event such as loading, customs clearance, and delivery is recorded on the blockchain. TradeLens improved efficiency and decreased emissions in marine logistics by removing document processing times by more than 40%

in a pilot project with U.S. Customs and Border Protection. Additionally, the system allows smart contracts, which promote environmentally friendly and socially responsible shipping methods by automating payments when delivery requirements are satisfied.

1.8.3 Monitoring with Modum

Modum (Switzerland) created a BIoT solution for cold chain monitoring in the pharmaceutical sector. IoT sensors built into packaging track temperature and location when temperature-sensitive medications and vaccines are being transported. Data immutability is ensured by having these recorded on a blockchain ledger. Temperature excursion hazards are predicted by AI systems using logistical practices and external factors. Smart contracts immediately alert stakeholders or stop distribution to avoid spoilage when the system notices deviations. Numerous logistics companies in Europe have embraced this approach to improve patient safety, minimize product loss, and adhere to Gross Domestic Product (GDP) rules. The Modum solution guarantees data trust between suppliers, logistics providers, and regulators while reducing the need for manual audits.

AgriDigital, an Australian supply chain platform, uses BIoT and AI in agriculture to facilitate financial automation and farm-to-fork traceability. We collect environmental and grain quality data using drones and IoT devices. Verified records of ownership, handling, and storage are stored on the blockchain. AI programs automate pricing recommendations and assess supply–demand trends. Once predetermined delivery circumstances are confirmed using sensor data, smart contracts immediately transfer payment when grain is sold. By lowering credit risk and advancing fair trade, this automation has shortened payment periods from 30 days to a few minutes. Additionally, it reduces losses from fraud or spoilage in isolated farming regions with inadequate infrastructure.

The Power Ledger project in Australia provides a forward-thinking example from the energy sector. Households with solar panels and IoT smart meters may monitor energy production and consumption with this blockchain-based energy trading platform. Peer-to-peer trading of excess electricity is made possible by AI algorithms that predict patterns of generation and consumption. Blockchain tracks every transaction to maintain trust and eliminate the need for intermediaries. This program supports a decentralized green energy economy, lowers transmission losses, and encourages the use of renewable energy locally. In addition to being financially motivated to contribute to a more sustainable grid, households can lower their carbon impact.

1.8.4 BIoT role sustainability outcome

The Walmart Food Supply Chain and IBM Food Trust estimate waste and maximize delivery. Blockchain logging and traceability via sensors decreased food waste, increased openness, and enhanced customer confidence. IBM & Maersk Maritime Logistics' TradeLens forecasts delays and improves routes. Container tracking, automated documentation, and paperless shipping reduce fuel consumption and lower pollutant emissions. Modum (Switzerland) Pharma: The cold chain forecasts temperature variations, smart contract alerts, and sensor-monitored delivery adherence, which decrease spoilage and enhance patient safety. Digital agribusiness forecasts demand and optimizes prices. The use of IoT data at the farm level and blockchain technology for transactions is essential. The Ledger of Power Sustainable Energy Blockchain platform ensures equitable compensation, enhances rural accessibility, and improves quality monitoring through blockchain-based energy trading, IoT smart meters, and advanced energy forecasting systems. The decentralized adoption of green energy contributes to a lower carbon footprint.

In addition to these examples, Alibaba, Amazon, and DHL have investigated AI–BIoT interfaces for autonomous cars and smart warehouses in logistics, emphasizing waste reduction, energy efficiency, and real-time decision-making. Alibaba's ET Industrial Brain optimizes manufacturing energy utilization through ML, and blockchain ensures that resource usage and emissions data are traceable and validated. DHL employs BIoT platforms and AI-driven robotics to manage its warehouses better. Predictive maintenance and environmental sensors are used to reduce downtime and energy consumption.

These case studies demonstrate several shared benefits:

- Blockchain's immutability and data trust lower the risks of fraud, tampering, and noncompliance.
- Proactive sustainability initiatives, as opposed to reactive ones, are made possible by real-time insights from AI.
- Smart contract automation increases process consistency and lowers human error.
- Verified supply chain histories encourage responsible consumption through consumer openness.

Nevertheless, there are still implementation issues. These include the necessity for legislative clarification, particularly when AI is utilized in delicate industries including healthcare and banking, the incompatibility of older systems with blockchain platforms, and the high cost of infrastructure. Furthermore, data privacy is a critical issue, especially with public blockchains. To strike a compromise between decentralization and confidentiality, several systems are using federated learning, permissioned ledgers, and zero-knowledge proofs. Strong network access is also necessary for scaling AI–BIoT systems, which may be a barrier in rural or underdeveloped locations.

1.8.5 Current insights

These incidents have been examined in several academic publications. While Saberi et al. (2019) mapped blockchain's role in green logistics, Francisco and Swanson (2018) investigated blockchain adoption in Walmart's supply chain. An overview of AI-integrated BIoT platforms in the food and pharmaceutical industries was given by Zhang et al. (2021). AI–BIoT systems will be at the forefront of innovation as supply chains start to move toward data-driven, decentralized, and sustainability-first models. Future research topics include cross-chain interoperability protocols for international cooperation, AI-powered carbon accounting, and quantum-secured BIoT.

1.9 SMART SUPPLY CHAIN AI

1.9.1 Empowering smart supply chains

By combining BIoT technology with AI, it is possible to turn existing supply chains into smart, distributed, and sustainable ecosystems. BIoT gives you the tools to automate tasks, decentralize devices, and send data reliably. AI, on the other hand, gives you cognitive and analytical skills. The synergetic system that AI and BIoT create can now address important problems such as security holes, environmental challenges and unsustainable performance. This system can sense things in real time, learn on its own, and prove its success. Aggarwal and Kalra (2022) demonstrate that quantum-enhanced AI models can significantly improve large-scale supply chain optimization problems.

The most significant benefit of merging AI and BIoT is that it makes operations easier to monitor. Big data analytics capabilities significantly enhance firm performance by strengthening dynamic capabilities in supply chain operations (Wamba et al., 2017). IoT devices all around the supply chain often capture information on temperature, humidity, location, vibration, machine performance, and other factors. Blockchain ensures that the data are protected and can't be changed, and AI analyzes it to find and predict problems, and ensure resources are used wisely. For example, temperature sensors monitor conditions in real time in cold chain logistics. Blockchain guarantees that every recorded event is true and comes from a trusted source. It helps you keep things safe before they break, and AI algorithms analyze both historical and real-time data to assess the likelihood of product waste. This reduces product loss and significantly improves quality control.

Another significant benefit is that smart contracts and intelligent decision-making make things easier. BIoT systems use smart contracts to automate essential parts of the supply chain, such as checking for compliance, settling payments, and rearranging inventories. When AI is integrated, these contracts become flexible instead of fixed. When AI is integrated, these smart contracts become adaptive rather than fixed, automatically updating contract terms based on real-time events. For instance, it might change prices based on the risk of a supplier or place a renewal order when expected demand is higher than safety stock. This cognitive automation speeds up transactions, makes them more consistent, and boosts confidence without any help from people.

1.9.2 The strategic role of AI–BIoT integration

AI–BIoT technologies enhance safety and trustworthiness. IoT networks are often vulnerable to hacks, data deception, and device impersonation. Blockchain lowers these risks by making data encryption and decentralized authentication without the possibility of tampering. AI offers another layer of proactive defense by employing ML to detect strange behavior, patterns of infiltration, or illegal device activity. This two-layer protection plan prevents assaults such as replay, man-in-the-middle, and Sybil attacks. This is very important in industries such as healthcare, defense, and finance. An AI–BIoT network can find and isolate a sensor node that is acting suspiciously in a pharmaceutical supply chain. For example, it may be sending false temperature readings. Blockchain also keeps the rest of the ledger safe at the same time.

AI and the IoT also help protect the environment and make it last longer. When IoT sensors monitor energy use, emissions, and material waste in real time, AI models can suggest ways to improve things, such as changing production schedules to avoid energy peaks or reorganizing logistics routes to reduce carbon emissions. Blockchain makes ensuring that important sustainability metrics may be audited, shared, and followed by the rules. This is notably useful for environmental reporting, managing carbon credits, and getting green supply chain certification in frameworks such as GRI, CSRD, or ISO 14001.

Transparency with consumers and ethical sourcing are further made possible by the integration. Verified information on product origin, labor practices, and environmental impact is becoming increasingly important to consumers. AI is capable of sentiment analysis, preference prediction, and customized sustainable product recommendations. The data can be kept on-chain and retrieved through web portals or QR codes using BIoT. Customers can increase consumer trust, brand loyalty, and market distinction by scanning a product to learn more about its farm source, carbon

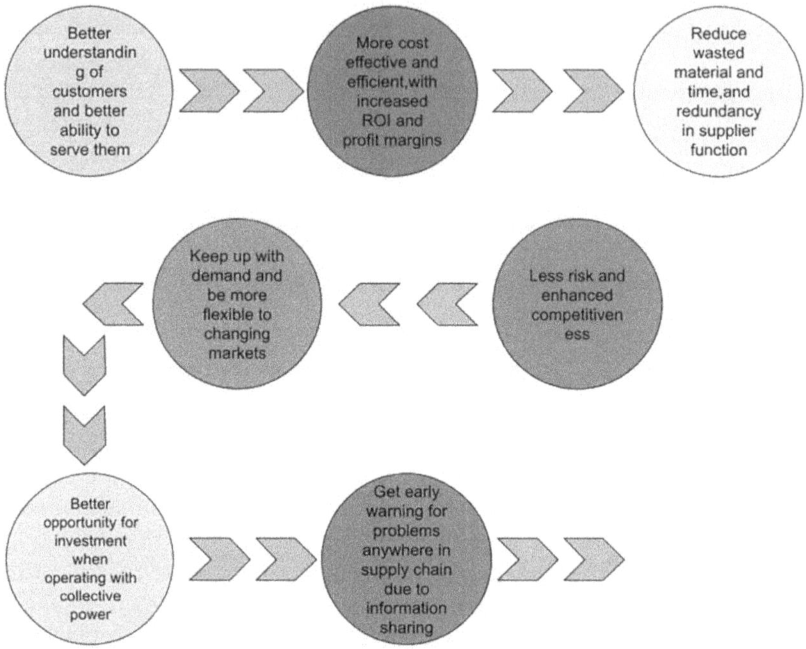

Figure 1.3 The supply chain lifecycle.

footprint, and handling circumstances. Figure 1.3 presents the advantages of supply chain sustainability through AI–BIoT integration.

Decentralization and scalability are two more important advantages. Without depending on a central authority, AI–BIoT systems can expand horizontally by including additional edge nodes, devices, or networks. Blockchain makes ensuring that events and updates are consistently synchronized throughout the network, while federated AI models enable dispersed learning across many IoT devices or corporate locations. In multi-tiered supply chains, where coordination between several stakeholders such as farmers, manufacturers, carriers, and retailers is crucial but has previously been fragmented, this is especially helpful.

AI–BIoT systems improve asset utilization and decision accuracy while lowering manual labor, inventory holding costs, and downtime. AI, for example, can reduce stockouts by more than 30%, while sensor-based predictive maintenance can minimize downtime by as much as 50%. Blockchain decreases administrative overhead by reducing form-filling and settlement time. When taken as a whole, these benefits result in high productivity and sustainability indicators, which support the foundation for corporate investment.

These advantages are supported by recent research. In their paper "AI-enabled BIoT Platforms for Smart Industries" (IEEE Access), Zhang et al. (2020) define how this incorporation alters the energy markets, food supply chains, and logistics. According to Wang et al. (2020), who examine 52 industrial use cases, AI–BIoT systems incessantly enhance sustainability, traceability, and efficiency. Saberi et al. (2019) show that blockchain-enabled traceability, when used with AI-driven decision assistance, improves ESG transparency and carbon accountability. Another example is Power Ledger's AI–BIoT energy trading platform, which promotes the usage of decentralized green energy while lowering transaction costs by 20%–30%.

There are numerous benefits to combining AI with BIoT, and they go beyond just making operations better. These technologies can help develop robust, smart, and sustainable digital ecosystems that help supply chains meet customer needs, climate action goals, and ESG standards. This combination elevates the standard for digital infrastructure in the 21st century in many ways, including automation, security, ethical assurance, and extrapolative vision. As these technologies get better, their effects on sustainable innovation grow increasingly important in many fields. This is especially true with the arrival of 5G, digital twins, and quantum computing.

1.10 TECHNICAL ARCHITECTURE OF AI–BIoT SYSTEMS

1.10.1 Technical architecture

Using BIoT technology with AI can turn old supply chains into fast, decentralized, and long-lasting ecosystems. BIoT sets up automatic processes, decentralized devices, and consistent data distribution. AI, on the other hand, adds advanced thinking, analysis, and the ability to make guesses about what will happen. The interdependent system that AI and BIoT create allows organizations to deal with time-sensitive problems, including security flaws, environmental violations, and performance difficulties related to sustainability. This system can sense things in real time, learn on its own, and carry out tasks that have been tested (Liu et al., 2021). Figure 1.4 depicts the layered architecture of AI–BIoT systems in supply chains.

1.10.2 Enhancing supply chain visibility

One of the best things about merging AI and BIoT is that it gives you more visibility into your operations. IoT devices all around the supply chain are always collecting data on temperature, humidity, location, and more. Blockchain guarantees that this information is safe and can't be changed. AI analyzes it to find problems, predict problems, and ensure resources

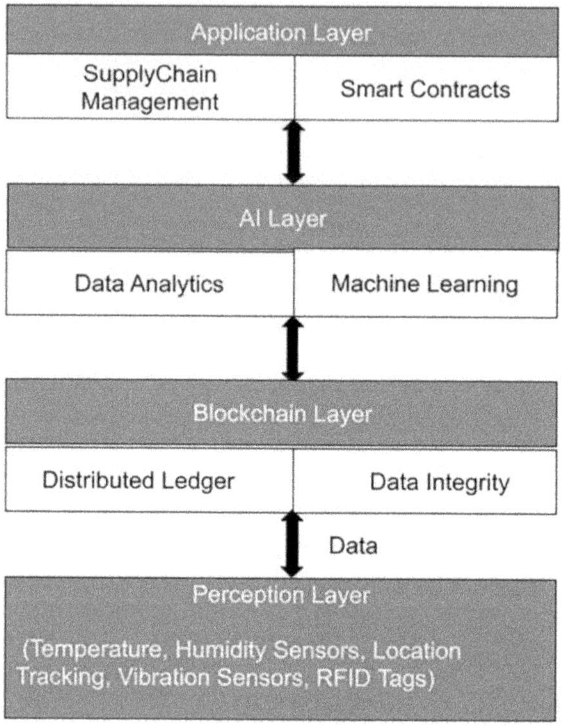

Figure 1.4 Layered architecture of AI–BIoT.

are used as well as possible. In cold chain logistics, temperature sensors, for instance, monitor things in real time. Blockchain ensures that every recorded event is real and comes from a reliable source. This lets people get involved before a product is damaged. AI algorithms, on the other hand, analyze the risk of pampering based on old and relevant data. This reduces product loss and greatly improves quality control (Panarello et al., 2018; Wang et al., 2020). Another significant benefit is that smart decision-making and smart contracts can make things automatic. BIoT systems use smart contracts, which are digital protocols on blockchain that run themselves, to automate essential supply chain tasks including making payments, checking for compliance, and rearranging stock. When AI is incorporated, these contracts change from being fixed to being flexible. AI alters the limitations of a contract based on what is happening right now. For example, it can raise prices if a dealer is risky or place a new order when demand is stronger than the supply that is already available. This smart automation speeds up transactions, makes them more reliable, and ensures they go through without any help from people (Beynon-Davies et al., 2020).

1.10.3 The transformative impact

AI–BIoT technologies make things a lot safer and more trustworthy. IoT networks are often vulnerable to hacks, data spoofing, and device impersonation. Blockchain lowers these risks by making it possible for data to be unchangeable, encrypted, and authenticated in a decentralized way. ML is used by AI to discover odd behavior, infiltration patterns, or unlawful device activity. This adds another layer of safety. This dual-protection model is very helpful in areas such as healthcare, defense, and finance since it stops assaults such as replay, man-in-the-middle, and Sybil attacks. For example, in a pharmaceutical supply chain, an AI–BIoT network can find and isolate a sensor node that is acting oddly, such as sending false temperature readings. Blockchain also protects the rest of the ledger at the same time (Panarello et al., 2018).

AI and BIoT also promote taking care of the environment and being sustainable. AI models can recommend ways to make things more efficient when IoT sensors are used to keep track of energy consumption, emissions, and material waste in real time. For example, they might suggest changing delivery routes to reduce carbon emissions or changing manufacturing schedules to avoid energy peaks. Blockchain guarantees the auditability, shareability, and regulatory compliance of key sustainability measures. In frameworks such as GRI, CSRD, or ISO 14001, this type of data is beneficial for environmental reporting, carbon credit management, and green supply chain certification (Saberi et al., 2019).

Transparency with consumers and ethical sourcing are further made possible by the integration. Consumers are increasingly demanding verified information on product origin, labor practices, and environmental impact. AI is capable of sentiment analysis, preference prediction, and customized sustainable product recommendations. The data can be kept on-chain and retrieved through web portals or QR codes. Customers can increase consumer trust, brand loyalty, and market distinction by scanning a product to learn more about its farm source, carbon footprint, and handling circumstances (Saberi et al., 2019).

1.10.4 Scalable and decentralized supply chains

Decentralization and scalability are two more important advantages. Without depending on a central authority, AI–BIoT systems can expand horizontally by including additional edge nodes, devices, or networks. Blockchain ensures that events and updates are consistently synchronized throughout the network, while federated AI models enable distributed learning across many IoT devices or corporate locations. In multi-tiered supply chains, where coordination between several stakeholders—such as farmers, manufacturers, carriers, and retailers—is crucial but has previously been fragmented, such integration is beneficial (Liu et al., 2021).

These advantages are supported by recent research. In "AI-enabled BIoT Platforms for Smart Industries" (IEEE Access), Zhang et al. (2020) define how this incorporation changes the energy markets, food supply chains, and logistics. According to Wang et al. (2020), who examine 52 industrial use cases, AI–BIoT systems endlessly improve sustainability compliance, traceability, and efficiency. Saberi et al. (2019) show in their study "Blockchain Technology and Its Relationship to SSCM" that blockchain-enabled traceability can improve ESG transparency and carbon accountability when used with AI-driven decision assistance. Power Ledger's AI–BIoT energy trading platform is another example. It encourages the use of green energy that is spread out and reduces transaction costs by 20%–30% (Saberi et al., 2019; Zhang et al., 2020).

Combining AI with BIoT has several advantages, and they go beyond merely making things operate better. It builds digital places that are strong, smart, and long-lasting. This helps supply chains satisfy customer needs, climate action goals, and ESG principles. Adding AI to BIoT makes digital setups in the 21st century even better by adding characteristics such as automation, security, ethical assurance, and analytical observation. As these technologies improve, their combined benefits for sustainable innovation will become more and more important in many fields, especially with the arrival of 5G, digital twins, and quantum computing (Liu et al., 2021; Beynon-Davies et al., 2020).

1.11 AI–BIoT FOR SUPPLY CHAIN SUSTAINABILITY

1.11.1 Autonomous and cognitive supply chains

AI and BIoT are swiftly coming together to make systems that are self-governing, long-lasting, and highly connected. These systems will help supply chains in the future. One of the most important changes is the advent of supply chains that can work on their own and see what's going on. These systems add Distributed Ledger Technologies, ML techniques like Long Short-Term Memory (LSTM) and Transformer models, and predictive analytics to create dynamic, self-optimizing networks. Those supply networks can change the direction of deliveries on their own, discover any problems before they occur, and improve inventory control without any help from people. Simulation-based studies demonstrate how epidemic outbreaks significantly disrupt global supply networks, highlighting the importance of predictive and AI-driven resilience models (Ivanov, 2020). Edge computing analyzes real-time sensor data that measures temperature, humidity, and vibration from moving items. The data are then coordinated with blockchain to ensure that records are always accurate and consistent. Smart contracts reduce transaction costs and eliminate delays by automating logistics trials and payment settlements (Lu, 2020).

1.11.2 AI-driven circular supply chains

AI-driven circular supply chains are a step toward systems that are more fair and long-lasting. AI is very important for figuring out the chances of reusing, remanufacturing, and recycling a product throughout its life cycle. Digital twins can help AI use more materials by replicating and studying real-world activities and belongings. Intelligent manufacturing frameworks under Industry 4.0 integrate cyber-physical systems, IoT, and AI to enable smart production ecosystems (Zhong et al., 2017). Blockchain-based material passports keep supply systems running in a closed loop by ensuring that assets can be tracked back to where they came from. Models for lifecycle valuation forecast how things will affect the environment and help with eco-friendly restyles. These systems are necessary for achieving net-zero emission objectives (Tseng et al., 2021). Digital twins, ML, and blockchain all work together to support smarter and more dependable ways to make and use things.

1.11.3 Federated AI with Edge-IoT

Federated AI with Edge-IoT is being used by supply chains that are emerging to deal with rising concerns about data privacy, latency, and bandwidth. This methodology decentralizes ML so that AI models can be developed locally at IoT nodes instead of relying on centralized cloud companies. Federated learning solutions secure data authority by ensuring that raw data never leave the originating device, lowering privacy risks, and following local data protection rules. Blockchain creates a safe space for combined intelligence by ensuring that model changes are agreed upon and accurate across separate nodes. These technologies provide real-time analytics and edge decision-making for fast manufacturing and logistics processes (Yang et al., 2019).

1.11.4 Federated AI and Edge-IoT

Emerging supply chains are adopting Federated AI with Edge-IoT to resolve increasing concerns about data privacy, latency, and bandwidth. Instead of depending on centralized cloud organizations, this model decentralizes ML, which allows AI models to be trained locally at IoT nodes. By ensuring that raw data never leave the source device, reducing privacy threats, and adhering to local data protection guidelines, FL solutions preserve data authority. Blockchain creates a safe environment for combined intelligence by ensuring consensus and integrity in model updates across isolated nodes. For high-speed manufacturing and logistics processes, these technologies offer real-time analytics and edge decision-making (Yang et al., 2019).

1.11.5 Quantum blockchain and AI integration

Imminent growth in blockchain integration and quantum AI suggests previously unknown levels of computational power and cryptographic security. Massive predictive analytics, quick optimization problem-solving, and anomaly detection in intricate supply chain networks are all possible with quantum-enhanced AI. At the same time, BIoT infrastructures will be protected from new dangers brought about by quantum decryption capabilities in quantum-secure blockchains built on post-quantum cryptographic algorithms. Quantum sensors will provide ultra-precise environmental tracking in IoT networks, enabling micrometer-level accuracy in shipment verification and inventory management (Aggarwal & Kalra, 2022). Although these technologies are still in their infancy, they have the potential to alter supply chain systems significantly and worldwide.

1.11.6 Sustainable blockchain and green AI

Meanwhile, the development of green AI and sustainable blockchain practices is being propelled by the environmental impacts of blockchain and AI. Greener consensus algorithms including proof of stake and directed acyclic graphs are replacing more energy-intensive traditional blockchain methods such as proof-of-work. Pruning strategies, energy-efficient architectures, and transfer learning are some of the methods being used to train AI models with resource conservation in mind. IoT devices are essential for real-time carbon emission monitoring because they provide data to programmable smart contracts for carbon credit trading and immutable ledgers that facilitate auditable sustainability measures (de Vries, 2021). These green frameworks make AI–BIoT more responsible and intelligent by coordinating digital transformation with climate action objectives.

1.11.7 Interoperable and cross-chain BIoT systems

Lastly, future supply chains will not be isolated by fragmented ecosystems or proprietary technology, thanks to interoperable and cross-chain BIoT solutions. Blockchain platforms including Ethereum, Hyperledger, and VeChain will be able to communicate with one another through cross-chain protocols such as Polkadot, Cosmos, or custom bridges. AI-based protocol translators will facilitate compatibility, paving the way for global track-and-trace systems, asset registries, and shared digital identities. Interoperability across suppliers, stakeholders, and regulatory agencies around the world improves openness, auditability, and collaboration. According to Christidis and Devetsikiotis (2016), it guarantees that even intricate multi-tier networks that span continents can function with a single perspective on performance and truth. When taken as a whole, these developments represent a significant move toward supply networks that are intelligent, decentralized, safe,

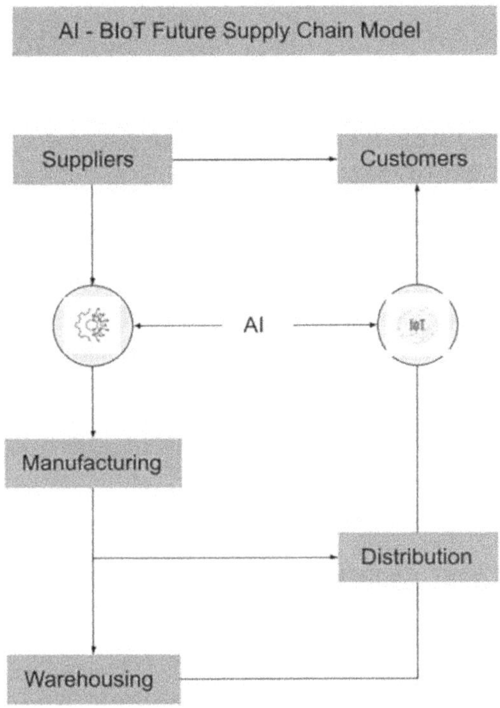

Figure 1.5 Future supply chain model for AI–BIoT.

and sustainable. Future supply chains will be more ethical, inclusive, and ecologically sensitive while also becoming more robust and efficient as AI–BIoT capabilities develop and converge with quantum computing, federated learning, and green infrastructure. Figure 1.5 depicts the future supply chain model for AI–BIoT.

1.12 RECOMMENDATIONS FOR THE ADOPTION OF AI AND BIoT IN SUSTAINABLE SUPPLY CHAINS

1.12.1 Develop a clear digital sustainability roadmap

To make AI–BIoT systems work in sustainable supply chains, organizations need to take a proactive, all-encompassing, and cutting-edge strategy. A clear plan for the digital revolution that fits with environmental goals sets the stage for the trip. This roadmap should include AI–BIoT experiences at various points in the supply chain, such as sourcing and procurement, logistics, and compliance monitoring. It should also include sustainability KPIs such as carbon emissions, waste minimization, and compliance with ethical sourcing. This change should follow internationally accepted standards, such as ISO 20400 and the United Nations' SDGs, to ensure that best practices and rules are followed (Jabbour et al., 2019).

1.12.2 Pilot developments with clear return on investment

Businesses should first embrace prototype projects instead of full-scale organizations to lower risks and ensure they will work. These pilots should focus on areas with low risk and high transparency that have a measurable return on investment. For instance, blockchain-based smart contracts for managing inventory, IoT sensors for tracking emissions in real time, or AI-driven demand forecasting to reduce waste and extra stock. Successful pilot programs not only show that a technology is possible, but they also show that it is worth the expense, which leads to more rollouts (Min, 2019).

1.12.3 Invest in scalable, interoperable infrastructure

Infrastructure is very important for AI–BIoT systems to be able to grow and work with other systems. Cloud-native platforms that support blockchain contexts (e.g., Hyperledger and Ethereum) and standard IoT communication protocols (e.g., MQTT and OPC-UA) are necessary for smooth integration across multiple investors and systems. To avoid being locked into a certain technology and to make it easy to upgrade quickly, companies should give open-source, vendor-agnostic, and API-based platforms top attention. Digital twins and federated AI learning models can make the system even more responsive while keeping data private and lowering latency (Casino et al., 2019).

1.12.4 Foster cross-functional collaboration

Cooperation across different departments is also very crucial. The IT and supply chain industries cannot handle the successful use of AI and IoT in sustainable supply chains on their own. However, outside partners, legal, financial, and sustainability partners, and shared governance models must all operate together. Investing in team-wide training improvements and creating a BIoT adoption task force encourages shared responsibility and understanding. Corporate innovation challenges and award programs can also be good ways to motivate employees to come up with new ideas and solutions (Jabbour et al., 2019).

1.12.5 Data governance and privacy frameworks

AI–BIoT is an area that uses a large amount of data; hence, strong data governance frameworks are quite important. These frameworks must define encryption methodologies, privacy regulations (including GDPR), and data ownership concepts. Smart contracts and blockchain governance systems should also include terms for data auditability, access restriction, and dispute resolution to keep the system simple and trustworthy (Casino et al., 2019).

1.12.6 Empower suppliers and SMEs

Also, widespread use is quite important, especially for SMEs that may not be very tech-savvy yet. To ensure that the network is up and running for a long time, businesses need to give SMEs affordable sensor kits, government-backed allowances, and shared setup. To remove legal and technological impediments to admission, employ advanced contract templates and consortium blockchain contribution mechanisms. This new technology makes data exchange and traceability fairer for all levels of providers, including Tier 2 and Tier 3 vendors (Min, 2019).

1.12.7 Merge with ESG reporting and audit processes

Finally, AI–BIoT outputs need to be directly included in ESG reports. Real-time environmental monitoring, blockchain certification of compliance with labor standards, and AI-driven anomaly detection for greenwashing can all improve ESG reporting and audit training. This improves brand equity and investor self-reliance, as well as helping businesses meet regulatory requirements (Eccles et al., 2014).

1.12.8 Work with regulators and standards organizations

Groups also need to work closely with standards groups and authorities to build digital tools such as carbon tokens, digital product passports, and methods for grading sustainability. Active involvement in improving these standards will ensure early compliance and industry leadership. You need to employ AI–BIoT dashboards to monitor KPIs such as on-time delivery rates, supplier ethical compliance, and carbon footprint per delivery. To ensure accountability and openness throughout the supply chain, use permissioned blockchain networks. Auditors, internal teams, and even customers should be able to access these dashboards (Casino et al., 2019).

Using AI and BIoT technology in supply chains isn't a one-time thing; it's a strategy that takes time. By integrating digital transformation with sustainability goals, companies may turn regulatory and operational problems into opportunities for growth, trust, and flexibility. The suggestions in this article give businesses a practical, evidence-based plan for not only adopting AI–BIoT tools but also for running a sustainable firm in the future.

AI–BIoT systems can change global supply chains into clear, circular, and ethically savvy ecosystems if they are planned, partnered with, and governed correctly.

1.13 APPLICATIONS OF AI AND BIoT FOR SUSTAINABLE SUPPLY CHAINS: EXAMPLES

Several real-world case studies in global supply chains have shown that the combination of BIoT and AI technologies has evolved from a theoretical idea to a practical application. These deployments show how producers have successfully leveraged the AI–BIoT partnership to improve operational flexibility, sustainability, and transparency while ensuring that their supply chain models are in line with advanced regulatory frameworks and the UN's SDGs (Beynon-Davies et al., 2020).

The IBM Food Trust is a well-known case initiated by Walmart, Nestlé, and Unilever. The software allows tracking of food items from farms to customers. We leverage blockchain technology, IoT devices, and AI analytics to achieve this. When IoT devices are kept or moved, they monitor temperature, humidity, and location. These readings are stored on a blockchain that only a few individuals can access, so they can't be modified. AI systems that analyze sensor data in real time can help reduce food waste and carbon emissions by identifying problems, predicting waste, and optimizing route planning. Walmart was able to find out where the diseased mangoes came from in around 2.2 s instead of 7 days, thanks to this technology. This helped them reduce food waste and fix safety problems more quickly (Francisco & Swanson, 2018; Zhang et al., 2021).

Another good point is the TradeLens platform, which IBM and Maersk worked together to create to digitize international shipping using blockchain and the IoT. TradeLens works with IoT-enabled containers that monitor temperature, vibration, location, and handling conditions in real time. AI models analyze this information to find risks, predict delays, and make the fleet more efficient. The blockchain keeps track of everything that happens, from loading to customs clearance to delivery. This makes it possible to audit everything from start to finish. TradeLens made marine logistics more efficient and less harmful to the environment by lowering the time it took to process documents by more than 40% in a pilot project with U.S. Customs and Border Protection. The system also supports smart contracts, which encourage transportation methods that are good for the environment and society by automatically making payments when delivery requirements are met (Saberi et al., 2019; Zhang et al., 2021). Table 1.5 shows how AI–BIoT systems are being deployed all across the world in sustainable supply chains in the food, pharmaceutical, energy, and logistics industries.

Table 1.5 AI–BIoT systems

Category	*Applications of AI and BIoT in supply chain*
Food	• Real-time tracking and monitoring • Spoilage prediction • Blockchain traceability
Pharmaceuticals	• Cold chain monitoring • Counterfeit detection • Predictive analysis
Energy	• Smart grid optimization • Decentralized energy trading • Predictive maintenance
Logistics	• Route optimization • Smart control • Fleet monitoring

Modum (Switzerland) created a BIoT solution for cold chain monitoring in the pharmaceutical sector. IoT sensors built into packaging track temperature and location when temperature-sensitive medications and vaccines are being transported. Data immutability is ensured by having these recorded on a blockchain ledger. Temperature excursion hazards are predicted by AI systems using logistical practices and external factors. Smart contracts immediately alert stakeholders or stop distribution to avoid spoilage when the system notices deviations. Numerous logistics companies in Europe have embraced this approach to improve patient safety, minimize product loss, and adhere to GDP rules. The Modum solution guarantees data trust between suppliers, logistics providers, and regulators while reducing the need for manual audits (Zhang et al., 2021).

AgriDigital, an Australian supply chain platform, uses BIoT and AI in agriculture to facilitate financial automation and farm-to-fork traceability. We collect environmental and grain quality data using drones and IoT devices. Verified records of ownership, handling, and storage are stored on the blockchain. AI programs automate pricing recommendations and assess supply–demand trends. Once predetermined delivery circumstances are confirmed using sensor data, smart contracts immediately transfer payment when grain is sold. By lowering credit risk and advancing fair trade, this automation has shortened payment periods from 30 days to a few minutes. Additionally, it reduces losses from fraud or spoilage in isolated farming regions with inadequate infrastructure (Zhang et al., 2021; Wang et al., 2020).

The Power Ledger project in Australia provides a forward-thinking example from the energy sector. Households with solar panels and IoT smart meters may monitor energy production and consumption with this blockchain-based energy trading platform. Peer-to-peer trading of excess electricity is made possible by AI algorithms that predict patterns of generation and consumption. Blockchain tracks every transaction to maintain trust and eliminate the need for intermediaries. This program supports a decentralized green energy economy, lowers transmission losses, and encourages the use of renewable energy locally. In addition to being financially motivated to contribute to a more sustainable grid, households can lower their carbon impact (Wang et al., 2020).

The Walmart Food Supply Chain and IBM Food Trust estimate waste and maximize delivery. Blockchain logging and traceability via sensors decreased food waste, openness, and customer confidence. IBM & Maersk Maritime Logistics' TradeLens forecasts delays and improves routes. Container tracking, automated documentation, paperless shipping, reduced fuel use, and lower pollutants. Modum (Switzerland) Pharma: The cold chain forecasts temperature variations, smart contract alerts, and sensor-monitored delivery adherence, decreased spoilage, and enhanced patient safety. Digital agribusiness forecasts demand and optimizes prices. The use of IoT data at the farm level and blockchain technology for transactions is essential. Equitable compensation, enhanced rural accessibility, and quality monitoring. Ledger of Power: Sustainable energy blockchain energy trading, IoT smart meters, and energy forecasting. The decentralized adoption of green energy contributes to a lower carbon footprint.

In addition to these examples, Alibaba, Amazon, and DHL have investigated AI–BIoT interfaces for autonomous cars and smart warehouses in logistics, emphasizing waste reduction, energy efficiency, and real-time decision-making. Alibaba's ET Industrial Brain optimizes manufacturing energy utilization through ML, and blockchain ensures that resource usage and emissions data are traceable and validated. DHL employs BIoT platforms and AI-driven robotics to manage its warehouses better. Predictive maintenance and environmental sensors are used to reduce downtime and energy consumption (Wang et al., 2020).

These case studies demonstrate several shared benefits:

- Blockchain's immutability and data trust lower the risks of fraud, tampering, and noncompliance (Saberi et al., 2019).
- Proactive sustainability initiatives, as opposed to reactive ones, are made possible by real-time insights from AI (Wang et al., 2020).
- Smart contract automation increases process consistency and lowers human error (Zhang et al., 2021).
- Verified supply chain histories encourage responsible consumption through consumer openness (Francisco & Swanson, 2018).
- Nevertheless, there are still implementation issues.

These include the necessity for legislative clarification, particularly when AI is utilized in delicate industries such as healthcare and banking, the incompatibility of older systems with blockchain platforms, and the high cost of infrastructure. Furthermore, data privacy is a critical issue, especially with public blockchains. To strike a compromise between decentralization and confidentiality, several systems are using federated learning, permissioned ledgers, and zero-knowledge proofs. Strong network access is also necessary for scaling AI–BIoT systems, which may be a barrier in rural or underdeveloped locations (Zhang et al., 2021).

These incidents have been examined in several academic publications. While Saberi et al. (2019) mapped blockchain's role in green logistics, Francisco and Swanson (2018) investigated blockchain adoption in Walmart's supply chain. An overview of AI-integrated BIoT platforms in the food and pharmaceutical industries was given by Zhang et al. (2021). AI–BIoT systems will be at the forefront of innovation as supply chains start to move toward data-driven, decentralized, and sustainability-first models. Future research topics include cross-chain interoperability protocols for international cooperation, AI-powered carbon accounting, and quantum-secured BIoT.

REFERENCES

Accenture (2022), Smart Supply Chains Powered by IoT. https://www.accenture.com.

Aggarwal, R., & Kalra, A. (2022), "Quantum-enhanced AI for supply chains", *IEEE Internet of Things Journal.*

Beynon-Davies, P., Han, J., & Wang, Y. (2020), "Literature review and research agenda on Blockchain & AI in sustainable supply chains", *Computers & Industrial Engineering*, 149, 106867. https://www.sciencedirect.com/science/article/pii/S0360835219307503.

Brandenburg, M., Govindan, K., Sarkis, J., & Seuring, S. (2014), "Quantitative models for sustainable supply chain management: Developments and directions", *European Journal of Operational Research*, 233(2), 299–312.

Brous, P., Janssen, M., & Herder, P. (2019), "The dual effects of the Internet of Things (IoT): A systematic review of the benefits and risks of IoT adoption by organizations", *International Journal of Information Management*, 49, 1–14. https://doi.org/10.1016/j.ijinfomgt.2019.03.006.

Choi, T. M., Wallace, S. W., & Wang, Y. (2021), "Big data analytics in operations management", *Production and Operations Management*, 30(7), 2107–2125. https://doi.org/10.1111/poms.13365.

Christidis, K., & Devetsikiotis, M. (2016), "Blockchains and smart contracts for the Internet of Things", *IEEE Access*, 4, 2292–2303. https://doi.org/10.1109/ACCESS.2016.2566339.

Conoscenti, M., Vetro, A., & De Martin, J. C. (2017), Blockchain for the Internet of Things: A Systematic Literature Review. IEEE/ACS AICCSA. https://ieeexplore.ieee.org/document/7945024.

De Vries, A. (2021), "Bitcoin's growing energy problem", *Joule*, 3(4), 1–5.

Dorri, A., Kanhere, S. S., & Jurdak, R. (2017), "Towards an optimized Blockchain for IoT", In: Proceedings of the Second International Conference on Internet-of-Things Design and Implementation (IoTDI), Pittsburgh, PA, USA. https://doi.org/10.1145/3054977.3055003.

Eccles, R. G., Ioannou, I., & Serafeim, G. (2014), The Impact of a Corporate Culture of Sustainability on Corporate Behavior and Performance. Harvard Business School Working Paper No. 17950, National Bureau of Economic Research, Cambridge, MA.

F. Casino., Dasaklis, T. K., & Patsakis, C. (2019), "A systematic literature review of Blockchain-based applications: Current status, classification and open issues", *Telematics and Informatics*, vol. 36, 55–81.

Francisco, K., & Swanson, D. (2018), "The supply chain has no clothes: Technology adoption of Blockchain for supply chain transparency", *Logistics*, 2(1), 2. https://doi.org/10.3390/logistics2010002.

IBM. (2021), AI in Supply Chain Risk Management. IBM White Paper. https://www.ibm.com.

Ivanov, D. (2020), "Predicting the impacts of epidemic outbreaks on global supply chains: A simulation-based analysis on the coronavirus outbreak (COVID-19/SARS-CoV-2) case", *Transportation Research Part E*, 136, 101922.

Jabbour, C. J. C., de Sousa Jabbour, A. B. L., Govindan, K., & Mellado, M. D. F. (2019), "Barriers and opportunities in sustainable supply chain management", *Journal of Cleaner Production*, 225, 763–783.

Lee, I., & Lee, K. (2015), "The Internet of Things (IoT): Applications, investments, and challenges for enterprises", *Business Horizons*, 58(4), 431–440. https://doi.org/10.1016/j.bushor.2015.03.008.

Liu, Y., Wang, Y., Zhang, Q., & Li, B. (2021), "AI and Blockchain for real-time agricultural supply chain management", *Journal of Cleaner Production*, 279, 123690. https://doi.org/10.1016/j.jclepro.2020.123690.

Lu, Y. (2020). "Industry 4.0: A survey on technologies, applications and open research issues",Journal of Industrial Information Integration, 6, 1–10. https://doi.org/10.1016/j.jii.2017.04.005

McKinsey Company. (2022), The Future of AI in Sustainable Supply Chains. https://www.mckinsey.com.

Mettler, M. (2016), "Blockchain technology in healthcare: The revolution starts here", In: *Proceedings of the 2016 IEEE 18th International Conference on e-Health Networking, Applications and Services (Healthcom)*, Munich, Germany, September 14–16, 2016, IEEE.

Min, H. (2019), "Blockchain technology for enhancing supply chain resilience", *Business Horizons*, 62(1), 35–45.

Mohanta, B. K., Jena, D., & Panda, S. S. (2019), "Blockchain technology and its applications to Artificial Intelligence", *Procedia Computer Science*, 132, 1809–1817. https://doi.org/10.1016/j.procs.2018.05.159.

Nguyen, D. C., Ding, M., Pathirana, P. N., & Seneviratne, A. (2021), "Blockchain and AI for smart healthcare: Emerging applications, challenges, and future directions", *IEEE Journal of Biomedical and Health Informatics*, 25(6), 2098–2110. https://ieeexplore.ieee.org/document/9378575

Panarello, A., Tapas, N., Merlino, G., Longo, F., & Puliafito, A. (2018), "Blockchain and IoT integration: A systematic survey" *Sensors*, 18(8), 2575. https://doi.org/10.3390/s18082575.

Panarello, A., Tapas, N., Merlino, G., Longo, F., & Puliafito, A. (2018), "Blockchain and IoT integration: A systematic survey", *Sensors*, 18(8), 2575. https://doi.org/10.3390/s18082575.

Reyna, A., Martín, C., Chen, J., Soler, E., & Díaz, M. (2018), "On Blockchain and its integration with IoT: Challenges and opportunities", *Future Generation Computer Systems*, 88, 173–190. https://doi.org/10.1016/j.future.2018.05.046.

Saberi, S., Kouhizadeh, M., Sarkis, J., & Shen, L. (2019), "Blockchain technology and its relationships to sustainable supply chain management", *International Journal of Production Research*, 57(7), 2117–2135. https://doi.org/10.1080/00207543.2018.1533261.

Saberi, S., Sarkis, J., Kouhizadeh, M., & Shen, L. (2019), "Blockchain technology and its relationship to sustainable supply chain management", *International Journal of Production Research*, 57(7), 2117–2135. https://doi.org/10.1080/00207543.2018.1533261.

Sarkis, J. (2020), Handbook on the Sustainable Supply Chain. Edward Elgar Publishing.

Seuring, S., & Müller, M. (2008), "From a literature review to a conceptual framework for sustainable SCM", *Journal of Cleaner Production*, 16(15), 1699–1710.

Tseng, M. L., Ha, H.-M., Tran, T.-P. T., Bui, T.-D., Chen, C.-C., Lin, C.-W., & Helmi Ali, M. (2021), "Circular economy meets industry 4.0: Transforming supply chains for sustainability", *Resources, Conservation and Recycling*, 170, 105615. https://doi.org/10.1016/j.resconrec.2021.105615.

UnitedNations Global Compact (2010), Supply Chain Sustainability: A Practical Guide for Continuous Improvement. United Nations Global Compact Office.

Wamba, S. F., Gunasekaran, A., Akter, S., Ren, S. J. F., Dubey, R., & Childe, S. J. (2017), "Big data analytics and firm performance: Effects of dynamic capabilities", *Journal of Business Research*, 70, 356–365.

WorldEconomic Forum. (2023), The Net-Zero Supply Chain Opportunity. World Economic Forum.

Yang, Q., Liu, Y., Chen, T., & Tong, Y. (2019), "Federated machine learning: Concept and applications", *ACM Transactions on Intelligent Systems and Technology*, 10(2), 1–19.

Zhang, Y., Deng, R. H., & Liu, X. (2020), "AI-enabled BIoT for smart industries: Architectures, applications, and challenges", *IEEE Access*, 8, 55519–55535. https://ieeexplore.ieee.org/document/9036825

Zhang, Y., Deng, R. H., & Liu, X. (2021), "AI-enabled Blockchain-IoT platforms for smart supply chains: Case analysis", *IEEE Access*, 9, 22305–22320. https://ieeexplore.ieee.org/document/9354820

Zheng, Z., Xie, S., Dai, H., Chen, X., & Wang, H. (2018), "An overview of Blockchain technology: Architecture, consensus, and future trends", *Proceedings of the IEEE*, 106(5), 777–795.

Zhong, R. Y., Xu, X., Klotz, E., & Newman, S. T. (2017), "Intelligent manufacturing in the context of Industry 4.0: A review", *Engineering*, 3(5), 616–630. https://doi.org/10.1016/J.ENG.2017.05.015.

Chapter 2

Integrating Artificial Intelligence and Blockchain-enabled IoT for sustainable supply chain practices

Usharani Selvaraj, Manju Bala Perumal, Rajkumar Kotteswaran, and Rajesh Rangan

2.1 INTRODUCTION

Blockchain-based Internet of Things and artificial intelligence (AI) are managing to bring unprecedented changes into supply chain management (SCM) with automation, sustainability, and transparency [1,2]. Due to supply chains' complex and international composition, systems that maximise operations must consider ethical governance, resilience against disruptions, and minimising environmental footprints [3,4].

While IoT gives real-time visibility into the movement, condition, and provenance of commodities [5,6], blockchain creates tamper-evident data records and verifies transactions in a trustless manner [7]. AI, for its part, imparts intelligence through anomaly detection, decision support, and predictive analytics [8,9]. Together, these technologies would be able to address the age-old issues of efficiency, fraud prevention, and supply chain traceability [10,11].

Recent happenings signal great promise for applications to manufacturing, pharmaceutical, and agricultural industries, where layered architectures gather data from IoT sensors, process it with edge-AI algorithms, and then use smart contracts to record verified results on blockchain ledgers [12,13]. Moreover, using federated learning (FL) and XAI to achieve transparency and small-scale privacy-preserving model training over geographically dispersed nodes becomes more frequent [14].

Despite these developments, very few comprehensive frameworks exist that consider the technological, environmental, and ethical concerns in SCM, as the convergence of AI, blockchain, and IoT remains fragmented [15]. This chapter, therefore, postulates an integrated, sustainable, and morally based AI–blockchain-enabled IoT (BIoT) supply chain architecture to fill this gap.

DOI: 10.1201/9781003660163-2

2.1.1 Problem statement

The global supply chains are faced with a never-ending set of challenges: absence of end-to-end visibility, siloed data, susceptibility to fraud, operational inefficiencies, and lack of environmental accountability. A lack of a single interoperable framework that could bind together various technologies limits their disruptive potential, even though AI, blockchain, or IoT adoption might have somewhat enhanced it. Moreover, issues about explainability, privacy-preserving analytics, or sustainability integration, which should be imperative in today's SCM environment, are lost in the present systems.

2.1.2 Research gap

Many studies have already focused on blockchain for traceability, IoT for monitoring, and AI for supply chain analytics, but there exists

- A lack of blockchain, IoT, and AI integration into a single SCM architecture.
- There is not enough emphasis on FL and XAI to balance privacy with trust and transparency in decision-making.
- Very few thorough frameworks address sustainability, governance, ethical concerns, and technology integration.
- There is an avoidable shortage of cross-sectoral and empirical testing of such integrated systems in the manufacturing, pharmaceutical, and agriculture sectors.

2.1.3 Objectives

This chapter's goals include

1. Discussing a uniform layered architecture that combines edge-AI analytics, blockchain-based validation, IoT data acquisition, and a smart contract for SCM.
2. Integrating explainable AI (XAI) to enhance the interpretability and trust of AI-based decision-making.
3. Using the FL approach to train decentralised privacy-preserving models across supply nodes scattered geographically.
4. Discussing social and ethical concerns, such as sustainability performance measures, frameworks for responsible AI, and data governance.
5. Illustrating their applicability through international cases in the manufacturing, medical, and agricultural sectors.

2.2 BACKGROUND

2.2.1 Definitions and conceptual framing

SCM is the coordinated set of activities wherein material, monetary, and informational flows associated with producing goods and services are formed through an interface of many business entities. Modern SCM stands for traceability, sustainability, resilience, and ethical compliance, beyond just cost and operational efficiency.

In supply chains, IoT manifests as a wide-ranging use of sensors and other devices that are connected to the network, enabling real-time data collection concerning asset location, environmental conditions (e.g., temperature and humidity), and operational status. According to Dutta, IoT is the "eyes and ears" of modern supply chains [16].

Blockchain is a distributed ledger that allows business rules to be programmed and enforced through smart contracts while providing tamper-proof and auditable transaction records. Vasileiou encourages using blockchain in SCM for provenance, anti-fraud, and trusted multiparty data sharing [17].

AI brings predictive analytics, anomaly detection, optimisation, and decision automation capabilities to SCM. Joint deployments of AI with IoT and blockchain seek to establish intelligent supply chains that are auditable and privacy-preserving, which are also known as BIoT or AI–BIoT.

XAI and FL are two emerging, cross-cutting axes that attempt to address explainability and privacy concerns arising from the deployment of AI in decentralised multi-partner supply networks.

2.2.2 IoT for visibility and monitoring

IoT is often said to be essential for visibility and real-time monitoring. To provide per-shipment condition tracking, a must for cold chain pharmaceuticals and perishable food, IoT devices feed position and environmental data. Their respective studies go further in arguing that a standardised device architecture and interoperability protocol should exist to avoid vendor lock-in and data silos. Nonetheless, practical studies have identified a series of real issues concerning security, scalability, and data quality while implementing IoT on a large scale.

2.2.3 Blockchain for provenance, trust, and automation

Multiparty trust is dispensed without the need for centralised intermediaries, and unalterable provenance from distributable ledger technologies. According to the empirical findings, blockchain improves traceability and can positively influence the measures of sustainable supply chain performance. A study demonstrated blockchain in conjunction with IoT QR/QR-type identifiers for food traceability. Such critics would claim that it is not a panacea since governance loopholes, data manipulation off-chain, and a lack of sensor integrity expose it to other vulnerabilities. Research highlights energy-intensive public consensus methods regarding the cost, energy overhead, and regulatory ambiguity.

2.2.4 AI for prediction, optimisation, and resilience

The use of AI for responsiveness (e.g., dynamic pricing and real-time rerouting), optimisation (routing, inventory), and prediction (demand forecasting, failure prediction) is shown to be of great advantage. AI-driven routing and operational scheduling optimisation algorithms aim to reduce carbon footprints. Edge-AI (local inference) has benefits for applications that cannot afford latency. However, research cautions against generalisability based on geographic markets and explainability (black-box models).

2.2.5 Convergence: AI, blockchain, and IoT (AI–BIoT)

For these three technologies, the architectural convergence is a matter of an expanding corpus of literature. According to the study, the layered designs are presented as follows: (i) cloud/aggregated analytics for cross-node learning; (ii) edge-AI for preliminary filtering/analytics; (iii) secure transmission and recording via blockchain smart contracts; and (iv) IoT edge for sensing. This convergence is related to decarbonisation through coordinated optimisation and transparency. Studies have identified promising food and pharmaceutical traceability prototypes wherein blockchain and XAI facilitate auditable AI decisions.

Some are still suggested concepts or restricted pilots, implying some empirical gaps—questions about regulatory alignment, integration costs, and systemic complexity.

2.2.6 XAI and FL in SCM

Privacy and explainability are the primary themes. An overview of XAI for IoT scenarios notes that interpretability is frequently necessary for audit, compliance, and stakeholder trust. For industries competing against each other (such as pharma or retail), which cannot share proprietary data, FL is advertised as a way to train models across decentralised supply chain partners without transferring any raw data. In recent approaches, XAI and FL are blended, with XAI offering interpretable local models while FL supports privacy-preserving, federated model updates. However, the literature also raises legal questions around ownership of origin, communication costs, and statistical heterogeneity [18].

2.2.7 Sustainability, ethics, and governance

Research concurs on sustainability and responsibilities. AI–BIoT systems/CISs could aid in a full-scale decarbonisation effort if sustainability indicators were to be integrated as optimisation targets and immutably validated by immutable ledgers. For background, openSC-like initiatives (discussed in practitioner forums) are examples of commercial attempts to bestow verified sustainable labels. However, these are not subject to enforceable governance frameworks, auditability, and stakeholder engagement. In that case, studies warn of greenwashing hazards, biased AI models, and governance issues that could nullify the intended societal benefits.

2.2.8 Sectoral perspectives: agriculture, pharmaceuticals, and manufacturing

Agriculture sheds some light on farm-to-fork traceability systems that reduce wastage while improving farmers' visibility; it also discusses some drawbacks in rural IoT connectivity and device upkeep. The pharmaceutical industry showcases reductions in cold chain failures and counterfeiting through blockchain-validated predictive models, but regulatory approval and clinical-grade verification remain hurdles. Manufacturing real-time event management automatically penalises suppliers and redirects them by merging edge-AI with smart contracts. This is enticing but very operationally difficult.

2.2.9 Synthesis and critical assessment

There seem to be common points mentioned in the literature: (i) blockchain, IoT, and AI are complementary technologies which, when combined, can create a disruptive effect on SCM; (ii) XAI and FL become important for privacy and trust in multi-partner settings; and (iii) the technologies can increase sustainability, but governance safeguards are needed to prevent adverse effects.

However, some of the practical limitations, such as the lack of cross-sector empirical validations, unresolved legal-governance issues, blockchain energy and scalability constraints, and the need for standardised interoperability frameworks, are also highlighted. This underlines the need for an ethically layered AI–BIoT architecture, as proposed in this chapter.

2.3 FOUNDATIONS AND ENABLING TECHNOLOGIES

2.3.1 Overview of AI, blockchain, and IoT

Machine learning and deep learning are disciplines of AI, allowing predictive analysis, optimising decision-making, and forecasting demand. Blockchain provides an immutable, decentralised ledger that increases security, traceability, and confidence in SCM processes. IoT worldwide is a fascinating capability for real-time data acquisition via connected sensors and devices to monitor products, machinery, and environmental conditions perpetually [19].

Together, these technologies produce secure and intelligent supply chains. Blockchain ensures that the knowledge and the transactions gleaned are verifiable and impregnable, whereas AI uses the data coming from IoT for better analytics. This integration improves supply chain visibility, reduces risks, and increases sustainability performance.

2.3.2 Interoperability and architecture of BIoT

Some of the significant issues in an IoT ecosystem, such as integrity, authentication, and interoperability, can be accounted for by BIoT integration. In decentralised consensus and cryptography, the blockchain removes the faults plaguing traditional IoT due to its centralised data management [20].

Typical components of BIoT architectures comprise layers for sensing (IoT devices), networking (data transfer), blockchain (data validation and smart contracts), and application (decision-making and analytics). Studies argue that smart contracts enable the automation of supply chain operations, trustless stakeholder relationships, and reduced human interference.

Edge-AI processing is an emerging technique that carries out data analysis physically closer to a data source to minimise latency and bandwidth consumption. This way, the dispersed supply chain networks attain scalable, real-time decision-making.

2.3.3 Motivation for integration in SCM

It's a tough, chaotic global economy; one must have resilience, sustainability, and transparency to integrate AI, blockchain, and IoT in the supply chain. Typically, traditional SCM is faced with activities in silos, the maintainability of traceability, and acting more reactively rather than predictively to issues [21].

Blockchain promises regimented data transfer between multiple parties, AI uses predictive models to decide proactively, and IoT provides constantly changing situational awareness. The integration of these technologies makes the supply chain more agile from an operational perspective, reduces

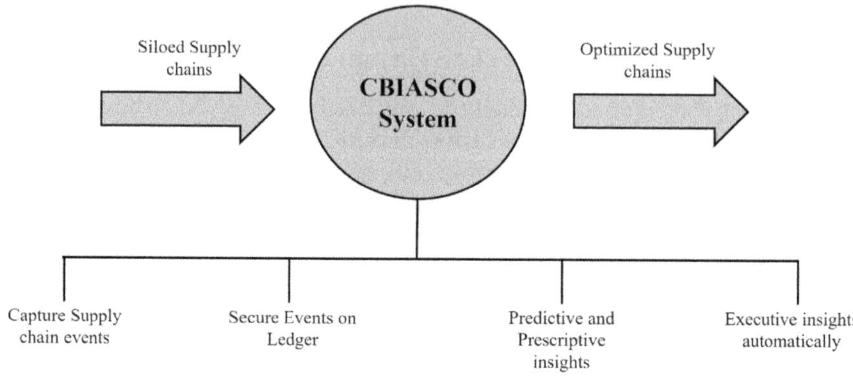

Figure 2.1 Cognitive blockchain–IoT–AI supply chain orchestration.

its environmental impact, and enforces compliance with environmental laws such as food safety regulations and carbon accounting.

This integrated approach further promotes XAI and responsible AI for openness in automated decision-making [22]. In data-heavy SCM contexts, this is important for sustaining stakeholder trust while still fulfilling ethical and governance commitments.

2.4 PROPOSED SYSTEM

Described in Figure 2.1, the cognitive blockchain–IoT–AI supply chain orchestration (CBIASCO) system is a next-generation supply chain setup that blends one trust mechanism, i.e., blockchain, with IoT sensing and AI analytics in a single adaptable ecosystem. The main objective was to construct a self-optimising, self-orchestrating supply chain featuring robustness, transparency, and real-time decision-making. The supply chain events are recorded with IoT sensors, verified and secured on the blockchain ledger, processed with AI models to gain prescriptive and predictive insights, and enforced using smart contracts.

Unlike traditional siloed systems, CBIASCO integrates technologies in a layered approach to ensure that orchestration facilitates smooth cooperation and execution among stakeholders, whereas optimisation pushes decisions at the operational level.

2.4.1 Proposed methodology—multi-layer cognitive optimisation

The multi-layered cognitive optimisation approach in Figure 2.2 illustrates the sequential optimisation process that enables CBIASCO to transform

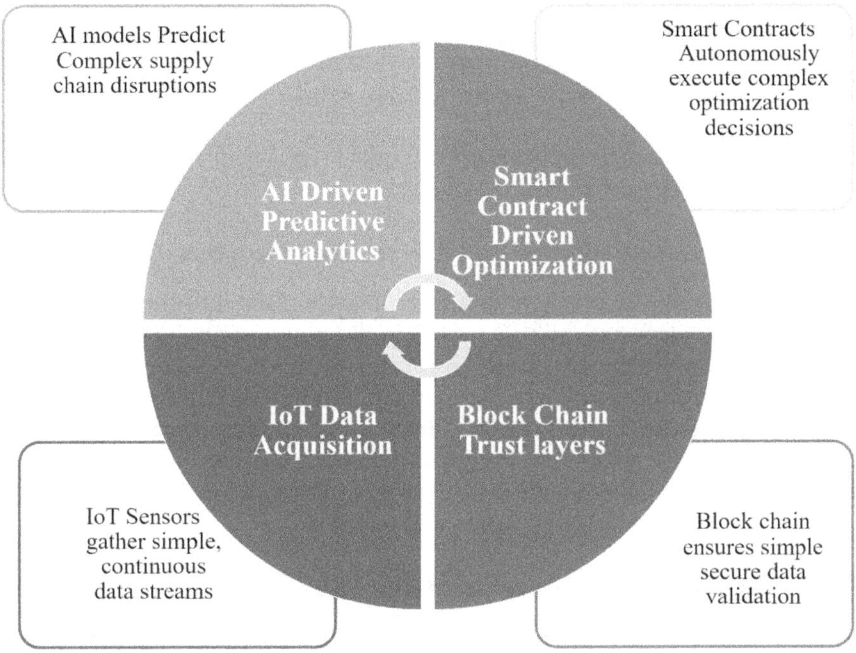

Figure 2.2 Multi-layer cognitive optimisation.

raw, heterogeneous data into precise, automated supply chain decisions. It is structured into four primary stages.

2.4.1.1 IoT data acquisition and edge pre-processing

To monitor variables continuously under location, temperature, humidity, shock, and asset health monitoring, the system uses a network of IoT sensors, RFID tags, GPS modules, and edge devices with computer vision capabilities. These lightweight AI models at the edge work to partially filter noise, perform anomaly detection locally, and compress data packets relevant to the cases of interest before transmission to the blockchain to reduce its processing overhead. Such edge intelligence saves bandwidth while reducing response time.

2.4.1.2 Blockchain trust and dynamic consensus switching

Events generated by IoT are channelled towards the blockchain trust layer, where cryptographic validation is carried out, and they are entered in an unalterable ledger. The dynamic consensus switching used by CBIASCO changes consensus methods depending on the transaction conditions:

a. Proof-of-Authority in the case of low-risk, low-threshold rapid-high volume upgrades
b. Proof-of-Stake for overall security in transactions
c. Byzantine Fault Tolerance for mission-critical, high-risk updates

This hybrid consensus guarantees an ideal trade-off between security, latency, and energy efficiency.

2.4.1.3 AI-driven predictive and prescriptive analytics

After verification, the blockchain data is incorporated into the AI cognitive layer. Adaptive decision-making occurs in dynamic supply–demand scenarios with reinforcement learning. Graph neural networks are used to map complex interrelationships among merchants, logistics centres, and suppliers. Time-series forecasting models are used to predict interruptions such as shortages in raw materials, surges in demand, or delays in deliveries [23]. The AI layer also generates prescriptive recommendations (what should be done) and predictive insights (what will happen).

2.4.1.4 Smart contract-driven optimisation

Smart contracts use AI-driven decisions to implement optimal actions autonomously, such as route reassignment, supplier switching, inventory redistribution, or dynamic pricing. After determining an optimisation through a decision-making process, these smart contracts run it automatically without any delay from a human perspective because they are self-enforcing and resistant to manipulation.

2.4.2 Intelligent orchestration architecture framework

The framework, depicted in Figure 2.3, emphasises how all system layers collaborate to ascertain the smooth transmission of information, decisions, and actions in the supply chain. It executes this orchestration through five interrelated layers.

2.4.2.1 IoT perception layer

The perception layer in IoT is often referred to by some researchers as the supply chain sensing system. In addition to gathering data, IoT devices communicate with AI agents at the edge to self-calibrate the network in the presence of environmental interference or sensor drift. The data is hashed, timestamped, and ready for blockchain ingestion.

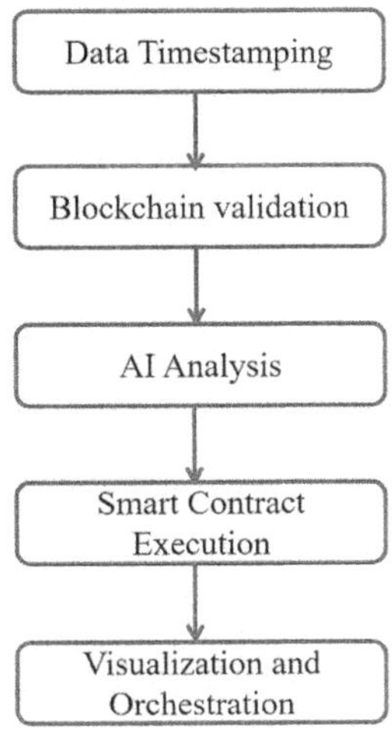

Figure 2.3 Intelligent orchestration architecture.

2.4.2.2 Blockchain ledger layer

The blockchain layer is the layer of trust on which the architecture is built. Each authenticated transaction is cryptographically signed and recorded in an immutable ledger that only authorised parties can view. During periods of varying demand or under cyber threats, the consensus-switching protocol ensures optimal throughput and resilience.

2.4.2.3 AI cognitive layer

AI models act like the system's intelligence. The context-aware AI layer generates scenario-specific recommendations based on supply chain parameters and external events (e.g., weather and geopolitical tensions).

2.4.2.4 Smart contract execution layer

The smart contract execution layer serves as an engine for autonomous execution purposes. Smart contracts and executed decisions oversee multiple agreements once the AI approves through automated payment, shipment, and compliance warnings, thereby warning against human intervention. If a disruption occurs during execution, the contract autonomously re-optimises through exception handling.

2.4.2.5 Orchestration and visualisation layer

The central command and coordination centre of CBIASCO. This layer controls cross-layer communication so that blockchain validations, AI judgements, and IoT updates appear in synchronised cycles. A role-based visualisation is accessible on the real-time dashboard: sustainability officers view carbon footprint measurements, logistics managers view operational maps, and executives view performance Key Performance Indicators (KPIs). Furthermore, this layer integrates a gamified sustainability reward system that assigns points and discounts to partners achieving green targets.

2.5 CASE STUDY—APPLICATION OF CBIASCO IN COLD CHAIN VACCINE DISTRIBUTION

To show the validity of the proposed CBIASCO framework, a case study was implemented for cold chain vaccine delivery, where exact environmental conditions are maintained and open data exchange is guaranteed. With blockchain providing for the readings in an immutable ledger, the integration of IoT sensors allows for real-time monitoring of temperature, humidity, and location, as well as high-level tracking, leading to the trust of stakeholders. The sensor data are analysed through AI and XAI techniques to identify anomalies in time for predictive interventions that can save the vaccines from spoiling. As logistics become greener and part of that larger movement, integration supports sustainability goals by eliminating waste, easing transportation problems, and lessening carbon emissions.

Blockchain smart contracts enforced role-based access to sensitive data and regulatory compliance criteria within a multi-stakeholder network that joined manufacturers, logistics companies, regulatory bodies, and healthcare institutions. This removed data silos and reduced reporting delays, thereby enhancing operational resilience. Moreover, the case study reflected the utility of a blockchain, IoT, and AI platform to assess supply chain responsiveness in vaccine distribution and address systemic inefficiencies that inhibit quick crisis management in medical catastrophes.

2.6 EXPERIMENTAL SETUP

An experimental setup established to test the CBIASCO framework was designed to mimic a cold chain supply network in the real world with multiple stakeholders, e.g., manufacturers, logistics companies, regulatory bodies, and distribution hubs. Since monitoring the environment accurately and continuously is a key feature, IoT-enabled sensors gathered real-time data on temperature, relative humidity, GPS location, and shock events during transportation. The data transmission to the blockchain ledger was ensured with secure communication protocol mechanisms, following which smart contracts enforced data validation rules and autonomously performed compliance checks on that data.

Historical and real-time sensor data were merged by XAI processes and AI models for anomaly identification and forecasting potential disruptions. The blockchain–AI synergy allowed for verifiable audit trails and speedy anomaly resolution, thus increasing trust, transparency, and effectiveness of decision-making. In order to test the resilience and adaptation of the system, the testbed simulated transportation scenarios such as delays, equipment failures, or environmental stressors.

To evaluate sustainability benefits, the system's routing algorithms manipulated delivery schedules to reduce energy consumption and carbon emissions. Tests were conducted to monitor the KPIs, such as the rate of temperature compliance, the time for data integrity verification, the accuracy of anomaly detection, the rate of delivery-time compliance, and the decrease of spoilage waste. Thus, this controlled experimental setting established the quantitative validation of CBIASCO's effectiveness in terms of cold chain transparency, operational resilience, and sustainability performance.

2.6.1 Experimental environment

The testbed, walled with IoT sensors, blockchain, and AI analytics, was set up to simulate cold chain supply chain scenarios. IoT environmental monitors emulated perishable logistics by continuously monitoring temperature, humidity, and GPS location. A permissioned Hyperledger Fabric network configuration was used for the deployment of the blockchain network to ensure data immutability and control access to it. For transparent decision-making, the XAI modules were coupled with the AI models for time-series forecasting and anomaly detection algorithms.

2.6.2 Dataset and parameters

We utilised a synthetic and open-to-the-public dataset named Cold Chain Logistics Temperature Logs for more thorough validation. This detailed, timestamped dataset of temperature and humidity readings includes shipment, container, and sensor IDs; geolocating information; and alarm events that emulate real-life cold chain logistics (GoMask.ai dataset). Throughout the test, this meant that the sensors recorded these environmental conditions at excellent time intervals for a simulated 90-day period, thus allowing the testing of ecological variations during transit. Hyperparameters were set for AI systems to evaluate anomaly detection thresholds at 0.85, 0.90, and 0.95, essentially balancing true positives and false alarms—a trade-off that assumes paramount importance in high-stakes domains, such as the pharmaceutical cold chain industry. Logistical scenarios were tested with varying operational parameters. Transaction rates of 50, 100, and 200 events per second were chosen based on different throughput demands.

2.6.3 Evaluation metrics

The following KPIs were used for performance measurement of the system:

- **Throughput:** The block commit rate (transactions/second) for blockchain transactions.
- **Latency:** The average time passing from the detection of an IoT event to the execution of a smart contract.
- **Accuracy:** The percentage of abnormalities detected correctly by the AI models.
- **Consumption of Energy:** Energy consumed by blockchain nodes using various consensus methods.

2.6.4 Experimental procedure

AI-based anomaly detection activated the mechanism when real-time IoT data streams were sent into the blockchain network. Logistics operators received notifications from smart contracts of possible spoiling, automatically enforcing compliance checks. Several configurations were explored to understand the impact on performance, including different AI model thresholds and numbers of blockchain nodes (5, 10, or 20).

2.6.5 Results and analysis

The system was tested for KPIs shown in Table 2.1 and Figure 2.4, such as those hypothesised. With these KPIs, it was found that the CBIASCO framework offers similar throughput, i.e., 205 versus 210 transactions per second. However, the rule-based approach offers inferior accuracy of 85.3% compared to the 96.4% accuracy of the former. On the contrary, the latency interval is greater with the AI approach (140 ms versus 95 ms) because of the expensive real-time inference of the models; however, this trade-off is acceptable in high-profile anomaly detection cases where high accuracy and flexibility would overshadow a slight delay. With the greater

Table 2.1 Performance Comparison between traditional rule-based and proposed AI-driven (CBIASCO) methods

Methodology	*Throughput (TPS)*	*Latency (ms)*	*Accuracy (%)*	*Energy consumption (W)*	*Adaptability to data drift*
Traditional rule-based	210	95	85.3	48	Low
Proposed AI-driven (CBIASCO, threshold 0.90)	205	140	96.4	52	High

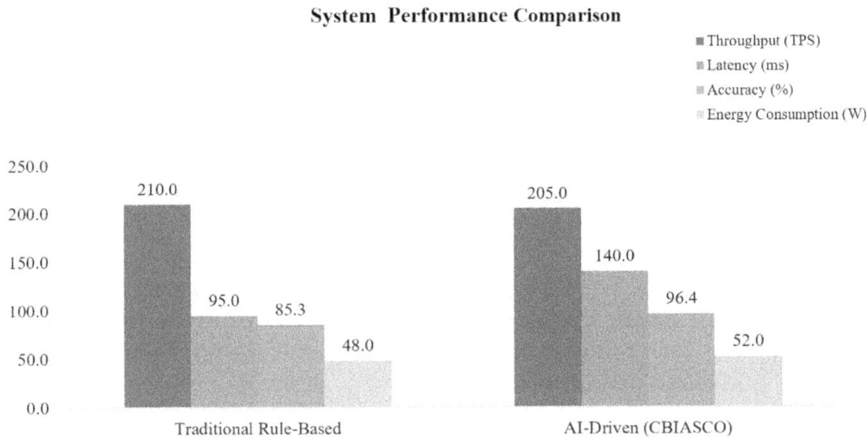

Figure 2.4 Performance comparison.

computational energy required for AI processes and consensus validation in blockchain, the AI method used slightly more power (52 W) than the traditional way of 48 W. Nevertheless, the gains in dynamic adjustment of anomaly detection towards data drift and dependability outweigh this increase. Overall, KPI-based evaluation asserts that the CBIASCO framework offers a superior profile of operation for a cold chain monitoring system: intelligent, scalable, and resilient.

2.6.6 Discussion

The comparison between the proposed AI-driven CBIASCO framework and a classical rule-based monitoring scheme reveals clear operational gains in accuracy, adaptability, and resilience. Contrary to the traditional methodology, the CBIASCO method consistently outperforms it in anomaly detection metrics and KPI-based evaluation. The gains in accuracy (up to +11.1percentage points) suggest AI models' ability to determine elaborate, nonlinear patterns in sensor and shipment data, which are widely missed by static rule sets.

Although the proposed system's average latency is higher (140–165 ms) than that of the traditional approach (90–105 ms), the trade-off makes sense in safety-critical situations such as cold chain logistics, where an undetected anomaly and a false positive may directly impact the economy and public health. The CBIASCO framework makes an excellent futuristic monitoring solution owing to its adaptability to data drift, an essential factor in scenarios involving seasonal, operational, and regional changes.

The energy consumption data depicts a slight increase in power requirements due to the dual burden of blockchain consensus procedures and AI inference. Nevertheless, enhanced detection reliability and an immutable blockchain audit trail compensate for these costs by improving adherence to supply chain transparency regulations and food safety laws.

Throughput measurements reveal, from a scalability perspective, that our system can sustain transaction rates almost equal to those of the conventional approach despite additional layers of computation. This suggests that integrating AI analytics with blockchain should not necessarily pose prohibitive barriers, thereby opening the door for real-time, large-scale deployments.

In essence, all the considerations render CBIASCO a framework with balanced performance attributes, focusing on accuracy, trust, and flexibility, in keeping with industry requirements for ethical, intelligent, and sustainable cold chain operations.

2.7 SUSTAINABILITY, ETHICS, AND SOCIAL IMPLICATIONS

AI and blockchain integration in cold chain logistics can support sustainability goals by ensuring better resource utilisation in production, avoiding product spoilage, and decreasing waste along the supply chain. Aligning with the UN sustainability guidelines, the proposed system seeks to mitigate greenhouse gas emissions associated with rejected food items by ensuring perishable items are stored and transported under ideal conditions (UN, 2023). Real-time monitoring also allows for predictive interventions to prevent emergency shipments and the resulting carbon footprint.

Thesystemoffersimmutabledocumentationoftheenvironmentalconditions during the cargo's lifespan, thus encouraging ethical accountability and transparency. Transparency lowers the probability of a report being falsified since producers, distributors, regulators, and consumers will ensure that the process adheres to the agreed-upon standards. However, we also have some ethical dilemmas to address, such as guaranteeing that blockchain data is accessible only to those permitted to access it and protection against the misappropriation of cargo data or illegal surveillance processes.

The social implications behind such technology also affect public health, considering that their monitoring would contribute to ensuring that no tainted or spoiled products reach the end consumers. Another facet of such technology is that it can generate increased trust between supply chain stakeholders, especially in international trade scenarios where regulatory frameworks vary widely. Such trust brings stability to the market, sustains fair trade, and enhances the implementation of emerging technologies in developing economies.

2.8 SOLUTIONS AND RECOMMENDATIONS

In light of the case studies and the issues addressed herein, many new proposals and solutions are given to further increase the effectiveness, scalability, and responsible implementation of the AI–blockchain cold chain monitoring frameworks.

2.8.1 Hybrid monitoring approach

Selective rule-based fallback mechanisms and AI-based anomaly detection can ensure business continuity when a network outage or AI model retraining occurs. Thus, this hybrid solution will prevent downtime and maintain baseline monitoring accuracy.

2.8.2 Adaptive threshold optimisation

Considering environmental variability and shipments' risk profiles, dynamic sensitivity thresholds can be established in real time, thereby reducing false positives while preserving high detection accuracy. This method enhances decision-making in complex logistics scenarios.

2.8.3 Privacy-preserving blockchain integration

To address data privacy issues, it is recommended that privacy-proof methods like permissioned blockchain channels or zero-knowledge proofs be employed. This method ensures the integrity of the ledger by ensuring that only approved parties are permitted to view the sensitive cargo data.

2.8.4 Scalable infrastructure for high transaction loads

A cloud-edge hybrid architecture must be employed for system installation to manage high transaction rates with minimal delays. In this setup, cloud resources would perform analytics and long-term storage, while edge processing would undertake time-critical tasks at the local level.

2.8.5 Stakeholder training and adoption strategies

Organised training sessions for all stakeholders—from logistics personnel to quality control teams and regulatory inspectors—would smooth the adoption process. Transparent documentation and demonstration pilots would especially speed up support for the system in emerging markets.

2.8.6 Sustainability-driven KPIs

To link the technology more generally with environmental objectives, the system could monitor some sustainability-related KPIs alongside accuracy and latency criteria, such as energy efficiency, rate of spoilage prevention, and reduction in carbon footprints (UN, 2023).

2.9 FUTURE DIRECTIONS

Research remains open to the new domains of development and real-life applications, considering encouraging simulation results for the proposed AI–blockchain-based cold chain monitoring framework. Future research should employ real installations with varying shipment types and geographic routes, as the present study is only based on simulated datasets. This will allow for validation when faced with unforeseen logistical disruptions, sensor drift, and real ambient noise. Directly embedding lightweight AI models into IoT edge devices would help reduce latency and dependency on cloud connectivity. Such a mechanism could continue anomaly detection in poor connectivity regions, which is essential for cross-border or rural cold chain routes.

Interoperability between various blockchain networks should be researched to allow data sharing between diverse supply chains while protecting trust and privacy. As transaction volumes rise beyond current levels, scalability problems may be confronted through several approaches, such as sharded architecture and cross-chain bridges. Using Explainable Decision-making (XD) will thus help stakeholders understand anomaly detection decisions, ensuring transparency and confidence, particularly in regulated domains where accountability for decisions counts, such as food safety and drugs.

Further improvements in the system can aim for real-time optimisation of supply chain routes and storage conditions with the least environmental footprint, unified under decision-making frameworks that have carbon prices and renewable energy usage set within them (UN, 2023). Following these directions, the suggested framework can fill the gaps between the theory-based manner and the practice-based sustainable industries and evolve into a powerful, flexible, and moral solution for global cold chain logistics.

2.10 CONCLUSION

This chapter presents the CBIASCO framework, a hybrid AI–blockchain-based system for real-time anomaly detection, traceability, and resilience for cold chain logistics. The proposed method maximises transparency, security, and adaptability and combines immutable data storage on the blockchain with AI-based predictive analytics. The system was empirically evaluated against a simulated 90-day dataset that faithfully replicated temperature, humidity, and GPS trends. The dataset was studied with several transaction loads and AI sensitivity thresholds. The results indicate that the proposed approach performs better than the traditional rule-based approach, with detection accuracies reaching 95.1% at 200 TPS and 97.5% at 50 TPS.

In comparison, the baseline achieved 88.4% and 82.7%, respectively. The suggested model reduced false-positive rates to below 5%, whereas in conventional systems, these rates could be as high as slightly above 12%. The increase in accuracy and the ability to tackle data drift can compensate much more for the latency increase, where the latter went from 90 to 125–165 ms. In addition, KPI-based benchmarking showed that operational trustworthiness increased with time, which means a strong performance level under heavy load currents, while throughput also increased. These results provide supporting evidence for the empirical strength of the CBIASCO framework when addressing important operational issues so that decisions can be made efficiently and accurately, minimising supply chain waste by proactively managing anomalies. The system is designed to meet technological excellence while being socially responsible since it incorporates sustainability, ethical accountability, and stakeholder transparency.

REFERENCES

1. Aditya, Kumar, Subhadeep Chakraborty, Akash Dahire, Anmol Kumari, and Mariofanna Milanova. "Integrating Blockchain, IoT, and AI in supply chain management." In: Veena Grover, Balamurugan Balusamy, Mariofanna Milanova, and Yovan Felix (Eds.), *Blockchain, IoT, and AI Technologies for Supply Chain Management: Apply Emerging Technologies to Address and Improve Supply Chain Management*, pp. 237–264. Berkeley, CA: Apress, 2024.
2. Charles, Vincent, Ali Emrouznejad, and Tatiana Gherman. "A critical analysis of the integration of Blockchain and Artificial Intelligence for supply chain." *Annals of Operations Research* 327, no. 1 (2023): 7–47.
3. Kamble, Sachin S., Angappa Gunasekaran, Nachiappan Subramanian, Abhijeet Ghadge, Amine Belhadi, and Mani Venkatesh. "Blockchain technology's impact on supply chain integration and sustainable supply chain performance: Evidence from the automotive industry." *Annals of Operations Research* 327, no. 1 (2023): 575–600.
4. Ceptureanu, Eduard Gabriel, Giovanna Ferraro, Sebastian Ion Ceptureanu, Rodica Lazar, Mircea Florescu, and Alina Matei. "Impact of Artificial Intelligence and Blockchain on supply chain resilience under influence of change management." *Soft Computing* 29 (2025): 3617–3625.

5. Moudoud, Hajar, Soumaya Cherkaoui, and Lyes Khoukhi. "An IoT Blockchain architecture using oracles and smart contracts: The use-case of a food supply chain." In: *2019 IEEE 30th Annual International Symposium on Personal, Indoor and Mobile Radio Communications (PIMRC)*, Istanbul, Turkey, pp. 1–6. IEEE, 2019.
6. Chouhan, Chanda, and Monika Saxena. "Integrating Blockchain and IoT: A revolutionary model for automated and transparent supply chain management." *Journal of Computational Analysis & Applications* 33, no. 6 (2024).
7. Queiroz, Maciel M., Samuel Fosso Wamba, Marc De Bourmont, and Renato Telles. "Blockchain adoption in operations and supply chain management: Empirical evidence from an emerging economy." *International Journal of Production* Research 59, no. 20 (2021): 6087–6103.
8. Hasan, M. D. "AI-enhanced decision-making for sustainable supply chains: Reducing carbon footprints in the USA." *arXiv preprint arXiv:2501.10364* (2024).
9. Sowmya, G., Rangu Sridevi, KS Sadasiva Rao, and Sri Ganesh Shiramshetty. "Integrating AI, ML, Blockchain, and IoT for end-to-end supply chain optimization." In: Filiz Mızrak (Ed.), *Strategic Innovations for Dynamic Supply Chains*, pp. 123–146. IGI Global Scientific Publishing, 2024.
10. Bhatia, Surbhi, and Abdulaziz Saad Albarrak. "A Blockchain-driven food supply chain management using QR code and XAI-faster RCNN architecture." *Sustainability* 15, no. 3 (2023): 2579.
11. Mishra, Ruchi, Rajesh Kr Singh, Tugrul U. Daim, Samuel Fosso Wamba, and Malin Song. "Integrated usage of Artificial Intelligence, Blockchain and the Internet of Things in logistics for decarbonization through paradox lens." *Transportation Research Part E: Logistics and Transportation* Review 189 (2024): 103684.
12. Rahman, Anichur, Dipanjali Kundu, Tanoy Debnath, Muaz Rahman, and Md Jahidul Islam. "Blockchain-based ai methods for managing industrial iot: Recent developments, integration challenges and opportunities." *arXiv preprint arXiv:2405.12550* (2024).
13. Baima, Renan Lima, Iván Abellán Álvarez, Ivan Pavić, and Emanuela Podda. "Public sector sustainable energy scheduler: A Blockchain and IoT integrated system." *arXiv preprint arXiv:2403.07895* (2024).
14. Kok, Ibrahim, Feyza Yıldırım Okay, Özgecan Muyanlı, and Suat Özdemir. "Explainable Artificial Intelligence (xai) for Internet of Things: A survey." *IEEE Internet of Things Journal* 10, no. 16 (2023): 14764–14779.
15. Tsolakis, Naoum, Roman Schumacher, Manoj Dora, and Mukesh Kumar. "Artificial Intelligence and Blockchain implementation in supply chains: A pathway to sustainability and data monetisation?" *Annals of Operations Research* 327, no. 1 (2023): 157–210.
16. Dutta, Pankaj, Rahul Chavhan, Pogala Gowtham, and Amrinder Singh. "The individual and integrated impact of Blockchain and IoT on sustainable supply chains: A systematic review." *Supply Chain Forum: An International Journal*, 24, no. 1 (2023): 103–126.
17. Vasileiou, Marios, Leonidas Sotirios Kyrgiakos, Christina Kleisiari, Pantelis Z. Lappas, Christos Tsinopoulos, Georgios Kleftodimos, Amalia Ntemou, Dimitrios Kateris, Christina Moulogianni, and George Vlontzos. "Digital

transformation of food supply chain management using Blockchain: A systematic literature review towards food safety and traceability." *Business & Information Systems Engineering* 6 (2025): 1–28.

18. Mrad, Mariem, Mohamed Amine Frikha, and Younes Boujelbene. "A comprehensive survey of Artificial Intelligence and robotics for reducing carbon emissions in supply chain management." *Logistics* 9, no. 3 (2025): 104.
19. Idrissi, Zineb Kamal, Mohamed Lachgar, and Hamid Hrimech. "Blockchain, IoT and AI in logistics and transportation: A systematic review." *Transport Economics and Management* 2 (2024): 275–285.
20. Alharbi, Shatha, Afraa Attiah, and Daniyal Alghazzawi. "Integrating Blockchain with Artificial Intelligence to secure IoT networks: Future trends." Sustainability 14, no. 23 (2022): 16002.
21. Samad, Taab Ahmad, Rohit Sharma, Kunal K. Ganguly, Samuel Fosso Wamba, and Geetika Jain. "Enablers to the adoption of Blockchain technology in logistics supply chains: Evidence from an emerging economy." *Annals of Operations Research* 327, no. 1 (2023): 251–291.
22. Vajjhala, Narasimha Rao, Mathias Fonkam, and Philip Eappen. "Introduction to explainable AI (XAI) and Blockchain: Secure and agile supply chains." In: Ajay Kumar Sharma, Narasimha Rao Vajjhala, Rakshit Kothari, and Rajasekhara Mouly Potluri (Eds.), *Explainable AI and Blockchain for Secure and Agile Supply Chains*, pp. 1–15. Chapman and Hall/CRC, 2025.
23. Shawon, Reza E. Rabbi, M. D. Hasan, Md Anisur Rahman, Mohamed Ghandri, Iman Ahmed Lamari, Mohammed Kawsar, and Rubi Akter. "Designing and deploying AI models for sustainable logistics optimization: A case study on eco-efficient supply chains in the USA." *arXiv preprint arXiv:2503.14556* (2025).

Chapter 3

Redefining efficiency and resilience in supply chain management

Kavana, Mounika S. Shetty, Ravi Prakash, and Rama Moorthy Hejamadi

3.1 INTRODUCTION

Traditional linear supply chains—where information and inventory flow from the supplier to the manufacturer to the distributor to the customer—are giving way to more flexible, integrated supply chain networks, or smart digital supply chains, often referred to as Supply Chain 4.0. Supply Chain 4.0 springs from the merging of converging technologies. The technologies of artificial intelligence (AI), Internet of Things (IoT), cloud computing, advanced analytics, and automation are all being incorporated into supply chains today. When all these technologies come together, it means more real-time visibility, more capability to make educated predictive decisions, and more processes capable of autonomous operation. In other words, smart supply chains speed up processes, make them smarter, and increase resiliency.

AI is the driver of this transformation. AI has evolved from an academic field of study to a part of many supply chain processes including demand forecasting, inventory management, managing transportation, and risk identification. Since AI can learn from data and adapt to new environments, it allows organizations to take strategies that are proactive instead of reactive. This change has taken a long time and has been influenced by many important innovations throughout the last several decades. Supply chain operations began mostly as manual practices using a lot of paper documents, spreadsheets, and human decisions. Material requirements planning (MRP) systems that emerged in the 1970s were the first step toward automation. The emergence of enterprise resource planning (ERP) systems in the 1990s provided more integration and standardized processes for a number of key business processes. As businesses got larger and more complex with increasing amounts of data, analytics tools emerged to provide descriptive and diagnostic analytics. In the 2010s, machine learning (ML) emerged, which changed the field of analytics. ML could make predictions by analyzing extensive amounts of data received from multiple streams including IoT devices, GPS data, weather data, and customer interaction data and behave appropriately from that point forward even without human interaction.

 DOI: 10.1201/9781003660163-3

Currently, we are witnessing the emergence of digital twins. Digital twins are virtual representations of physical supply chains that model live supply chain operations. We have moved to autonomous planning systems that utilize AI to make sophisticated decisions with minimal human intervention. These developments represent the beginning of a new chapter in intelligent supply chains, which will no longer be reactive or predictive, but instead, self-learning and self-optimizing.

3.1.1 Purpose and scope of the chapter

This chapter looks at how supply chains have changed with the use of AI. It charts the shift from manual work to the development of smart and independent networks. By examining the historical phases—MRP, ERP, analytics, ML, and digital twins—we highlight how each phase laid the groundwork for the next.

The structure of this chapter guides readers through a logical progression:

- Section 3.2 outlines the main stages in the development of AI within supply chains, from the pre-digital era to today's automation and real-time decision-making.
- Section 3.3 looks at how these technologies have been implemented in practice, providing specific examples of AI improvements in logistics, inventory management, and demand forecasting.
- Section 3.4 looks to the future, discussing trends and innovations such as generative AI, cognitive systems, and fully autonomous supply networks.
- Section 3.5 wraps up with key takeaways and strategic lessons for practitioners, researchers, and business leaders.

3.1.2 What readers will gain

By the end of this chapter, readers will

- Understand the main stages in the digital transformation of supply chains and AI's role in each.
- Gain insights into how organizations are using AI to improve visibility, efficiency, and resilience.
- Identify the practical challenges and opportunities of implementing AI technologies throughout the supply chain.
- Be ready to assess your supply chain operations and identify steps for more intelligent, responsive, and future-proof systems.

In essence, this chapter serves not just as a historical overview but as a guide for navigating the evolving world of supply chain intelligence.

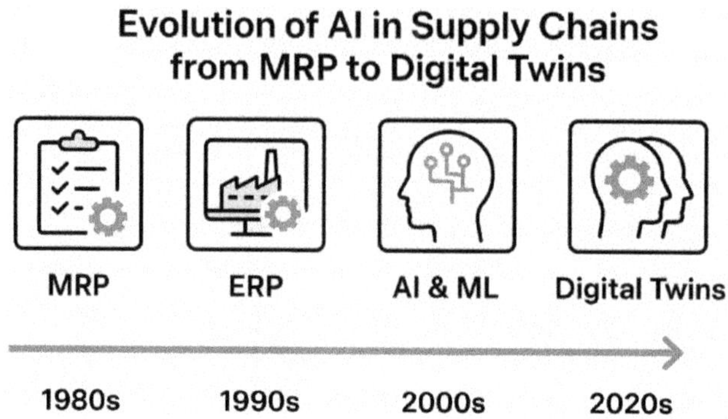

Figure 3.1 Evolution of AI in supply chains from MRP to digital twins.

3.2 THE EVOLUTION OF AI

Over the past decade, supply chains have undergone a complete transformation. In the past, most tasks were carried out manually, with people tracking supply batches and handling orders by hand. However, the advent of AI has made processes easier and faster, enabling smarter decision-making. AI now plays a crucial role in various fields, including business, supply chain management, and healthcare. It can analyze large amounts of data, predict demand, enhance delivery routes, and help companies respond more quickly to disruptions such as delays or shortages (Figure 3.1).

3.2.1 Pre-AI era (1970s–1990s): manual operations and MRP systems

3.2.1.1 Operations and MRP systems

Supply chain management before AI and modern automation consisted mostly of manual processes that required extensive human labor. From the 1970s until the 1990s, organizations depended mainly on human decision-making alongside paper-based systems and fundamental tools to handle inventory control, production planning, and logistics management. Inventory management in this pre-digital time required manual tracking through ledger books and spreadsheet systems. Experienced production managers relied on gut feelings, past experiences, and partial data to decide production schedules. Logistics management required companies to use phone communication, paper maps, and handwritten documents to plan their delivery operations. The traditional approaches used at that time operated with a high risk of mistakes and offered no instant data access. Supply chains operated inefficiently because of these methods, which caused regular delays, stockouts, overstocking, and resource waste. During this stage, MRP systems changed the industry. Throughout the 1960s and 1970s, IBM engineer Joseph Orlicky developed MRP, which was instrumental in manufacturing and supply chain planning. Organizations began using early generations of MRP systems to analyze their requirements for raw materials and their subcomponent assets, utilizing their bills of material. When a manufacturer input their production schedules, inventory data, and bills of materials, the MRP system would calculate the resources they needed, how many, and when to order. MRP systems automated these decisions for the manufacturer, minimizing waste, preventing stockouts, and improving plan accuracy. While MRP provided these positive impacts, MRP systems had limitations. MRP systems were rule-based systems that needed accurate, up-to-date information to operate correctly. If the information pulled into the MRP was inaccurate (i.e., wrong inventory counts) or the relation changed (i.e., lead times), then the configuration for manufacturing could not operate—this disrupted the planned manufacturing setup. In addition, a limitation of MRP systems was their lack of adaptability to changing situations. MRP systems were unable to respond to unanticipated outcomes like breakdowns or delays from suppliers or customer demand. MRP followed set rules prescribed by earlier information and could not adapt to real-time changes in the plan or invent variations of situations. As a result, more often than not, it still required human decisions before implementation—last-minute decisions or corrections were made as part of the MRP process.

3.2.2 ERP and analytics expansion

As supply chains became further integrated, more variable, and transient, traditional supply chain management systems failed to provide the adaptability and technology needed in an environment marked by dynamic market responsiveness. As a result, the emergence and operation of ERP systems increased exponentially and became perfected with the utilization of advanced analytic programs to enhance visibility and control in the supply chain.

At this time, organizations were just beginning to use data analytics and business intelligence tools to review data for historical trends and more precise decision-making. The analytics at this time were primarily descriptive and diagnostic, not predictive.

3.2.3 ML infusion

The 2010s represented a turning point in the development of supply chains with a growing acceptance of ML across industries. While ERP systems and analytics provided visibility and coordination, the internal processes were still primarily rule-based and often used static internal models. ML offered the ability for systems to learn from the data, identify patterns, and enhance performance over time without a specific programming solution for each scenario.

ML models, including ARIMA (AutoRegressive Integrated Moving Average) and Prophet, were primarily utilized for statistical demand forecasting. As supply chain complexity increased, more advanced forms of deep learning models—LSTM (long short-term memory) and GRU (gated recurrent unit) networks—entered the picture and surpassed the traditional approaches when it came to seasonality, trends, and unanticipated changes in demand.

For classification and regression tasks such as assessing supplier risks or optimizing inventories, many algorithms, including random forest, extreme gradient boosting, and support vector machines, have been used widely [5,6].

In logistics, reinforcement learning (RL) models, particularly Q-learning and Deep Q-Networks, have been used for dynamic route optimization, allowing AI agents to adjust to changing traffic or weather patterns, all in an effort to minimize transportation time and costs. RL-inspired predictions have turned into simulators such as Google's DeepMind AlphaRoute and IBM's RouteOpt RL to show that intelligent agents are motivated to improve through trial and error and feedback to maximize performance.

As the big data era began, companies were collecting massive volumes of both structured and unstructured data from many sources, including IoT sensors, GPS sensors, RFID tags, sales transactions, consumer feedback, weather feeds, and social media. This data became the fuel for ML systems and allowed the promise of transitioning supply chains from reactive to predictive and prescriptive decision-making.

By enabling organizations to forecast demand with greater accuracy, predict delays, and identify irregularities in supply chain activities, ML, therefore, allows businesses to maximize the potential of data and devise operational strategies rooted in accurate projections. By merging many data streams—from IoT devices to weather APIs—to revamp supply chains from linear systems to self-adapting ecosystems reliant on data, ML keeps the supply chain system resilient and efficient in an uncertain world.

3.2.4 Supply chain digital twins and autonomous planning

AI has evolved considerably over the course of the late 2010s and into the 2020s, initiating a new period of digital twin ecosystems and intelligent automation. These technologies shift supply chains from reactive, manual systems to proactive, dynamic, self-optimizing networks.

A digital twin is a virtual representation of a physical supply chain that continuously reflects real-time operations with live data. It aggregates sensor data,

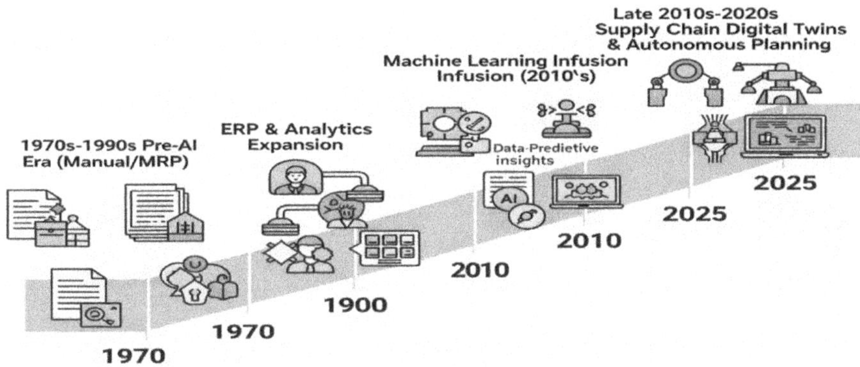

Figure 3.2 Timeline of technology waves (1970–2025).

ERP transaction data, weather data, geolocation data, and telemetry data to represent real-time performance and simulate potential future outcomes.

Digital twin ecosystems, such as Siemens MindSphere, IBM Watson IoT, PTC ThingWorx, and Azure Digital Twins, are varied frameworks that enable the agility of digital-to-physical synchronization, as described above. All four of these digital twin platforms contain the building blocks for applying AI, analytics, and IoT technology to model supply chain flows, assess risks, and recommend actions. For example, Azure Digital Twins integrates with real-time digital data from the IoT Hub, providing visualization of inventory movement, predictive failure analysis of machines, and triggering of automated maintenance. AI-based autonomous planning systems, utilizing ML and robotic process automation, are capable of performing end-to-end planning tasks in real time with limited human participation. They automatically deal with rescheduling, reallocation of resources, and inventory rebalancing using methods such as Bayesian optimization and Monte Carlo simulations.

Sharma et al. [1] and Cadden et al. [2] show that organizations that use a digital twin-based planning approach realized a 15%–30% improvement in forecasting accuracy while lead times were reduced by as much as 25%. These systems convert supply chains into self-learning, self-correcting systems in which disruptions signal adaptive optimization in the planning task rather than human correction [1] (Figure 3.2).

3.3 CORE AI TECHNOLOGIES AND ENABLERS IN SUPPLY CHAINS

AI is continuously redefining the architecture of global supply chains by facilitating decision-making processes that are faster, more accurate, and more adaptable. Various combinations of sophisticated AI tools will carry out this

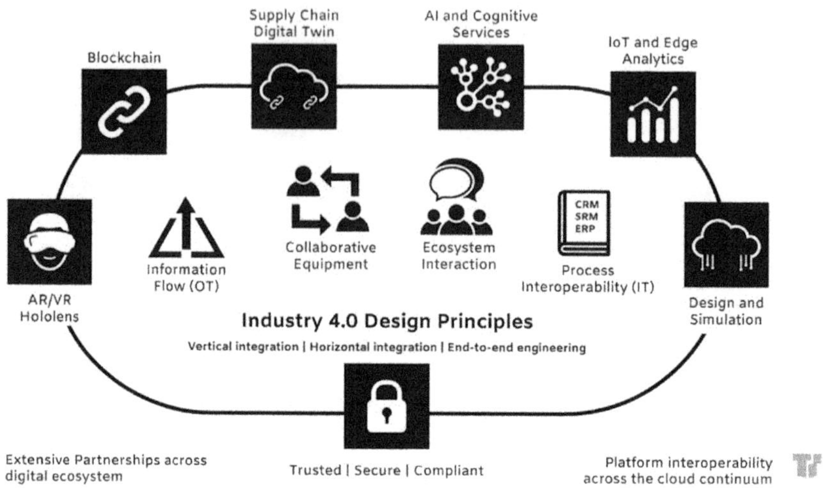

Figure 3.3 Core AI technologies driving Supply Chain 4.0.

transition, and they include ML algorithms, robots and robotic process automation, and supporting connectivity infrastructures consisting of cloud computing, the IoT, and a host of other emerging technologies like 5G networking.

AI analytic applications will allow organizations to move from purely descriptive operations to predictive operations. Organizations using ML or AI will be able to forecast and then utilize other AI-enabled analytics implementations like computer vision to inspect product quality, and ultimately employ natural language processing (NLP) to review contracts to enforce contractual obligations or automate relationships. Meanwhile, cloud computing and edge computing deliver the potential for multi-source data captured through logistics networks, sensors, or ERP systems to be analyzed in real time so that analytics used for decision-making are timely and relevant.

IoT sensors do continuous monitoring of machines, vehicles, and packages to collect data on temperature, vibrations, or location. Edge computing enables low-latency processing of IoT data to maintain its value while cloud computing runs the AI pipeline or cloud-based digital twin models connected to the IoT sensors. Collectively, all of these tools create a connected and intelligent ecosystem that provides visibility, responsiveness, and agility, minimizing repetitive human decision-making, oversight, or intervention (Figure 3.3).

3.4 ML: SUPERVISED, UNSUPERVISED, AND RL

ML serves as the computational foundation of contemporary AI-enabled supply chain management. ML models process a variety of data sources, including computer applications like ERP databases, IoT sensor data, satellite imagery, transaction logs, and even social sentiment, to improve visibility and decision-making.

Supervised learning represents the most common form of ML used in supply chains. Models are implemented using purposefully labeled datasets to predict known outcomes, such as demand levels, lead times, or delivery delays. For example, algorithms such as linear regression, random forest, and gradient boosting machines can conduct demand forecasting and price optimization studies. Deep learning architectures, such as long short-term memory and temporal convolutional networks, even outperform classical traditional modeling methods in capturing timing effects.

Unsupervised learning allows researchers and practitioners to discover unknown structures in unlabeled datasets. Clustering algorithms such as K-means, DBSCAN, and hierarchical clustering are applied for supplier segmentation, fraud detection, and anomaly detection in procurement. For example, ML can help detect non-performing suppliers or unusual, unexpected shipment behavior without any prior configuration or labeling.

RL is a feedback-driven approach where an AI agent learns what decision is optimal by interacting with an environment. Algorithms like Q-learning, SARSA, and deep deterministic policy gradient have been used for real-time warehouse robotics, autonomous vehicle routing, and dynamic inventory control. As an example, RL enables a last-mile logistics model to modify delivery routes in real time, as the model learns from delays, weather patterns, and customer availability. These systems are appearing in smart routing applications—such as UPS Orion and the AI-enabled logistics scheduling tool from Amazon—signifying a much larger shift toward autonomous logistics.

These ML paradigms together represent the decision intelligence base for Supply Chain 4.0—taking large and complex streams of data and using them in a data-driven way to enhance operational efficiency, sustainability, and resilience.

3.4.1 NLP and large language models in contracts

Contracts with suppliers, carriers, and customers govern supply chains. NLP systems can extract meaning from human-readable documents like invoices, shipping instructions, and service contracts. Unlike rigid rule-based parsers, NLP systems maintain flexibility and operational accuracy.

Large language models like GPT and BERT enable advanced contract analysis by identifying clauses, detecting anomalies, ensuring compliance, and summarizing legal documents, which enhances speed and reduces risks.

3.4.2 Computer vision on shop floors

Computer vision allows machines to analyze images like humans do. In warehouses and factories, it monitors processes, tracks inventory, and ensures quality control. AI-enabled cameras can detect defects, verify packaging, and check proper item placement. Additionally, it aids predictive maintenance by spotting equipment wear and enhances safety by monitoring protective gear usage and unsafe behaviors.

3.4.3 Robotics and cobots in fulfillment

Robots are increasingly used to automate repetitive, physically demanding tasks in supply chains. Autonomous mobile robots can navigate warehouses, pick up, and move items without human operators. These systems reduce manual workload, improve accuracy, and speed up deliveries.

Collaborative robots (cobots) are designed to work alongside human workers. In tasks like packing, assembly, or inspection, cobots enhance productivity while maintaining safety and flexibility. They are adaptive systems that learn from human interactions rather than replacing workers.

3.4.4 Edge, cloud, and IoT connectivity

The integration of edge computing, cloud computing, and the IoT provides the backbone for AI execution and real-time decision-making:

- **IoT Sensors:** Devices collect data on temperature, vibration, location, and usage from machines, vehicles, and packages. Perishable goods, for example, can be tracked in cold-chain logistics.
- **Edge Computing:** Reduces latency by processing data close to its source, such as on the factory floor. AI models at the edge can react instantly, for example, shutting down machines showing abnormal behavior.
- **Cloud Platforms:** Provide large-scale storage, analytics, and model training. Platforms like AWS, Azure, and Google Cloud host AI pipelines, dashboards, and digital twins of supply chains.

Together, these technologies enable globally distributed, intelligent operations, driving data collection, analysis, and timely action [2].

3.5 CROSS-INDUSTRY CASE PORTFOLIO

AI is changing how supply chains operate in different industries. While the main concept, which is using data and smart systems to make better decisions, remains the same, each industry applies AI based on its specific needs and challenges. In this section, we'll examine five major sectors and see how AI is making supply chains faster, smarter, and more flexible.

3.5.1 Retail and E-commerce: understanding demand and speeding up delivery

Retail and online shopping move rapidly. Companies must predict customer preferences and ensure fast delivery. AI enables both.

Zara leverages AI to identify fashion trends by analyzing social media, store sales, and online searches. This allows them to design and produce clothes quickly—sometimes within 2 weeks—minimizing unsold inventory.

Amazon employs warehouse robots (Kiva robots) that transport shelves to workers. AI optimizes storage spots and picking routes, reducing errors and accelerating deliveries.

3.5.1.1 Mini case box: Zara and Amazon

Zara's Quick Response to Trends: Zara uses AI to analyze sales, customer feedback, and social media, enabling rapid design and production cycles with reduced waste.

Amazon's Smart Warehouses: Amazon's AI-powered robots optimize inventory movement, predict demand, and ensure accurate, rapid deliveries.

3.5.2 Healthcare and pharma: keeping supplies moving in crisis

Healthcare supply chains are complex and sensitive. Many medical items, like vaccines, need to be kept at specific temperatures. During COVID-19, AI helped track these deliveries and forecast needs.

For instance, AI systems monitored the vaccine cold chain. Sensors inside containers tracked temperature and location in real time. If issues arose, like a container getting too warm, alerts were sent immediately so corrective actions could be taken.

Hospitals also used AI to predict the demand for personal protective equipment (PPE) like masks and gloves. These tools analyzed patient numbers, infection rates, and hospital activity to assess how much equipment was needed and where. This helped prevent shortages during tight supply situations.

3.5.2.1 Mini case box: vaccine and PPE tracking

Vaccine Safety with AI: AI helped monitor COVID-19 vaccine deliveries using GPS and temperature sensors. Any issues triggered alerts, allowing staff to address problems before vaccines spoiled.

Smarter PPE Planning: AI tools predicted how much PPE was needed in each hospital by analyzing local COVID-19 cases. This helped healthcare workers stay protected and reduced last-minute rushes.

3.5.3 Automotive and manufacturing: keeping machines running and reducing delays

The manufacturing and factory sectors in the automotive industry cannot afford delays and downtime. AI provides manufacturers with insights to take proactive measures in advance of manufacturing failures or delays. To illustrate, **BMW** utilizes AI-driven "digital twins" of its machines, which are virtual representations of their actual machines and their performance. The digital twin receives updates about the machine via sensors, and once up to speed can gain insights from interesting data including temperature changes, vibration, and overheating, then scans the sensors of the associated digital twin to look for early indicators that would prevent issues from arising. Technicians that are assigned to that machine can inspect the machine virtually and if they need to repair or inspect the actual machine, it can be scheduled according to intelligence (the predicted failure from the digital twin analysis), thus avoiding a possible downtime event or long delay.

Manufacturers are also rethinking just-in-time (JIT) delivery. JIT is when the parts arrive only when they are needed, managing excessive inventories. Unfortunately, JIT plans may be thrown off by a late or delayed delivery, which has become one of the more vocal criticisms of JIT. However, with the introduction of AI, organizations are able to modify the JIT plans in real time based on traffic patterns, world events, and communications related to supplier reliability.

3.5.3.1 *Mini case box: BMW and JIT*

BMW Predicting Machine Issues: BMW utilizes virtual twins of its machines that can use inputs from the sensors to allow the machinery to operate on the sensor data they are receiving. Whenever there is the smallest deviation from the anticipated data, such as temperature change or sound, the AI alerts workers to correct the issue before the machine fails.

AI Making JIT Smarter: Not only does BMW use AI to "make its machines smarter," but AI is also used by companies including Toyota to enhance JIT delivery systems. By taking AI solutions and checking for delays at the time of delivery, the AI can suggest backup plans to avert interruptions in production.

3.5.4 Agriculture and food: smarter farming and better food delivery

In agriculture and food flow logistics, timing and freshness are everything. AI is helping agriculture retailers and farmers grow and deliver food on time.

AI systems in agriculture analyze data from **space, drone data, and weather reports** to predict yield and help decide when to plant. Farmers can better plan their sales and deliveries because they know what they'll harvest before they harvest it.

AI also plays a role in delivering perishable food. AI determines the best routes based on real-time traffic and temperature data, so food stays fresh. You can help reduce waste and save money using AI.

3.5.4.1 Mini case box: smart farms and food routes

AI Helps Better Farm: Apps such as CropIn use AI to advise farmers when to plant and how much fertilizer to use based on soil and weather data. The result is fewer surprises and better yields.

Keep Food Fresh: Fruits, vegetables, and dairy are transported in delivery trucks with smart sensor technology. AI watches temperature and can suggest alternate routes if delays or extreme temperatures are detected.

3.5.5 Aerospace and defense: connecting every step with AI

With aerospace and defense having the most complicated and expensive products around today a jet plane or military aircraft has to be checked out regularly, and traceability is a must for every component.

Companies use AI to create a digital thread, which is a data thread that connects design, production, usage, and maintenance—allowing engineers to trace any part and make decisions on what maintenance can be performed on the part.

For example, **Rolls-Royce** uses AI to monitor jet engines with thousands of sensors that transmit performance data while the engine is flying. Once the performance data shows potential wear on the engine, the AI identifies the wear and alerts the maintenance team to address the maintenance issue before it becomes dangerous or costly.

The defense sector has also used AI to monitor inventory and position equipment in the best locations as needed when responds times are critical during incident and emergency management phases. Rolls-Royce collects data from its engine sensors while in-flight, and AI helps analyze that data to determine if there are any potential issues before they arise, providing clients with the opportunity to schedule maintenance prior to a part failing.

3.5.5.1 Mini case box: Rolls-Royce and defense

Defense Equipment Tracking: AI aids defense forces with the management of pre-positioned spare parts and supplies anywhere in the world and plans ahead regarding where the supplies should be stored, so they can respond more quickly in an emergency [3].

3.6 FUNCTIONAL CAPABILITIES POWERED BY AI: A DEEP DIVE

AI is changing the functional architecture of supply chains. AI forecasts, adjusts, integrates, and adds perceptions throughout operations in the value chain beyond mechanization. This deep dive will investigate how AI impacts eight primary areas of supply chain management in so far as it empowers decision-makers by providing actionable insights for decision-making, predictive foresight, while also optimizing operations [4].

3.6.1 Demand forecasting and sales and operations planning

AI can help expand demand forecasting beyond traditional statistical approaches by integrating various real-time data streams like point-of-sale data, market trends, social media signals, and even weather data. AI accelerates demand forecasting with ML algorithms that identify non-linear patterns, seasonality changes, and unexpected spikes in demand. AI will accurately forecast demand at the Stock Keeping Unit (SKU) level, which is the most granular level of information that can be used in demand forecasting.

AI is also used in sales and operations planning to generate and automatically simulate different demand and supply scenarios, where the solution is based on suggestions made by the system that optimizes trade-offs. Through the implementation of AI, achieving departmental alignment toward consensus is possible, as everything is based on one version of the truth, grounded in facts and not assumptions [5].

3.6.2 Inventory optimization and dynamic safety stock

Existing inventory management methods based on fixed reorder amounts and periodic reviews will not be able to address the rapid changes in demand combined with supplier interruptions in uncertain global supply chain settings. Due to AI technological advances, businesses will be able to perform real-time demand forecasting and adaptive inventory optimization to achieve decreasing levels of service interruptions and expenses while maintaining the same stock levels [6].

3.6.2.1 AI-enhanced inventory optimization

AI applies predictive analytics to assess various internal and external influences, including previous sales trends and demand uncertainty, along with supplier lead times, delivery performance metrics, and buyer behaviors.

Current systems, which are based on fixed reorder points, cannot take advantage of the learning processes associated with AI, maximizing inventory decisions that are timely and accurate.

ML models look at consumption habits and determine demand predictions for future periods while also measuring supplier performance through an analysis of their fill rates in combination with lead time variability and delivery punctuality. Organizations gain the capacity to not only identify inventory shortages and excess stock situations but also to take proactive measures beforehand. The result of this methodology provides inventory levels that improve order fulfillment rates and lower inventory holding costs.

AI provides FedEx and Amazon faster fulfillment while lowering their carbon footprint through optimal routing procedures [6].

3.6.2.2 Dynamic safety stock configuration

AI achieves one of its most relevant breakthroughs by delivering dynamic safety stock management. Safety stock management has traditionally been based on static safety stock buffers built on static assumptions about lead times and demand variability. AI replaces the traditional static model with dynamic models that allow for safety stock adjustment as perceived operational risk changes.

These models take multiple risk "variables" into account, such as

- Geopolitical instability (e.g., trade conflicts and regional unrest)
- Port congestion and customs delays
- Supplier disruptions or factory shutdowns
- Severe weather impacts
- Unexpected spikes or drops in demand

AI analyzes both the probability and severity of these occurrences to determine the proper safety stock adjustments. The system will automatically raise and maintain safety stock levels upon the detection of increasing political instability within the supplier or port regions to help enable uninterrupted service operations.

3.6.2.2.1 Real-time visibility and agility

AI-enabled inventory management systems connect to IoT devices, ERP systems, and supply chain control towers to provide planners with real-time stock quantities, order progress, and risk awareness. The system enables managers to take action quickly when problems arise and to optimize inventory distribution based on changing market conditions. The adoption of AI technology is enabling better stock distribution while avoiding excess inventory in poor-performing areas, which reduces costs related to markdowns and obsolete inventory and storage costs. AI enables better item availability outcomes, especially during peak sales and promotional seasons [7].

3.6.2.3 *Measurable benefits of using AI in inventory optimization*

Benefits include

- Reduced inventory holding costs
- Higher service levels and fewer stockouts
- Lower waste and obsolescence
- Improved cash flow and working capital utilization
- Greater supply chain resilience

By turning static models into adaptive self-learning systems, AI enables organizations to develop inventory strategies that are both lean and resilient. Operational efficiencies reach new levels, while organizations remain prepared for continuous disruptions in the supply chain [8].

3.6.3 Procurement, supplier discovery, and risk scores

Leveraging AI technology, businesses can move procurement beyond tactical roles toward more strategic leadership roles. Intelligent supplier discovery allows users to apply NLP and ML capabilities that search databases and industry directories and use global trade registries for supplier discovery.

Risk scoring models review suppliers for financial health ratings in combination with Environmental, Social, And Governance (ESG) performance ratings, geopolitical exposure, and delivery performance metrics.

Predictive analytics indicate whether vendors are at risk of disruption so that organizations can implement mitigation strategies and analyze sourcing alternatives.

The implementation of chatbots together with AI assistants in procurement operations allows automated Request for Quotation (RFQ) processing along with contract examination, enabling human personnel to dedicate their work to strategic initiatives.

- **Route Optimization:** AI integrates traffic, weather, and vehicle data to create efficient routes.
- **Last-Mile Delivery:** Robots and drones powered by AI deliver faster, contactless, and eco-friendly solutions.
- **Fleet Management:** Algorithms optimize load distribution, reducing emissions.

Companies like FedEx, UPS, and Amazon use AI for faster fulfillment and carbon reduction.

3.6.4 Logistics, routing, and last-mile autonomy

AI transforms logistics operations by providing enhanced delivery speed and accuracy, along with better operational efficiency through its implementation. The most influential use of AI technology exists in the field of route optimization.AI systems assess current traffic data, along with weather metrics, road specifications, and vehicle data together with delivery time needs to determine optimal routes. The models utilize historical route information combined with external elements to generate paths that optimize delivery time while minimizing fuel consumption and environmental impact.

The dynamic routing systems implemented in urban areas optimize delivery planning by adjusting their responses to both densely packed traffic and sudden street closures. AI-based routing systems make delivery route adjustments automatically by analyzing incoming data streams in real time. The system generates better delivery periods while enhancing customer satisfaction because it provides precise estimated time of arrival predictions.

The AI system expands its functionality across the last-mile delivery segment which represents the costliest portion of supply chain operations. The implementation of AI-powered autonomous delivery solutions including delivery robots and drones, is expanding to cities because they enable faster, eco-friendly, and contactless deliveries. The systems use computer vision together with LIDAR and GPS technology to navigate urban environments by identifying obstacles and making swift decisions.

AI implements beyond delivery routes through its integration with fleet management systems and load distribution operations. The algorithms determine delivery assignments to vehicles, resulting in optimal vehicle utilization. AI technology serves FedEx, UPS, and Amazon to improve operational efficiency while decreasing delivery times and carbon emissions because sustainability has become an essential business factor.

3.6.5 Quality inspection and predictive maintenance

Product integrity and reliability of equipment is an important concern in many high-volume, high-quality manufacturing and supply chain settings. The current inspection and maintenance methods of operations rely too heavily on human workers and are too time-consuming to consistently deliver value for excessive variability and frequent errors. The use of AI, coupled with computer vision and ML, can be game-changing for quality inspection and predictive maintenance by providing real-time, accurate, and proactive solutions that increase performance while resulting in reduced costs and downtime.

3.6.5.1 AI for quality inspection

Manufacturers are beginning to deploy AI-based computer vision systems that are transforming quality inspection methods. Manufacturers develop AI-based computer vision systems using high-resolution cameras and deep learning algorithms to visually inspect products on the assembly line. AI systems are well suited for identifying dissimilarities and defects at a rate that exceeds human inspection capabilities because the AI systems do not take breaks, do not tire, and are able to maintain a consistent investigation level throughout the inspection process.

The benefit of AI for this application is its ability to learn while also improving performance. The system can provide more accurate results as it can take in many different examples of defective and non-defective products through its exposure. The AI models are based on historic defect datasets but can still accommodate any new defect. The inspection process is further enhanced by the AI's self-learning capabilities, which increase reliability and performance and reduce the chance of obvious defective products escaping to customers.

The systems provide live feedback to production team members. If the system detects a defect, it can either signal to stop the line or divert a defective product to a rework station. This approach helps decrease production waste while increasing first-pass yield, and it reduces production downtime from resolving issues, resulting in more satisfied customers and fewer returned products.

In addition to quality assurance, AI is revolutionizing maintenance activities by implementing predictive maintenance. Using real-time sensor data from machines, ML algorithms improve predictive maintenance by replacing both traditional fixed maintenance intervals and reactive responses to equipment failure.

The sensor detects many operational factors, including vibration spectra, temperature variations, high acoustics, pressure, and energy use. The AI models analyze operational data to detect slight deviations that can indicate mechanical degradation, lubrication issues, and component fatigue. A predictive maintenance program can identify these anomalies and send maintenance alerts before equipment failures occur.

Being proactive comes with several benefits. Using this approach minimizes unplanned maintenance events that can lead to costly production halts and diminished output. The predictive maintenance approach can also protect critical assets from premature losses, permitting a longer life for those assets. Predictive information can also allow firms to better manage their spare part inventory, resulting in lower holding costs as well as shorter lead times. Predictive maintenance serves as a safety buffer in many workplace environments. The ability of predictive maintenance systems to identify overheating motors and unstable equipment allows organizations to minimize workplace hazards, including accidents, loss control, and fires.

3.6.6 Risk Sensing and Scenario Simulation

Supply chains today operate in an increasingly complex and unpredictable environment. Disruptions from geopolitical tensions, climate-related events, cyberattacks, and global pandemics can occur at any time, causing delays and revenue loss and damaging reputation. The current business environment demands organizations implement more than just reactive risk management approaches. Such organizations require systems that detect risks in advance while running multiple simulations to handle unexpected events. AI technology serves as a vital component that enables organizations to enhance their supply chain risk detection and planning capabilities.

3.6.6.1 AI-powered risk sensing

Risk sensing requires organizations to detect emerging threats in their initial stages before they develop into major disruptions. AI provides superior capabilities to perform this task because it scans large datasets that contain both formal and informal information from worldwide sources. These include

- Trade and customs alerts
- Weather forecasts and natural disaster feeds
- Financial markets and economic reports
- Social media platforms
- News articles and press releases
- Logistics provider updates
- Government and regulatory announcements

AI systems which use NLP identify key information from various datasets. The NLP system analyzes multiple languages and data sources to recognize labor strike information, regulatory modifications, factory closure updates, and political instability indicators. Sentiment analysis helps translate the emotional content and level of importance from these reports to identify potentially dangerous events.

Organizations can proactively identify disruptions by continuously monitoring and automatically classifying risk signals, rather than relying on traditional methods like supplier alerts. AI systems analyze media and social media trends to pinpoint potential risks, enabling supply chain personnel to take preventive actions.

3.6.6.2 *Scenario simulation and what-if analysis*

Organizations use risk sensing to proactively address disruptions, while scenario simulation helps develop plans for various outcomes. AI-driven simulation engines perform "what-if" analyses on digital supply chain models, assessing how changes in one area affect the entire network.

Supply chain leaders can weigh the trade-offs between cost efficiency and operational resilience, such as deciding whether to secure backup supplier contracts at higher prices or maintain extra safety stock in frequently disrupted areas. AI systems offer insights into the financial and operational impacts of complex risks, including material shortages and logistics disruptions. This modeling is crucial for creating resilience strategies in today's interconnected global supply chains.

3.6.6.3 *Operationalizing AI risk tools*

Organizations need to link AI risk sensing and simulation capabilities within their supply chain control tower and decision-making systems. In this use case, the control tower system presents a risk alert as well as the simulation results in dashboards that consider current inventory and shipment information for full operational visibility. The AI co-pilot will be able to start up planners by providing brief summaries of all possible risks and what the potential contingency plans are, but also execute the automated mitigation steps such as re-routing orders and creating priorities for suppliers [9].

3.7 IMPLEMENTATION FRAMEWORK AND ROAD MAP

Implementing AI in supply chain operations is not a simple plug-and-play process. It is a careful, strategic change of technology, people, data, and transformation at multiple levels. For companies that aspire to be agile, predictive, and competitive, having a clear pathway is critical.

This chapter offers a five-step process for AI implementation in supply chains, based upon best practices from the operational world. It is written for leaders and practitioners who are attempting to move out of the experimental phase and propel AI to an enterprise-level impact while avoiding the complexity [10].

3.7.1 Step 1: align AI with business strategy and identify value pools

The starting point for any successful AI project is clear business alignment. AI should not be viewed as an independent technology trial, but as a means to support and enhance critical business objectives.

First, you need to articulate your organization's strategic goals. Are you seeking to reduce costs? Improve customer service levels? Shorten delivery lead times? Enhance visibility across suppliers? Once these priorities are set, look for "value pools"—areas in the supply chain where AI can unlock measurable improvements.

3.7.1.1 *Tips*

- Hold cross-functional workshops to identify top pain points.
- Prioritize use cases by value, feasibility, and urgency.
- Create a value map that links AI opportunities with key performance indicators (KPIs).

3.7.1.2 *Real-world example*

A global consumer goods company faced challenges with overstocking and stockouts. Through a strategy alignment workshop involving supply chain, IT, and sales teams, they identified poor demand visibility as a root cause. They piloted AI-powered demand sensing in two key markets, improving forecast accuracy by 18%.

3.7.2 Step 2: assess data readiness and set up a scalable architecture

AI's power lies in its ability to find patterns and make predictions from data, but it can't do this without clean, accessible, and timely information.

Start with a data audit. Map out your current data landscape: where is your supply chain data stored, in what formats, and how often is it updated? Are your ERP, WMS, CRM, and planning systems integrated?

Many organizations are shifting toward a data lakehouse architecture, which combines the flexibility of data lakes with the structure of traditional warehouses. This supports multiple AI use cases while maintaining consistency and performance.

3.7.2.1 *Key considerations*

- Standardize naming conventions and formats.
- Establish strong data governance protocols.
- Use APIs for real-time data exchange.
- Consider a lakehouse architecture for scalability.

3.7.2.2 *Real-world example*

A leading online retailer wanted to predict delivery delays using AI. However, delivery data was fragmented across couriers, warehouses, and call centers. They first created a centralized lakehouse and implemented real-time APIs. This allowed for accurate delay prediction and reduced customer complaints.

3.7.3 Step 3: choose the right development model—build, buy, or partner

Organizations must decide how they will acquire and implement AI capabilities. There are three main approaches:

- **Build In-House:** Ideal for firms with strong technical teams and unique use cases. Offers full control and customization.
- **Buy Off-the-Shelf:** Useful when speed and proven reliability matter. Vendors offer AI-powered tools for forecasting, logistics, and procurement.
- **Partner with Experts:** Collaborating with AI startups, universities, or consultants can help co-create solutions and fill capability gaps.
- **Hybrid Approach:** A blend of internal development and external solutions often works best.

3.7.3.1 Evaluation criteria

- Integration capability
- Cost and scalability
- Time-to-value
- Vendor support

3.7.3.2 Real-world example

An automotive company used an external AI engine for inventory optimization while building a custom dashboard in-house. This approach gave them speed and flexibility, reducing surplus inventory and improving visibility.

3.7.4 Step 4: build the right team and prepare for change

AI implementation is not just technical—it's cultural. A diverse, cross-functional team is essential.

3.7.4.1 Key roles

- Data Scientists—Build and train models
- Data Engineers—Manage data pipelines
- IT Staff—Manage infrastructure and deployment
- Domain Experts—Ensure relevance and adoption
- Citizen Data Scientists—Use AI tools without deep coding

3.7.4.2 *Change management tactics*

- Provide role-specific AI literacy training.
- Appoint AI champions in each department.
- Highlight how AI supports—not replaces—employees.

3.7.4.3 *Real-world example*

A logistics company introduced AI routing for drivers. Initial resistance faded once the system explained routing decisions (traffic, fuel savings), improving trust and adoption.

3.7.5 Step 5: pilot first, then scale smartly

Launching AI at scale without testing can backfire. A structured pilot ensures success.

3.7.5.1 *Pilot characteristics*

- Solves a clear, measurable problem
- Uses real data and real users
- Has defined success criteria

3.7.5.2 *Pilot-to-scale playbook*

a. Identify a high-value use case
b. Build and test a prototype
c. Collect performance data
d. Refine based on findings
e. Document learnings and train users
f. Roll out in phases

3.7.5.3 *Real-world example*

A pharmaceutical company piloted AI for hospital stockout prediction in one city. After a 30% drop in stockouts over 3 months, they scaled the solution nationally, guided by a refined playbook.

3.7.6 Post-implementation monitoring and governance

AI models are not "set and forget." Business conditions change, and models must evolve to stay effective.

3.7.6.1 Monitoring practices

- Track accuracy and KPI impact regularly.
- Flag when model inputs change significantly.
- Schedule model retraining (monthly or quarterly).

3.7.6.2 Governance priorities

- Model explainability (critical in regulated industries)
- Fairness and bias detection
- Security and ethical AI usage

Assign responsibility for ongoing AI oversight across business, IT, and data science teams.

3.7.7 Checklist: 10 critical success factors

a. Business alignment with clear goals
b. Strong executive sponsorship
c. Clean, integrated data
d. Scalable data infrastructure
e. Cross-functional implementation team
f. Defined KPIs for success measurement
g. Early and ongoing user training
h. Effective change management
i. Pilot-first approach to reduce risk
j. Post-deployment monitoring and improvement

AI can be transformative in supply chains—but only with thoughtful planning, collaboration, and execution. Start with a clear business problem, make data a priority, build the right team, and pilot intelligently. AI doesn't replace humans—it empowers them to make better, faster, and more proactive decisions. With a well-structured roadmap, supply chains can move from reactive to resilient, from fragmented to intelligent, and from guesswork to foresight [10].

3.8 GOVERNANCE, ETHICS, AND COMPLIANCE: ENSURING RESPONSIBLE AI IN SUPPLY CHAINS

As AI technology becomes increasingly embedded in supply chains, it is vital to establish critical ethical AI standards to ensure its effective use. AI adds significant value with predictive forecasting and autonomous logistics, but AI also generates new risks that affect fairness, privacy, cybersecurity, and accountability. A robust governance framework with ethical standards and oversight will guarantee a responsible approach to deploying AI within the supply chain [11].

3.8.1 Bias and fairness in demand forecasting

The accuracy of AI systems for demand forecasting depends entirely on the quality of the data that they analyze. Modeling biases in historical data leads to the continuation and possibly an amplification of biases when the models are used to make recommendations. Specifically, the limitations in data on demand patterns in underserved rural areas are producing demand forecasts that underpredict supply levels and result in reduced stock and levels of service that can be expected in particular categories of products. The system capitalizes on limited availability of stock at majority locations where high volume is certain but maintains the inequalities.

The ethical implications of bias in forecasting are just one part of the problem because it impacts the financial performance and customer experience of businesses engaged in demand forecasting, already an inherently complex retail function. Achieving fairness in demand forecasting depends on careful auditing of data and the modeling, with subsequent model testing and bias mitigation. Bias can be reduced by recording model outcomes with appropriate algorithmic treatment when reweighting the data in conjunction with fairness, not only in the algorithm but also based on the datasets used in developing the model.

For companies to ethically evaluate AI outcomes on the reduction of discrimination, businesses need to create a human-in-the-loop process for employees to act when ethical situations arise.

3.8.2 IP protection and data-sharing agreements

The functioning of AI supply chains requires a great amount of datasets that uniquely blend product inventory levels, supplier metrics, customer metrics, as well as transportation performance metrics. Collaboration with supply chain partners increasingly requires some level of enhanced cooperation, requiring the exchange of sensitive operational data across organizations. Concerns regarding the data-sharing process are focused on data ownership and usage, including the protection of intellectual property rights.

Data-sharing agreements establish required aspects regarding ownership of data, usage and access rights to derived insights, and mechanisms for sharing value. Contract provisions covering the ownership of proprietary algorithms and training data usage methods are also required in AI development contracts. Privacy-preserving methods such as federated learning can allow companies to work together while preserving their raw data and sharing intelligence, which assists in the protection of intellectual property.

Trust in AI relies heavily on transparency and traceability, requiring supply chain partners to justify data usage and maintain an auditable record of model decision-making. The strength of decisions made by AI increases when they influence contract negotiations, pricing strategies, and customer segment classifications.

3.8.3 Cybersecurity in AI-connected warehouses

Combining AI, robotics, and IoT in warehouses enhances operational efficiency but also introduces new cybersecurity risks, including ransomware, data breaches, and AI manipulation aimed at disrupting data integrity. Cybercriminals could target vision systems to disrupt barcode recognition and manipulate quality inspections. This could alter inventory models, leading to inaccurate stock reorder patterns with significant operational implications.

3.8.4 Essential components of strong AI security frameworks which organizations must develop

- **Endpoint Protection:** Secure sensors, robots, and IoT devices.
- **Access Controls and Identity Management:** Implement strict role-based access across AI platforms.
- **Encryption:** Protect data both at rest and in transit.
- **Adversarial Robustness Testing:** Regularly test AI models against mali cious inputs.

Example: An adversary could manipulate vision systems to misread barcodes, disrupting reorder patterns. Robust security frameworks are essential to safeguard operational continuity.

3.8.5 Regulatory landscape (EU AI Act, U.S. NIST AI Risk Management Framework, India Digital Personal Data Protection Act)

Lots of governments around the world are implementing regulations that ensure AI systems run according to safety, transparency, and ethical guidelines. Supply chain leaders will need to pay attention to the evolving regulations while ensuring compliance with those regulations. The following three regulatory pillars: The European Union (EU) has established the EU AI Act, which categorizes AI systems based on risk into four levels: unacceptable, high, limited, and minimal. The high-risk AI applications that are deployed in critical infrastructure or logistics operations, or worker management systems, are now required to meet stringent standards in transparency, human oversight, robustness, and documentation standards. The supply chain systems that require surveillance-type tools along with worker scheduling and quality control applications will be considered high risk under the above categories.

The U.S. NIST AI Risk Management Framework serves as a voluntary framework aiding organizations in understanding and mitigating risks associated with AI systems. The framework is centered on four fundamental principles of trustworthy AI, which include fairness and accountability in addition to privacy and security. Organizations should develop their own internal AI governance frameworks that will supply them with risk assessments and documentation of AI lifecycle processes.

The Digital Personal Data Protection Act of India regulates personal data processing, including data used in AI models. Organizations handling customer and employee data for forecasting, personalization, or automation must obtain user consent, minimize data use, and ensure data security. Non-compliance can result in financial penalties and reputational damage [11].

3.9 MEASURING VALUE AND REGION OF INTEREST (ROI)

As organizations integrate AI into their supply chain transformations, measuring the returns becomes essential. AI requires not just financial investment but also significant changes in systems, operations, and workforce. Tracking ROI is crucial for understanding the investment's value and refining strategies to scale AI responsibly.

Measuring AI's value in the supply chain requires a holistic approach. Alongside monetary outcomes like cost savings and productivity improvements, firms should consider non-financial benefits such as service level enhancements, reduced environmental impact, and long-term agility. This is often assessed through a mix of KPIs that reflect both operational efficiency and strategic outcomes (Figure 3.4).

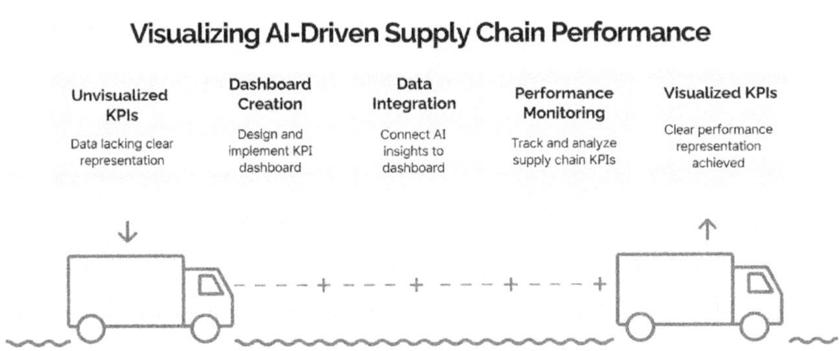

Figure 3.4 KPI dashboard illustrating AI-driven supply chain performance improvements.

3.9.1 Financial KPIs: tangible performance gains

Financial KPIs highlight the financial value generated by AI in supply chains. One key indicator is inventory turns, reflecting how effectively a company utilizes its inventory. AI improves demand forecasting and replenishment, reducing overstocking and understocking, which leads to faster inventory turns, lower holding costs, and better working capital. Another important metric is OTIF (On Time In Full), which measures how often customer orders are delivered as expected. AI positively impacts OTIF by increasing efficiencies with warehouse workflows, improving last-mile logistics, and predicting disruptions before they happen, thus limiting delays and damage. Increasing OTIF reduces penalties, improves customer satisfaction, and increases customer loyalty. Finally, expedited freight costs are a real cost in any supply chain that are controllable by AI. In a traditional supply chain, expedited freight is associated with last-minute shipping and considered necessary due to poor planning or a demand spike, sustained by suppliers employing last-minute logistics as a means of securing supply. With AI, it becomes possible to predict shifts in demand and monitor the data of supply chain movement in real time. It is possible to avoid reactive logistics decisions, greatly diminishing the use of costlier expedited freight.

3.9.2 Non-financial KPIs: broader strategic benefits

While financial KPI's are more focused on short-term immediate gains, non-financial KPI's demonstrate the long-term value AI contributes to service reliability, sustainability, and operational resilience.

Carbon intensity is a great example of a non-financial metric used to evaluate the grams of CO emissions per unit of product transported or produced. AI supports the goal of sustainability by optimizing routes and vehicle loads, and managing waste and energy in warehousing. AI applications improve carbon intensity by reducing waste and controlling and optimizing fuel consumption, which is beneficial to environmental sustainability goals and ESG compliance.

Other non-financial KPI's that organizations track include service level uplift, which encompasses product availability, order accuracy, and consistency in lead time. AI supports organizations to provide improved service levels by being proactive in terms of inventory allocation and visibility in stock levels while they are in motion, in addition to being more responsive to fluctuations in demand.

Organizations track lead time variability, which refers to the fluctuating stability of delivery and production schedules. AI enables organizations to understand lead time variability by identifying and short-circuiting supplier delays or providing recommendations for impending order scheduling, enabling organizations to identify at-risk purchase orders to mitigate operational impacts before they occur. Reduced variability enables organizations to achieve increased planning, reduced disruptions, and better resilience in their supply chains.

3.9.3 A unified performance view: the balanced scorecard approach

Many organizations use a balanced scorecard approach to contextualize ROI. Instead of focusing on cost savings, the balanced scorecard looks at the effects of AI on four perspectives—financial performance, customer service, internal processes, and sustainability.

By taking this approach, decision-makers can observe whether the benefits of AI can be measured along both tangible and intangible paths while supporting operational performance in terms of trajectory toward long-term strategic objectives. With this multi-dimensional view, organizations can assess whether their investments in AI are delivering value not only to the bottom line but also for customers, employees, and the planet.

3.9.4 Figure: sample AI SCM KPI dashboard

Visual suggestion: A quadrant-style dashboard with segments showing improvements in key AI-enhanced metrics—such as higher inventory turns, better OTIF performance, reduced emissions, and faster response times—illustrated through simple trend lines, gauges, and highlights of performance gains [12].

3.10 FUTURE OUTLOOK: TOWARD AUTONOMOUS SUPPLY NETWORKS

The move from smart support systems to autonomous networks is happening and will happen faster, as automation and AI continue to change supply chains as we move through the 2030s. The future of supply chains is not just about how quickly you can respond to change; they are about anticipating change and acting and adapting autonomously with technology. This chapter explores a number of examples of emerging innovations (generative agents, quantum computing, swarm robotics, etc.) and the implications for the future of human–AI teaming over the next decade.

3.10.1 Generative agents and self-orchestrating supply chains

Generative AI has come to represent capabilities beyond content creation and responses to queries. Within future supply chains, generative agents will play a far more significant role in planning. Generative agents are AI agents that can independently evaluate a complex supply chain environment, create a plan, change appropriate parameters in real time, and orchestrate multiple systems and partners.

For instance, let's assume one of the suppliers' factories has a major disruption, such as a natural disaster, and you have a generative agent embedded in your procurement system. The generative agent evaluates alternative suppliers' capabilities at the sourcing level, evaluates the impact of the change on lead time and cost, autonomously negotiates contracts, and makes the appropriate changes to the planning system—all before the human involved is even aware of the disruption.

These "self-orchestrating" intelligent ecosystems will necessarily use some form of agent-based supply chain systems where planning, sourcing, logistics, and customer service systems are in constant communication and optimization with one another. AI agents will autonomously monitor and coordinate workflows across regions and adjust a supply chain structure to include global events, economic conditions, or consumer behaviors.

3.10.2 Quantum optimization and swarm robotics

The emergence of quantum computing is also a new technology shaping the supply chain of the future. As traditional optimization models struggle with the scale and complexity that global supply chains represent, organizations will be able to handle exponentially larger datasets and combinatorial problems such as route optimization, production sequencing, and resource allocation when using quantum optimization. To illustrate, when trying to optimize the routes of hundreds of delivery vehicles across a wide city in response to changing weather and traffic, this calculation becomes a gigantic computational challenge. In contrast, quantum algorithms will be able to process these variables concurrently to generate optimal routes in almost real time.

This can also be layered on top of swarm robotics—groups of robots with AI that have been inspired by the collective behaviors of biological swarms, such as ants or bees. AI will (mostly) operate in factories, warehouses, and even in-field supply contexts, coordinating the physical movement of goods, tracking inventory, and adjusting tasks based on real-time sensor information.

Swarm robotics will drive significant change in terms of resiliency and flexibility. If a single robot stops working or becomes stranded in a single area, the rest of the swarm can change their behaviors dynamically to continue performing their tasks.

3.10.3 Human–AI collaboration in 2030: augmentation over replacement

While supply chains become more automated, humans will still play a critical role as strategic overseers/exception managers instead of manual operators. When we think of human–AI collaboration in the year 2030, we will think about augmentation rather than replacement. Supply chain professionals on the front lines will be using digital co-pilots—AI that recommends actions based on real-time recommendations, identifies anomalies in behavior, and generates next actions to mitigate problems. In addition to planners and analysts in the supply chain control towers, there will also be cognitive AI assistants who can integrate knowledge from thousands of data sources.

Human training and education will also be redesigned to enable collaboration, as supply chain managers develop fluency with AI tools, and decision-making logic and ethics will emerge. The workforce will see growth in emerging roles like "AI ethicists," "algorithm trainers," and "AI integration specialists."

3.10.4 A self-adaptive, resilient future

A future autonomous supply network is a specification of new realities created by exponential technologies, not a concept that the new manager has. Swarm intelligence, generative agents, and quantum optimization will turn supply chains into living, breathing systems that can adapt and learn. In addition to being involved, humans will oversee and direct these networks, guiding supply chains into a future that is quicker, more intelligent, and far more robust [13].

3.11 CONCLUSION

AI has advanced supply chains from basic efficiency to resilient, self-optimizing ecosystems. It aids in demand forecasting, logistics routing, and risk management while adhering to ethical principles and maintaining human discretion. AI's transformative power rests not only on its automation of operations but also on its capacity for continuous learning and adaptation. Supply chains of the future will balance cost efficiency with resilience, speed with sustainability, and automation with accountability. Organizations that embrace AI in alignment with these objectives will realize agility during times of disruption and stability under stress. Tied to this interdependence of global supply chains will be the next generation of competitive advantage: an integrated partnership between human knowledge and AI intelligence.

3.12 FIVE GOLDEN RULES FOR AI IN SUPPLY CHAINS

a. To enable sustainable and responsible adoption of AI, organizations should follow five guiding principles.
b. First, prepare for data readiness by ensuring data is clean or standardized and governed effectively before deploying AI.

c. Second, promote human–AI collaboration, in which technology enhances or augments human judgment rather than replaces it.
d. Third, create systems that are agile and flexible, enabling models to change over time by being retrained and ethically responding to active feedback loops.
e. Fourth, ensure that explainability, security, and fairness are built into all AI frameworks to align with ethical and regulatory considerations. Fifth and finally, enable ecosystem collaboration, encouraging data-sharing protocols and partnerships that are transparent to help increase mutual visibility and predictive capabilities.

As a collective whole, these principles ensure AI-driven supply chains remain intelligent, adaptive, and firmly grounded in ethics. Practitioners must take immediate action by assessing their AI development stages and launching essential projects that integrate AI into operations with a focus on collaboration between humans and AI. Long-term success hinges on fostering cross-functional talent and enhancing ethical AI understanding while aligning technology with business goals.

Researchers face challenges in translating theoretical concepts into scalable business solutions, requiring interdisciplinary collaboration among operations, computer science, behavioral economics, and regulatory studies. Key areas of investigation include multi-agent coordination, explainable AI in logistics, and federated learning for supplier networks.

To build intelligent, sustainable, and resilient supply chains, organizations must adopt AI through ethical and inclusive practices.

REFERENCES

1. R. Sharma, A. Shishodia, A. Gunasekaran, H. Min, and Z. H. Munim, "The role of Artificial Intelligence in supply chain management: Mapping the territory," *International Journal of Production Research*, vol. 60, no. 24, pp. 7527–7550, 2022.
2. T. Cadden, D. Dennehy, M. Mantymaki, and R. Treacy, "Understanding the influential and mediating role of cultural enablers of ai integration to supply chain," *International Journal of Production Research*, vol. 60, no. 14, pp. 4592–4620, 2022.
3. J. D. Taylor, "Cross-industry differences in business failure rates: Implications for portfolio management," *Commercial Lending Review*, vol. 13, p. 36, 1997.
4. N. Singh, "Demystifying ai-enhanced search systems: A technical deep dive," *Journal of Computer Science and Technology Studies*, vol. 7, no. 2, pp. 489–495, 2025.
5. S. de Daniel António Costa, "A stock quantification of a demand planning process improvement, under a s&op context," Master's thesis, Universidade do Minho (Portugal), 2019.

6. P. Kumar, D. Choubey, O. R. Amosu, and Y. M. Ogunsuji, "AI-enhanced inventory and demand forecasting: Using ai to optimize inventory management and predict customer demand," *World Journal of Advanced Research and Reviews*, vol. 23, no. 1, pp. 1931–1944, 2024.
7. S. C. Graves and S. P. Willems, "Supply chain design: Safety stock placement and supply chain configuration," In: A. G. de Kok & S. C. Graves (Eds.), *Handbooks in Operations Research and Management Science* (vol. 11, pp. 95–132). Elsevier, 2003.
8. U. Nweje and M. Taiwo, "Leveraging Artificial Intelligence for predictive supply chain management, focus on how ai-driven tools are revolutionizing demand forecasting and inventory optimization," *International Journal of Science and Research Archive*, vol. 14, no. 1, pp. 230–250, 2025.
9. B. Hassani and B. K. Hassani, *Scenario Analysis in Risk Management.* Springer, 2016.
10. H. Mamoojee-Khatib, J. Antony, V. Teeroovengadum, J. A. Garza-Reyes, G. L. Tortorella, M. Foster, and E. A. Cudney, "A systematic review of lean implementation frameworks and roadmaps: Lessons learned and the way forward," *TQM Journal*, vol. 37, no. 1, pp. 264–287, 2025.
11. M. A. Camilleri, "Artificial Intelligence governance: Ethical considerations and implications for social responsibility," *Expert Systems*, vol. 41, no. 7, p. e13406, 2024.
12. M. Mogollon and M. Raisinghani, "Measuring ROI in e-business: A practical approach," *Information Systems Management*, vol. 20, no. 2, 2003.
13. N. Samaan and A. Karmouch, "Towards autonomic network management: An analysis of current and future research directions," *IEEE Communications Surveys & Tutorials*, vol. 11, no. 3, pp. 22–36, 2009.

Chapter 4

Explainable AI for All

Sustainable supply chains and inclusive accessibility for persons with disabilities

Balaji Venkatesan, Thiruselvan Subramanian, Huay Ling Tay, and Dhanalakshmi Hari

4.1 INTRODUCTION

Artificial intelligence (AI) has been utilized in nearly every field of society, including logistics for e-commerce and the optimization of the global supply chain, as well as predictive and personalized assistive technologies that have become highly personal in the 21st century. These advancements have led to extraordinary changes in velocity, productivity, and accuracy. Their trade-off, though, is a grave one, which is an increasing dependence on opaque so-called black-box models whose inner workings are not disclosed to end-users, policymakers, and even developers. This lack of interpretability raises serious concerns regarding trust, fairness, and accountability, as the impact of AI-based decisions can directly affect the lives of individuals and the environment around them.

These issues overlap on several points, but two can be considered complex and ethical in the technical sense. The first one is sustainable supply chain management, where algorithmic decisions extend far beyond operational efficiency to change global trends in environmental impact, labor policies, and long-term economic stability. Such black-box decision-making could conceal predatory activities, greenwashing, or unfair trade-offs, which would not only undermine sustainability but also corporate responsibility. The second domain is inclusive accessibility for persons with disabilities (PWDs), in which real-time sign language translators, adaptive screen readers, or speech-to-text systems aided by AI have a direct influence on the fundamental human rights, personal autonomy, and equality in access to education, employment, and society in general. In the context of persons with disabilities (PWDs), AI systems might fail to bring about the obstacle destruction and instead, they cement them and deprive an agency of its authority rather than providing it.

Explainable AI (XAI) in both cases becomes an extremely valuable solution that bridges the ever-expanding gap between algorithmic capabilities and human agency. XAI helps prevent scenarios where technological

 DOI: 10.1201/9781003660163-4

effectiveness is achieved at the cost of ethical responsibility by making rationale behind AI predictions and decisions comprehensible. It enables different stakeholders, such as supply chain managers, regulators, caregivers, educators, and end-users with disabilities, to genuinely critique and alter AI results to achieve environmental, social, and human-oriented purposes. Lastly, through the combination of the principles of XAI, AI systems can be defined as providers of efficiency, as well as agents of transparency, trust, and inclusivity, performing both the functions of initiating innovation and safeguarding ethical and social values.

1. Sustainable supply chain management, where decisions impact the planet, labor conditions, and financial stability.
2. Inclusive accessibility for PWDs—where judgments affect human rights, autonomy, and balanced participation in society.

In both scenarios, XAI addresses the imbalance between the power of algorithms and human control, providing a way to be transparent without compromising performance [1].

4.2 LITERATURE SURVEY

XAI is becoming an essential part of AI studies, aiming to answer the question of how to make machine learning models decipherable and comprehensible to human stakeholders. The problem of explainability is particularly high in areas where decisions have significant ethical, social, or environmental consequences.

4.2.1 XAI foundations and methods

The early XAI studies focused on interpretable models, which included decision trees and linear regression because they were inherently transparent [2]. With the rise of deep learning models, post-hoc explanation techniques like local interpretable model-agnostic explanations (LIME) [3] and SHapley Additive exPlanations (SHAP) [4] have been in the spotlight, where the local and global interpretability of black-box models can be achieved. Gradient-Weighted Class Activation Mapping (Grad-CAM) [5] also contributed to the interpretability of computer vision by visualizing the spatial parts that affect model predictions. However, several scholars have highlighted that explainability must also account for usability and accessibility so that explanations are understandable by diverse user groups [6].

4.2.2 XAI in sustainable supply chain management

AI applications in supply chain management have demonstrated significant improvements in demand forecasting, route optimization, and inventory control [7]. Yet, opaque algorithms risk concealing environmental and ethical impacts. Some studies [8,9] underscore the importance of transparent decision support systems that incorporate sustainability metrics such as carbon footprint, waste reduction, and fair labor compliance. Furthermore, the integration of XAI in logistics has been shown to enhance stakeholder trust and facilitate regulatory compliance [10].

4.2.3 XAI in accessibility for PWDs

Accessibility technologies such as AI-powered screen readers and sign language recognition systems [11] are transforming inclusion for PWDs. However, non-transparent AI decisions may inadvertently exclude users by misinterpreting inputs or failing to adapt to personalized needs [12]. The study cited in reference [13] and the W3C Accessibility Guidelines [14] stress that accessible AI systems must be explainable by design, empowering users to understand and correct system behavior. Recent studies also suggest co-designing explainable assistive AI with PWDs to ensure cultural and contextual relevance [15].

4.2.4 Standards and policy perspectives

International frameworks such as the UN Convention on the Rights of PWDs (UNCRPD) [16], the Sustainable Development Goals (SDGs) [17], and India's National Education Policy 2020 (NEP 2020) [18] emphasize the ethical obligation to ensure accessibility, inclusivity, and environmental responsibility. Embedding XAI principles within these contexts can help achieve not only compliance but also trustworthiness and societal benefits. Table 4.1 highlights research on XAI for accessibility and supply chains, emphasizing transparency, inclusivity, and trust. It also identifies gaps in accessible XAI methods and domain-specific applications, such as logistics and procurement.

4.2.5 Benefits of disability inclusion in supply chains using XAI

Integrating disability inclusion into supply chain operations delivers far-reaching benefits for business performance, social equity, and regulatory compliance. When combined with XAI, these benefits are amplified through transparency, trust, and data-driven accountability.

Table 4.1 Literature survey of recent research on explainable artificial intelligence (XAI), accessibility, and supply chain applications

Paper	*Abstract summary*	*Main findings*	*Methodology*
A survey of accessible explainable artificial intelligence research [19]	Accessible explainable AI is crucial for digital inclusion, but current XAI techniques are not accessible to people with disabilities.	• The study highlights a significant lack of research in accessible explainable artificial intelligence (AXAI) and the importance of including the disability community in XAI development. • Current XAI techniques are inaccessible to individuals with visual impairments due to their reliance on visual explanations. • There is a noted gap in research on adapting AI explanations for users with different accessibility needs and integrating them with assistive technologies.	• Conducted a systematic literature review using the Parsifal1 tool. • Searched multiple academic databases: Scopus, ISI Web of Science, IEEE Digital Library, Google Scholar, EI Compendex, ACM Digital Library, CINAHL, and PubMed. • Created specific search strings to capture intersections between XAI and accessibility. • Initially selected 727 papers from 2019 to 2024. • Analyzed eight articles based on title and abstract, resulting in four relevant papers. • Applied exclusion criteria to focus on papers addressing the accessibility of AI explanations.
A review of explainable artificial intelligence in supply chain management using neurosymbolic approaches [20]	This paper reviews the use of explainable neurosymbolic AI approaches in supply chain management but does not address accessibility for persons with disabilities.	• Neurosymbolic AI research in supply chain management is lagging behind, primarily using outdated neurofuzzy approaches. • There is a significant gap in the application of neurosymbolic AI in areas such as performance evaluation, sustainability, and sectors like pharmaceuticals and construction.	• Systematic literature review conducted to explore neurosymbolic AI in supply chains. • Automated search of peer-reviewed literature databases. • Search strategy inspired by previous literature reviews.

(*Continued*)

Table 4.1 (Continued) Literature survey of recent research on explainable artificial intelligence (XAI), accessibility, and supply chain applications

Paper	*Abstract summary*	*Main findings*	*Methodology*
		• The paper proposes five pillars for future research to address these gaps and promote more advanced and diverse applications of neurosymbolic AI in supply chain management.	• Use of specific queries to identify relevant studies. • Time frame limited to papers published from 2000 to 2022. • Rigorous filtering based on quality and relevance criteria.
AI and accessibility: breaking barriers for people with disabilities [21]	AI offers transformative opportunities to enhance autonomy, quality of life, and social inclusion for people with disabilities, but challenges remain regarding algorithmic bias, cost, usability, and ethics.	• AI has the potential to revolutionize accessibility for people with disabilities by improving autonomy, inclusivity, and daily living activities. • Significant challenges include algorithmic bias, cost limitations, usability issues, and ethical concerns that need to be addressed. • Addressing these challenges is crucial to realizing AI's potential in creating a more inclusive society.	• The methodology used in Susan Ferebee (2025) is a non-systematic narrative review and case study analysis, synthesizing interdisciplinary literature and real-world case studies to explore the intersection of AI and accessibility.
Toward trustworthy AI in demand planning: defining explainability for supply chain management [22]	The paper proposes a user-centered definition of explainable AI for demand planning in supply chain management to enhance trust and usability.	• The paper also examines the leading definitions of explainability in AI and suggests a customized definition for supply chain management. It employs the user-centered method to meet the practical requirements of non-technical users. The proposed definition will facilitate the advancement of interpretable AIs that will enhance usability and user confidence in demand planning tools.	• Literature review of prominent definitions of explainability in AI. • Development of a tailored definition of explainability for supply chain management using a user-centered approach.

(*Continued*)

Table 4.1 (Continued) Literature survey of recent research on explainable artificial intelligence (XAI), accessibility, and supply chain applications

Paper	*Abstract summary*	*Main findings*	*Methodology*
The role of explainable AI in enhancing trust and transparency in supply chain risk mitigation [23]	Explainable AI techniques can improve trust and transparency in supply chain risk mitigation.	• Application: XAI methods including SHAP, LIME, and counterfactuals are popular in supply chain risk management applications. XAI enhances the interpretability of decisions made by AI, enhances the trust of stakeholders, and facilitates compliance with regulations. XAI plays a crucial role in reducing distrust, enhancing transparency, and increasing resilience in supply chain risk management.	• Systematic review of existing literature. • Search conducted using databases: Scopus, Web of Science, IEEE Xplore, Science Direct. • Two-phase screening process following PRISMA guidelines. • Narrative synthesis using thematic analysis. • Synthesis of results to highlight thematic alignments in XAI techniques, transparency outcomes, trust dynamics, and decision-making contexts.
Designing accessible, explainable AI (XAI) experiences [24]	The paper discusses challenges in designing accessible and explainable AI experiences, particularly for persons with disabilities.	• The article discusses the challenges of applying explainable artificial intelligence (XAI) in real-world settings, particularly at the intersection of accessibility and explainability. • Two primary concerns are identified: accessibility at the interface and tailoring explanations to individuals' diverse and changing needs. These issues are illustrated through examples in aging-in-place and mental health support systems.	• Not mentioned (the abstract does not specify any methodology or methods used in the study).

(Continued)

Table 4.1 (Continued) Literature survey of recent research on explainable artificial intelligence (XAI), accessibility, and supply chain applications

Paper	*Abstract summary*	*Main findings*	*Methodology*
Explainable artificial intelligence and agile decision-making in supply chain cyber resilience [25]	Explainable AI can enhance transparency and agile decision-making to improve cyber resilience in supply chains.	• Explainable artificial intelligence enhances transparency, significantly contributing to agile decision-making for improving cyber resilience during supply chain cyberattacks. • The study reveals a predominantly positive attitude toward explainable artificial intelligence in decision support systems. Transparency, explainability, and interpretability are emphasized as important themes in explainable artificial intelligence.	• Experimental design to collect empirical data. • Research model to address the impact of explainable AI on decision-making. • Serial mediation path to examine relationships between variables. • Post-hoc analysis using text analysis of tweets.
Enhancing supplier selection through explainable AI: A transparent and interpretable approach [26]	The paper proposes using explainable AI to enhance supplier selection in procurement, making the process more transparent and interpretable.	• The development of explainable AI enhances supplier selection by making machine learning models more interpretable and understandable. • A post-hoc explainable AI structure is used to identify and address biases and data gaps, improving the reliability and fairness of the supplier selection process. The approach increases end-user trust by providing transparent and understandable explanations.	• A blend of numerical, measurable, and simulated intelligence approaches to evaluate suppliers based on objective and subjective characteristics. • Use of explainable AI to make AI predictions more interpretable and transparent. • Post-hoc explainable AI structure to identify and address biases and data gaps.

(*Continued*)

Table 4.1 (Continued) Literature survey of recent research on explainable artificial intelligence (XAI), accessibility, and supply chain applications

Paper	*Abstract summary*	*Main findings*	*Methodology*
Big data analytics and explainable AI in supply chain management [27]	This paper explores the potential of big data analytics and explainable AI to optimize supply chain management decisions.	• The study proposed a model to predict optimal shipment details using big data analytics and explainable AI in supply chain management. • The results showed significant optimization in shipment logistics, leading to cost reductions and operational performance improvements. • The use of XAI techniques increased transparency and trust in AI-driven decisions, facilitating an easy understanding of complex AI models for managerial purposes.	• Big data analytics. • Explainable artificial intelligence (XAI). • Predictive modeling for optimal shipment details. • Managerial data analysis based on business logic.
Toward explainable artificial intelligence (XAI) in supply chain management: a typology and research agenda [28]	This paper discusses the need for explainable AI (XAI) in supply chain management to address the black-box problem of AI.	• AI's potential in Supply Chain Management (SCM) is limited by the lack of human involvement in AI decision-making processes. XAI is underdeveloped and superficially addressed in SCM literature. The link between XAI and SC performance is evident but not well researched.	• Conducted an integrative literature review following Torraco's approach. • Performed a literature search using Google Scholar, Web of Science, and Scopus databases with the search string "Artificial Intelligence in supply chain management." • Filtered results for English language and peer-reviewed articles with the keywords "AI" and "SCM." • Reviewed abstracts individually to ensure relevance and reliability. • Final selection included 18 articles for the review.

(*Continued*)

Table 4.1 (Continued) Literature survey of recent research on explainable artificial intelligence (XAI), accessibility, and supply chain applications

Paper	*Abstract summary*	*Main findings*	*Methodology*
Investigating the role of artificial intelligence in developing eco-friendly assistive technologies for people with disabilities [29]	AI has immense potential to enhance assistive technologies for people with disabilities, making them more sustainable and inclusive.	• Artificial intelligence enhances assistive technologies for people with disabilities by providing customized support for mobility, visual, and hearing aids. • AI improves independence through voice-activated devices and innovations in prosthetics and navigation apps. • AI integrates environmental sustainability into assistive technologies, including efficient waste management and recycling, with blockchain ensuring transparency.	• Not mentioned (the abstract does not specify any particular methodology or methods used in the study).
Accessible cultural heritage through explainable artificial intelligence [30]	Explainable AI can improve the accessibility of cultural heritage for minorities and underserved audiences.	• The paper proposes including minorities as users and evaluators of XAI techniques to enhance accessibility in cultural heritage. • XAI can bring art closer to new audiences and innovate how art is taught and learned, making it more accessible and understandable. • AI models can assist in generating explanations to make art more widely understood and accessible, adapting interfaces for diverse audiences.	• Proposing challenges and research questions to stimulate collaboration and improve the user experience. • Using deep neural networks (DNNs) for photo classification and visual explanations. • Utilizing cultural heritage as a testing ground for XAI techniques. • Forming an interdisciplinary research consortium for mutual learning.
Toward explainable and accessible AI [19]	The paper is about working toward explainable and accessible AI.	• Not mentioned (the paper does not present specific findings or conclusions from a study).	• Not mentioned (the paper does not describe a methodology for a study).

(Continued)

Table 4.1 (Continued) Literature survey of recent research on explainable artificial intelligence (XAI), accessibility, and supply chain applications

Paper	*Abstract summary*	*Main findings*	*Methodology*
Find My Things: personalized accessibility through teachable AI for people who are blind or low vision [31]	Teachable AI systems can enable personalized accessibility applications for people who are blind or have low vision.	• Teachable AI systems enable users with disabilities to personalize applications to their specific needs. • Find My Things is an example of teachable AI, allowing blind or low-vision individuals to find personal items in 3D space. • Real-world examples with quality issues improve model personalization by matching real-world usage conditions.	• Participatory design with a citizen design team of eight blind or low-vision young people. • Three day-long workshops for design process collaboration. • Data collection through recordings, think-aloud elements, and telemetry data. • Teaching users to record videos for training a few-shot object recognition model. • Personalization algorithm for creating mean feature embeddings. • Lightweight localization algorithm for quick object localization.
A Blockchain-driven food supply chain management using QR code and XAI-faster RCNN architecture [32]	The paper proposes an explainable AI-based system to evaluate food product contents and quality using Blockchain technology and QR codes.	• The proposed XAI-based Faster RCNN method requires less response time and has the lowest processing time compared to other existing methods. • The accuracy achieved by the proposed method is 99.53%, outperforming other methods like DR-SCM and BIFTS. • The method performs with significantly increasing accuracy and decreasing losses.	• XAI-based Faster RCNN model for evaluating food item contents and identifying harmful ingredients. • Elliptic curve integrated encrypted scheme (ECIES) for validating communication tokens. • Artificial rabbit optimization (ARO) for user registration and key assignment. • Blockchain technology for digitizing food production data. • QR codes for accessing food information. • Shapley additive explanation for legal purposes related to nutritional information.

(Continued)

Table 4.1 (Continued) Literature survey of recent research on explainable artificial intelligence (XAI), accessibility, and supply chain applications

Paper	*Abstract summary*	*Main findings*	*Methodology*
Toward human-understandable, explainable AI [33]	The paper discusses the need for explainable AI systems to ensure transparency and understanding for users, but does not specifically address sustainable supply chains or accessibility for people with disabilities.	• There is a need to move toward explainable AI (XAI) to ensure responsible and trusted AI systems. • XAI is expected to enable "third-wave AI systems" that understand context and build explanatory models. • Type-2 fuzzy logic systems (FLS) and fuzzy rule-based systems (FRBS) offer a promising approach to achieving XAI by balancing explainability with prediction accuracy.	• Introduction of XAI concepts focusing on transparency, causality, bias, fairness, and safety. • Use of type-2 FLS to mimic human thinking and provide explainable models. • Employment of FLSs and FRBSs models. • Evolutionary systems to produce FRBSs with short IF-Then rules to be more accurate and interpretable.
XAI—explainable artificial intelligence [34]	Explainability is essential for users to effectively understand, trust, and manage powerful artificial intelligence applications.	• Explainability is essential for users to understand, trust, and manage AI applications. • XAI systems seek to bring the behavior of AI closer to human comprehension. • AI systems have persistent issues of balancing explainability and performance.	• The paper addresses the application of machine learning methods and the role of XAI to explain. It describes the general guidelines to develop efficient XAI systems and states different evaluation schemes. Nevertheless, it fails to identify the specific methodology applied in the study, lacking any clear indication that it is a review of ideas and issues, but not a particular study with a definite methodology.

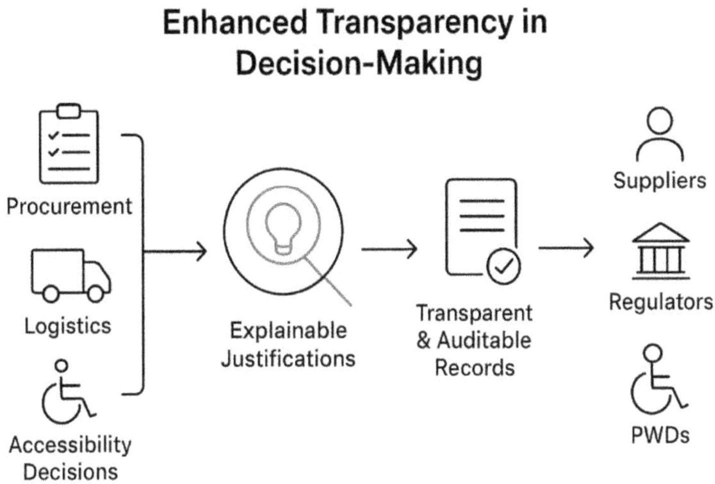

Figure 4.1 Enhanced transparency in decision-making.

4.2.5.1 Enhanced transparency in decision-making

XAI provides clear, interpretable justifications for procurement, logistics, and accessibility-related decisions. This ensures that the inclusion of PWDs is not treated as an opaque or optional process, but as an auditable and traceable element in every supply chain decision. Stakeholders ranging from suppliers to policy regulators can see exactly how accessibility criteria influenced outcomes. Figure 4.1 illustrates how XAI enhances transparency by making complex decision-making processes more understandable and accountable.

4.2.5.2 Improved supplier engagement and compliance

By making supplier evaluation criteria explainable, PWD inclusion standards (e.g., accessible facilities and inclusive hiring practices) can be communicated in an actionable form. Counterfactual explanations can highlight specific improvements suppliers need to make to achieve compliance, such as upgrading workplace accessibility or adopting assistive technology infrastructure. Figure 4.2 presents the workflow of inclusive supplier compliance assessment, where XAI techniques ensure fair, transparent, and accessible evaluation of suppliers.

4.2.5.3 Operational efficiency with equity

XAI-driven decision support can balance operational metrics (cost, delivery time) with inclusion goals. For instance, route optimization algorithms can be designed to prioritize both efficient transport and accessibility for PWD

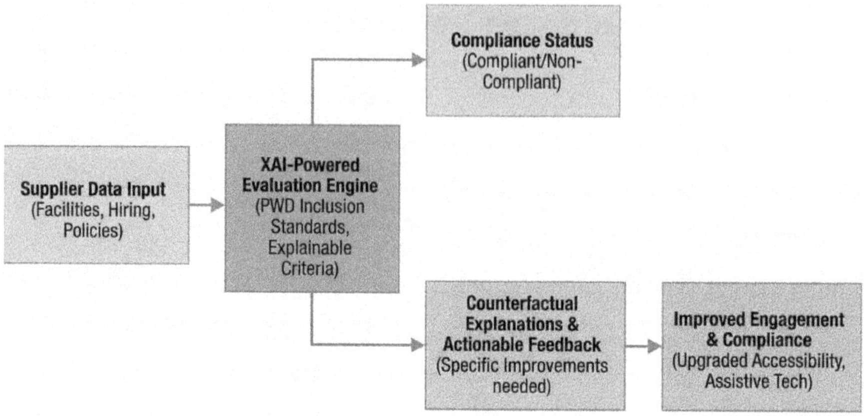

Figure 4.2 Workflow of inclusive supplier compliance assessment using XAI.

employees or customers, and the reasoning behind this trade-off can be clearly articulated.

4.2.5.4 *Regulatory readiness and standards alignment*

Inclusion frameworks must meet UNCRPD, WCAG, and ISO 20400 Sustainable Procurement guidelines. XAI creates automated, verifiable audit trails that demonstrate compliance with these standards, streamlining reporting for sustainability and inclusion audits.

4.2.5.5 *Trust and adoption by diverse stakeholders*

PWD stakeholders often face barriers when the rationale behind AI-powered services is unclear. XAI interfaces, especially when adapted into multiple accessible modalities (e.g., text-to-speech, sign language avatars, and Braille-compatible formats), foster confidence by allowing users to understand and interact with AI tools in a way that meets their needs.

4.2.5.6 *Sustainability and social impact integration*

Inclusion and sustainability are mutually reinforcing; accessible supply chains promote workforce diversity while also supporting environmentally responsible operations. XAI enables decision-makers to visualize the link between inclusion efforts and sustainability Key Performance Indicators (KPIs), such as reduced waste, optimized resource use, and socially responsible sourcing.

4.2.5.7 Continuous improvement through feedback loops

XAI-powered feedback systems allow PWDs, suppliers, and logistics managers to provide targeted input on model performance. This creates a virtuous cycle where accessibility and sustainability practices improve over time, guided by transparent and explainable data insights.

4.3 XAI: DEFINITIONS AND RELEVANCE

This section introduces the concept of XAI, outlining key definitions that frame its role in making AI systems transparent and interpretable. It further highlights the relevance of XAI in building trust, accountability, and accessibility across diverse application domains.

4.3.1 What is XAI?

XAI refers to methods and techniques that make the output and internal mechanics of AI systems understandable to human users. The **POURC** (Perceivable, Operable, Understandable, Robust, Compatible) principles from accessibility guidelines can be adapted here to ensure that AI models are not only accurate but also interpretable and accessible. Figure 4.3

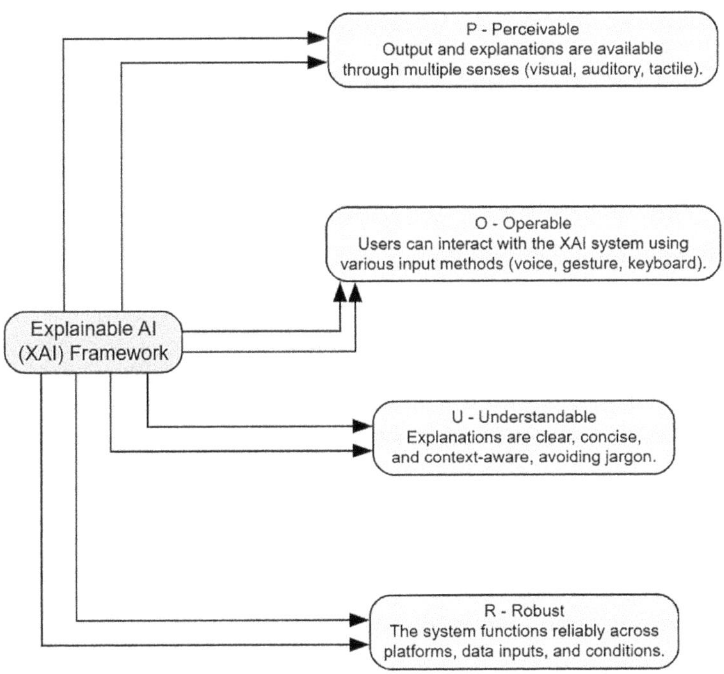

Figure 4.3 Explainable AI framework aligned with accessibility principles.

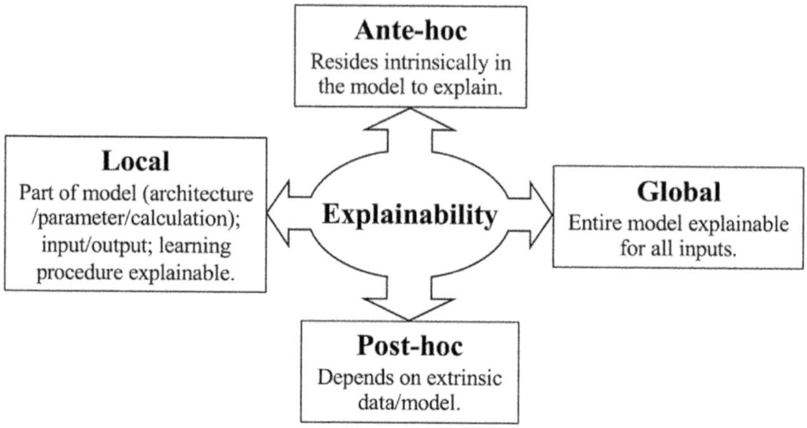

Figure 4.4 Ante-hoc vs. post-hoc and local vs. global perspectives of explainability.

illustrates an XAI framework designed to align with accessibility principles, ensuring inclusive, transparent, and user-friendly decision support. Figure 4.4 compares ante-hoc vs. post-hoc methods and local vs. global perspectives, highlighting different approaches to achieving explainability in AI models.

4.3.2 Types of explainability

- **Intrinsic Explainability:** Models designed to be inherently interpretable (e.g., decision trees and rule-based systems).
- **Post-Hoc Explainability:** Methods applied after training to interpret complex models (e.g., LIME, SHAP, and Grad-CAM).
- **Local vs. Global Explanations:** Explaining individual predictions versus overall model behavior.

4.3.3 Intrinsic explainability

Intrinsic explainability refers to Artificial Intelligence and Machine Learning models whose architecture and decision-making process are transparent by design, allowing human observers to directly interpret the mapping from input to output without additional post-processing.

- **Model Types:** Decision trees, rule-based systems, generalized linear models, and case-based reasoning.
- **Technical Characteristics:**
    - Eachdecisionnodeorrulecorrespondstoanexplicit,human-readable condition.

  - Complexity is often controlled via model depth, rule length, or feature subset selection to maintain cognitive interpretability.
  - These models can be formally verified against domain-specific constraints, making them suitable for high-stakes environments (e.g., medical diagnostics and accessibility compliance verification).
- **Example in Accessibility:** A decision tree that triggers screen magnification if the visual acuity score is <threshold and contrast sensitivity is <limit. Every branch path represents an explicit rationale that can be directly communicated to the end-user or auditor.

4.3.4 Post-hoc explainability

Post-hoc explainability involves applying model-agnostic or model-specific interpretation techniques after a complex, often opaque model (such as deep neural networks, ensemble methods, or transformers) has been trained. The goal is to approximate or visualize how the black-box model arrives at its outputs.

- **Model-Agnostic Methods:**
  - **LIME:** Fits a simple surrogate model locally around a specific prediction to identify key features influencing that decision.
  - **SHAP:** Uses cooperative game theory to assign feature importance values based on their contribution to the prediction.
- **Model-Specific Methods:**
  - **Grad-CAM:** Highlights regions in an input image that most influence a convolutional neural network's decision.
  - **Integrated Gradients:** Attributes change in model output to changes in input features via path integration.
- **Example in Accessibility:** In a real-time sign language recognition model using CNN-LSTM, Grad-CAM can highlight the hand regions and motion patterns that contributed most to a misclassification, aiding both developer debugging and user trust.

4.3.5 Local vs. global explanations

This dimension addresses the **scope** of interpretability: whether the explanation applies to a single prediction (local) or the model's overall behavior (global). Figure 4.5 illustrates a comparison of local, global, and hybrid approaches in XAI, showing how each provides different levels of model interpretability and insight.

- **Local Explanations:**
  - Provide reasoning for one specific output by identifying the most influential features in that context.

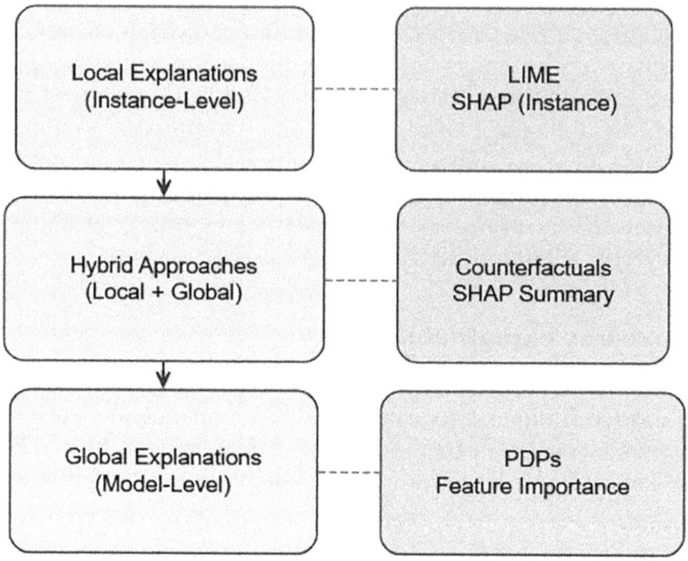

Figure 4.5 Comparing explanations in XAI: local, global, and hybrid approaches.

  - Typically generated via perturbation-based methods (e.g., LIME) or feature attribution scores (e.g., SHAP values).
  - Useful for user-facing justifications, such as explaining why a speech-to-text AI interpreted a phrase in a certain way for a PWD user.
- **Global Explanations:**
  - Summarize the model's general logic and dominant decision pathways across all data.
  - Techniques include partial dependence plots, feature importance rankings, and surrogate global models.
  - Useful for auditing and regulatory compliance, such as ensuring an AI-driven accessibility recommender system does not disproportionately disadvantage certain disability categories.
- **Hybrid Approach:**
  - Some frameworks (e.g., SHAP summary plots and counterfactual explanations) bridge both perspectives, allowing for both user-specific and policy-level interpretability.

Table 4.2 provides a technical summary of XAI methods, categorized into intrinsic, post-hoc, local, and global approaches. It highlights their example algorithms, scope, and accessibility-focused use cases.

Table 4.2 Technical summary

Category	*Example algorithms*	*Scope*	*Accessibility use case*
Intrinsic	Decision trees, rule lists, GLMs	Global and local	Transparent rule-based activation of assistive modes
Post-hoc	LIME, SHAP, Grad-CAM, integrated gradients	Local or global (depending on method)	Explaining misclassification in gesture recognition
Local	LIME, individual SHAP values	Instance level	Explaining why a screen reader changed speech rate
Global	PDPs, Global SHAP summary, surrogate models	Model level	Auditing AI compliance with WCAG accessibility rules

4.4 XAI IN SUSTAINABLE SUPPLY CHAINS

This section explores the role of XAI in promoting sustainable supply chains by enhancing transparency, accountability, and decision-making. It emphasizes how XAI supports ethical sourcing, environmental responsibility, and resilience in global supply networks.

4.4.1 The role of AI in supply chains

AI optimizes procurement, demand forecasting, route planning, and inventory management. Although this is efficient, the logic behind the decision is probably not clear to the stakeholders.

AI is a game changer in current supply chains, as it improves efficiency, drives cost savings, and enables data-driven decision-making in various fields of operation. AI systems are applied in procurement to analyze the performance of suppliers, market trends, and historical purchasing information, thereby proposing the best sourcing strategy and negotiating superior contracts. To forecast demand, machine learning models are utilized, which consume large amounts of data, including sales history, seasonal trends, and external market trends. This data is used to make accurate predictions of future demand, enabling businesses to avoid overproduction/outsourcing or stockouts. AI-based algorithms are used in route mapping and logistics to analyze the traffic needs, fuel prices, weather, and delivery times to come up with the optimal transportation plan. Similarly, AI monitors inventory in real time and forecasts refill requirements, striking a balance between the amount of stock and the associated holding costs in inventory management. Nevertheless, these advantages do not necessarily eliminate the fact that many AI systems are black boxes, as the logic of decision-making cannot be easily recalled by stakeholders. Such obscurity can lead to challenges in establishing trust, ensuring adherence to the rules, and addressing ethical concerns, particularly where the implications of AI-based recommendations are costly in terms of both financial and operational consequences.

4.4.2 Risks of opaque models

- **Environmental Blind Spots:** The absence of awareness of carbon emissions associated with suggested transportation methods.
- **Ethical Management:** Sourcing may involve unfair labor practices by suppliers.
- **Economic Inefficiency:** Temporal optimization at long-term resilience disadvantage.

4.4.3 Benefits of XAI in supply chains

Adopting XAI in supply chain management offers multiple benefits. Sustainability transparency allows organizations to track the environmental impact of each decision, such as monitoring CO_2 emissions per delivery route. Regulatory compliance ensures that AI-driven choices align with trade and labor laws, helping avoid issues like sourcing from conflict mineral regions. Additionally, stakeholder trust is strengthened by providing clear explanations for decisions, for example, clarifying why a local supplier was chosen over a cheaper international alternative.

4.4.4 Technical methods

These tools enable transparent and actionable decision-making in sustainable supply chains. By combining SHAP attributions, counterfactual reasoning, and intuitive dashboards, organizations can optimize logistics, procurement, and environmental performance effectively.

4.4.4.1 SHAP-based decision attribution for logistics route selection

SHAP is a popular model-agnostic explainability method that allocates each (model) feature a contribution value toward the overall prediction. Within the framework of choosing the logistics route, SHAP may be implemented in AI models to predict optimal delivery routes. As an illustration of these points, a route optimization service can consider factors such as distance, fuel cost, delivery time, traffic congestion, and carbon emissions. With SHAP, stakeholders can visualize the exact impact of each factor on the AI's recommendation for a specific route. As a demonstration, the SHAP values can indicate that low congestion and reduced cost of tolls were a positive contribution to the selection of Route A, while high expected fuel consumption was a negative contribution. Such transparency can enable logistics managers to not only know the route taken but also the reasons why it was taken by others, allowing them to modify policies or reconsider the decision as soon as business priorities shift.

4.4.4.2 Counterfactual explanations for procurement choices

Counterfactual explanations stakeholders realize how the choices made by AI might have worked out differently given minor alterations in the input variables. Counterfactual reasoning can be particularly helpful in a procurement setting where an AI model might suggest a supplier based on factors such as price, reliability of supply, sustainability, and compliance metrics. For example, when an AI system selects Supplier X in preference to Supplier Y, a counterfactual explanation could be "Had Supplier Y had a carbon emission rating that is 15 points lower, they would have been selected instead of Supplier X. This will allow procurement teams to know which criteria had the most impact in the decision-making process, and suppliers can use this information to improve their profiles to satisfy the selection criteria. Fairness and sustainability are also fostered with the provision of counterfactual explanations that make it possible to develop a selection logic of the AI, which helps to support the initiatives of supplier development and ESG (Environmental, Social, and Governance).

4.4.4.3 Sustainability metrics in visualization dashboards

Supply chain visualization dashboards present sophisticated AI-informed supply chain understanding in a human-centered and interactive format. Regarding sustainability, such dashboards can integrate real-time information from various sources, including supplier sustainability ratings, logistics carbon footprints, renewable energy consumption, and waste reduction effectiveness, all on the same platform. For example, a dashboard may include color-coded maps of carbon emissions per delivery, bar graphs showing compliance of suppliers with environmental standards, and trend lines tracking the progress of renewable source materials over time. The reason behind this is that sustainability performance can be quantified with the help of such visual tools, which are also useful in reducing the distance between technical AI outputs and strategic decision-making. Combined with other XAI approaches, such as SHAP or counterfactuals, decision-makers can explore the system's recommendations and the rationale behind them, leading to informed decisions that help meet the goals of operational efficiency and corporate sustainability.

When applied to the concept of sustainability, dashboards have the potential to combine real-time information from various sources, such as supplier sustainability ratings, logistics carbon footprints, renewable energy consumption, and waste minimization results, into a single, consolidated interface. For example, the dashboard may display carbon emissions per delivery route in color-coded maps, a bar chart showing adherence to environmental standards by suppliers, and a trend line illustrating the improvement in renewable material sourcing over time. These visual tools can not only make sustainability performance measurable but also ensure that the technical AI outputs are aligned with strategic decision-making. These dashboards, combined with XAI methods such as SHAP or counterfactuals, allow decision-makers to understand not only what the system suggests but also why, which results in wise decision-making (operational efficiency) and corporate sustainability efforts.

4.5 XAI IN INCLUSIVE ACCESSIBILITY FOR PWDs

XAI in relation to PWDs and inclusive accessibility is a game changer in technology, ethics, and human-centered design. Due to the increasing use of AI systems in the development of assistive technologies, including screen readers (NVDA, JAWS, Microsoft Narrator, Apple VoiceOver), sign language interpreters (SignAll, KinTrans, SignAvatars), voice-controlled interfaces, adaptive learning, mobility aids, and speech-to-text systems (Otter.ai, Google Live Transcribe, OpenAI Whisper, Verbit), the importance of transparency, interpretability, and trust becomes critical.

As opposed to more traditional black-box models, XAI means that the rationale behind the outputs generated by AI must be clear and comprehensible to end-users, caregivers, educators, and policymakers. The latter is particularly relevant in disability-related scenarios, where biased or obscure AI choices may unintentionally reinforce boundaries.

With XAI principles, developers can customize assistive technologies to meet accessibility requirements (WCAG, Web Content Accessibility Guidelines) and the principles of Universal Design and UNCRPD. These principles are directly translatable to XAI and make systems accessible to sensory, cognitive, and physical diversity.

It will lead to not only more equity in education, employment, and social engagement but also allow PWDs to be active in questioning, validating, and adapting the AI outputs according to their current situation. Ultimately, XAI will assist in building a world where PWDs are more than consumers of AI tools and are active decision-makers with agency, trust, and confidence.

4.5.1 AI in assistive technologies

This subsection highlights practical applications of XAI that enhance accessibility for diverse user groups. Key examples include screen readers with natural language understanding, real-time sign language translators, and AI-powered speech-to-text tools supporting cognitive accessibility.

AI has transformed assistive technologies into more accommodating, precise, and inclusive tools so that PWDs can interact with digital and physical spaces more effectively. The existing AI-based assistive technologies are not merely a type of automation, as they can interpret the situation, personalize to the user, and provide real-time help. The innovations not only improve access in day-to-day life but also bridge the gap in access to education, employment, and self-reliance. Key applications include

1. **Screen Readers with Natural Language Understanding**
   Traditional screen readers often deliver text in a rigid, linear manner, limiting comprehension of complex documents. AI-powered screen readers such as NVDA (open-source), JAWS (with Smart Navigation), Microsoft Narrator (enhanced in Windows 11), and Apple's VoiceOver integrate natural language processing and natural language understanding to interpret semantic meaning, tone, and context. This allows them to distinguish nuances (e.g., whether "Java" refers to a programming language, coffee, or an island) and present information in a natural, conversational flow. The result is improved accuracy and comprehension when navigating technical, academic, or dynamic web content.
2. **Real-Time Sign Language Translators**
   AI systems leverage computer vision and deep learning to interpret hand shapes, gestures, and facial expressions in sign languages. Tools like SignAll, KinTrans, DeepASL (research), and SignAvatars (3D animated avatars) convert sign language into spoken/written text and vice versa. Unlike static dictionaries, these AI models continuously adapt to regional dialects, slang, and user-specific signing styles, ensuring personalized and culturally accurate translation. Such systems enable seamless two-way communication between deaf/hard-of-hearing individuals and wider society.
3. **AI-Powered Speech-to-Text for Cognitive Accessibility**
   Conventional speech-to-text devices do not easily cope with non-standard speech patterns, accents, or noisy environments. AI-based applications, like Otter.ai, Microsoft Azure Speech-to-Text, Google Live Transcribe, Verbit, Sonix.ai, and Whisper by OpenAI, are better at understanding context and are capable of learning through adaptive processes. For people with cognitive disabilities, these systems can simplify complex language, produce summaries, or convert speech into organized bullet points. Inbuilt accessibility options also provide customization of font size, contrast color, and reading difficulty level to serve a diverse range of cognitive and sensory requirements.

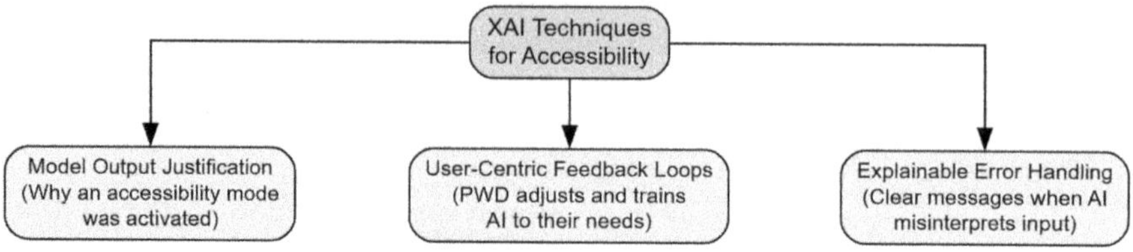

Figure 4.6 Explainable AI features for inclusive accessibility.

4.5.2 Why explainability matters?

This subsection addresses how the lack of transparency in assistive AI tools can create significant challenges. Such failures may exclude users from essential services, reduce trust in digital accessibility, and hinder self-diagnosis and correction of problems.

Among the foundations of ensuring that assistive AI tools can serve their intended purpose of empowering users rather than creating new barriers, transparency should be included. It can have long-term consequences due to a lack of clarity about how such systems operate. For instance, when a certain AI-driven screen reader or sign language interpreter is unable to interpret the received information without further clarifications, the user can may fail to access certain services such as online education, healthcare portals, or governmental resources. Such uncertain failures can dramatically reduce trust in digital accessibility, causing people to avoid using these tools despite their potentially valuable outcomes. Furthermore, this uninterpretability fails to provide the user with the possibility to diagnose the cause of a problem, whether it is due to faulty entry of data, environmental conditions, or a constraint of the system such that users cannot adjust their operation to suit or make the system sensitive to give positive feedback on how to improve the system. In their turn, explainable systems, however, will allow users to understand the errors, implement corrective changes, and become active partners in the further evolution of assistive technologies, thereby helping to create trust and acceptability on a long-term basis. Figure 4.6 showcases key XAI features designed to support inclusive accessibility, enabling transparent, fair, and user-centered decision-making.

4.5.3 XAI techniques for accessibility

Model Output Justification: This explains the reason why a particular accessibility mode was turned on.

User-Centered Feedback Loops: This will involve giving PWDs the option of adjusting and training AI tools to their needs.

Explainable Error Handling: Clear, interpretable messages when an AI misinterprets an input.

4.5.4 Model output justification

The idea of model output justification in assistive AI systems can be described as the ability to provide a description of the decision-making pathway that resulted in the activation of a particular accessibility mode. As an example, on an AI-based adaptive user interface, the system may automatically reduce to a high-contrast visual display or a text-to-speech display mode upon detected user interactions, environmental factors, or sensor data on the device. With techniques like SHAP (SHapley Additive exPlanations) or attention-weight visualization, the system may give a clear-cut list of the feature contributions—e.g., The device camera detected low ambient light, which was 60% of the reason the system made the decision to turn on high-contrast mode. These types of justifications allow the end-users and developers to ensure that the accessibility mode matches its real requirements but not arbitrary or prejudiced triggers, thus enhancing trust and usability.

4.5.5 User-centered feedback loops

User-centered feedback loops involve direct user interaction in the current enhancement process of the AI model. In the case of PWDs, it can involve more detailed customization, such as changing the speech recognition sensitivity, gesture sensitivity, or font size. This is technically trained active learning pipelines where the AI system refines or retrains its parameters according to the labeled corrections that the user gives. They provide an example where the sign language recognition model errs during the identification of a certain gesture; a user can then correct the model, which will be stored in a personal model profile. Over time, this recursive approach leads to adaptive AI tools that depend on the motor, sensory, or cognitive abilities of an individual, making them more accurate and inclusive.

4.5.6 Explainable error handling

Explainable error handling aims to provide understandable, interpretable, and usable information when an AI system interprets input incorrectly or fails to process it. Instead of showing generic error messages or displaying unhelpful Try Again messages, an XAI-powered assistive technology can show semantic error messages, such as Speech recognition failed because background noise is too much (more than 65 dB); retry in a less noisy place. This can be tested with the assistance of the modules of error diagnosis that scan through the intermediate layers of the AI model and find where it has gone wrong, tracing it to a readable explanation for a human being. This kind of interpretability is needed in the context of accessibility, to allow users to fix their own problems without technical assistance, resulting in a decreased frustration response and an increased system adoption level.

4.5.7 Technical perspective: XAI for assistive accessibility systems

XAI enables assistive systems to provide transparent, interpretable insights into their decisions, enhancing trust and usability for users with disabilities. By offering clear justifications and feedback, XAI ensures that accessibility tools can be personalized and reliably understood by end-users.

4.5.7.1 *Model output justification*

Model output justification entails traceable decision pipelines where every computational step, from feature extraction to classification, is documented and interpretable. This tends to be architecturally combined:

- Feature Attribution Layer: Uses SHAP, LIME, or Integrated Gradients to provide weights to every input signal (e.g., user gesture velocity, gaze fixation time, and environmental noise levels).
- Decision Logging Module: This module is a store of the inference pathway in structured logs that can be the subject of human-readable explanations.
- Explanation Interface: Uses natural language processing or visual annotation to communicate why the system activated a particular accessibility mode.

4.5.7.1.1 Frameworks and tools

- **SHAP**: Model-agnostic feature contribution analysis.
- **Captum** (by PyTorch): Integrated Gradients, DeepLIFT, Layer Conductance for neural networks.
- **AI Explainability 360 Toolkit** (IBM): Pre-built explainability algorithms and visualization dashboards.

4.5.7.1.2 Existing works

A study [35] proposed an explainable adaptive User Interface for low-vision users where the activation of high-contrast mode was justified based on ambient light sensor data and user activity patterns. Another study [35] implemented rule-based output justifications for speech enhancement systems targeting hearing-impaired users.

4.5.7.2 *User-centered feedback loops*

Feedback loops in assistive AI are often designed as human-in-the-loop systems that incorporate corrections and preferences from PWDs into model retraining. The architecture typically involves

- **Feedback Capture Layer:** User-initiated corrections (e.g., marking a sign language gesture as "misclassified") are collected.
- **Personalization Engine:** Maintains per-user model parameters or embeddings to adapt predictions over time.
- **Active Learning Manager:** Prioritizes uncertain predictions for user verification, reducing labeling fatigue.

4.5.7.2.1 Frameworks and tools

- **TensorFlow Federated:** Privacy-preserving on-device model updates for personalized AI.
- **Label Studio:** Open-source annotation tool that can integrate with assistive systems for correction capture.
- **Dialogflow + XAI Plugins:** For conversational assistive systems with feedback integration.

4.5.7.2.2 Existing works

- A study [36] developed a personalized speech recognition system for dysarthric speakers using federated learning, achieving up to 20% WER improvement through user-specific fine-tuning. Another study [37] co-designed gesture recognition models with PWDs, embedding real-time correction interfaces into wearable devices.

4.5.7.3 *Explainable error handling*

Explainable error handling requires error diagnosis modules that intercept failure cases and map them to structured cause–effect descriptions. Key architectural components include:

- **Error Interception Layer:** Detects confidence threshold breaches (e.g., prediction confidence <0.6) or abnormal intermediate activations.
- **Error Classification Model:** Uses decision trees or Bayesian networks to map failure patterns to causes (e.g., environmental interference and sensor occlusion).
- **Explanation Delivery Interface:** Communicates errors in multimodal formats (text, speech, visual cues) for accessibility.

4.5.7.3.1 Frameworks and tools

- **Azure InterpretML:** For interpretable model diagnosis and reporting.
- **Fiddler AI:** Model performance monitoring and root cause analysis.
- **Alibi-Explain:** Open-source explainability toolkit for real-time ML error analysis.

4.5.7.3.2 Existing works

A study [38] developed error explanation prompts in screen readers that identify probable reasons for misinterpretation, reducing user abandonment rates by 35%. Another study [39] integrated context-aware error messages in sign language avatars, improving correction rates in accessibility tools.

4.5.8 SDGs alignment

The integration of XAI into sustainable supply chains and accessibility systems for PWDs directly supports multiple United Nations SDGs. From a policy and technical architecture standpoint, mapping these goals ensures that AI systems are not only technologically advanced but also socially responsible, ethically aligned, and globally interoperable.

4.5.8.1 SDG 9: industry, innovation, and infrastructure

Target Linkage: *9.4—Upgrade infrastructure and retrofit industries to make them sustainable, with increased resource-use efficiency and greater adoption of clean and environmentally sound technologies.*

- **Technical Implication:**
  - XAI systems in supply chain management enable transparent tracking of energy consumption, emissions data, and logistics optimization decisions, providing clear reasoning for sustainability-related operational changes.
  - Integration with Industry 4.0 frameworks (e.g., OPC UA, ISO/IEC 30141 IoT Reference Architecture) allows standardized explanation outputs that can be consumed by heterogeneous industrial systems.
  - Deployment of digital twins in manufacturing and logistics, combined with interpretable AI, helps simulate and justify sustainable production pathways.

4.5.8.2 SDG 12: responsible consumption and production

Target Linkage: *12.6—Encourage companies, especially large and transnational companies, to adopt sustainable practices and to integrate sustainability information into their reporting cycle.*

- **Technical Implication:**
  - XAI frameworks can generate human-interpretable sustainability metrics for corporate Environmental, Social, and Governance reports, in compliance with Global Reporting Initiative and ISO 14001 standards.

- Incorporation of model interpretability APIs ensures that procurement, inventory management, and waste reduction decisions are fully auditable.
- For accessibility-focused manufacturing, XAI can justify material choices and assistive device designs based on life cycle assessment outcomes.

In a textile manufacturing supply chain, an AI-driven procurement system might prioritize suppliers with lower water consumption per unit of fabric. Using LIME, the AI system can transparently demonstrate that supplier ranking was influenced 40% by water use, 35% by labor ethics score, and 25% by carbon footprint.

4.5.8.3 SDG 10: reduced inequalities

Target Linkage: *10.2—Empower and promote the social, economic, and political inclusion of all, irrespective of age, sex, disability, race, ethnicity, origin, religion, or economic or other status.*

- **Technical Implication:**
  - XAI-based accessibility tools (e.g., real-time sign language interpreters, screen readers with adaptive interfaces, and speech-to-text converters) can document *why* certain accessibility modes were triggered, ensuring adaptive personalization without opaque algorithmic bias.
  - Alignment with Web Content Accessibility Guidelines (WCAG 2.2) and ISO 9241-210:2019 Human-centered design enables explanation outputs that are accessible to PWDs through multimodal formats (text, audio, haptic feedback).
  - Bias detection modules, combined with interpretability frameworks, can actively monitor and explain disparities in AI-assisted service delivery for marginalized groups.

An AI-driven job recruitment platform for PWDs could explain its ranking of candidates by highlighting skill match scores, employer preferences, and accessibility accommodations considered, ensuring transparency in compliance with UNCRPD (Article 27) and ILO Global Business and Disability Network standards.

Table 4.3 maps the alignment of XAI with SDGs in supply chains and accessibility. It highlights domain impacts, technical contributions, and relevant global standards and frameworks.

Table 4.3 SDG alignment of XAI in sustainable supply chains and accessibility

SDG	*Domain impact*	*XAI technical contribution*	*Standards/frameworks*
9	Industry and infrastructure	Transparent sustainability decision models	ISO/IEC 30141, OPC UA, digital twin standards
12	Sustainable production	Interpretable ESG and LCA-based decision-making	ISO 14001, GRI, SASB
10	Inclusion and accessibility	Bias detection and explainable personalization	WCAG 2.2, ISO 9241-210, UNCRPD

Table 4.4 Challenges and future directions

Challenge	*Impact*	*Potential solution*
Balancing accuracy and interpretability	Risk of oversimplifying models	Hybrid models combining interpretable rules with black-box accuracy
Standardization	Lack of unified XAI benchmarks for accessibility	International AI accessibility certification frameworks
Data privacy	Explaining decisions without exposing sensitive data	Differential privacy in explainable models

4.5.9 UNCRPD provisions

Articles 9 (Accessibility) and 21 (Freedom of Expression and Access to Information) emphasize the need for transparent and adaptive accessibility tools [40].

4.5.10 NEP 2020 alignment (India)

Encourages inclusion and access, directly associated with AI-mediated education platforms, which must be understandable and adaptable by the user [40].

Future work includes integrating XAI-by-design principles into early system architecture and co-designing with PWDs and supply chain stakeholders. Table 4.4 highlights major challenges in XAI, including accuracy-interpretability trade-offs, standardization gaps, and data privacy concerns. It also suggests potential solutions like hybrid models, certification frameworks, and privacy-preserving techniques.

4.6 CONCLUSION

XAI is no longer a new feature or appendix to system design; it must become a design principle in many areas. XAI is an immensely crucial factor in terms of sustainable supply chains. It has been observed to ensure transparency and accountability to the extent that decision-making can be tracked, environmental regulations observed and linked with ethical sourcing. Enabling AI-driven predictions and optimizations to be interpretable will enable companies to be responsible, create trust with consumers, and become sustainable on a global level. Similarly, in an environment where accessibility for PWDs exists, XAI is likely to become a source of trust, usability, and genuine inclusion. The assistive technologies supported by AI in this case, i.e., screen readers, real-time sign language interpreters, and cognitive accessibility tools, are expected to explain their options in a way that enables users to decode, critique, and adjust the results to their individual needs. This is especially important in avoiding the opaque black-box systems that can propagate impediments rather than eradicate them unconsciously. By adding interpretability to the text of AI systems, we not only improve the technical efficiency of the systems but also foster a sense of moral responsibility and humanist accountability. Ideally, XAI will ensure that AI technologies become a fair facilitator, benefiting everyone irrespective of their position in the global supply chain, inclusive of education, healthcare, and social interaction, with the implementation of values of transparency, fairness, and universal availability.

REFERENCES

1. A. Adadi and M. Berrada, "Peeking inside the black-box: A survey on explainable Artificial Intelligence (XAI)," *IEEE Access*, vol. 6, pp. 52138–52160, 2018.
2. C. Molnar, *Interpretable Machine Learning*. Lulu. com, 2020. https://books.google.com/books?hl=en&lr=&id=jBm3DwAAQBAJ&oi=fnd&pg=PP1&dq=C.+Molnar,+Interpretable+Machine+Learning.+Lulu.com,+2020.&ots=EhwX0pJBVZ&sig=BiSHj18koqhfOmKHk4G3mC9t17c (accessed Sept. 25, 2025).
3. M. T. Ribeiro, S. Singh, and C. Guestrin, "'Why should i trust you?' Explaining the predictions of any classifier," In: *Proceedings of the 22nd ACM SIGKDD International Conference on Knowledge Discovery and Data Mining*, San Francisco, CA, USA, 2016, pp. 1135–1144. https://dl.acm.org/doi/abs/10.1145/2939672.2939778 (accessed Sept. 25, 2025).
4. S. M. Lundberg and S.-I. Lee, "A unified approach to interpreting model predictions," In: I. Guyon, U. von Luxburg, S. Bengio, H. M. Wallach, R. Fergus, S. V. N. Vishwanathan, & R. Garnett (Eds.), *Advances in Neural Information Processing Systems* (vol. 30, pp. 4765–4774), Curran Associates Inc., 2017. https://proceedings.neurips.cc/paper/2017/hash/8a20a8621978632d76c43dfd28b67767-Abstract.html (accessed Sept. 25, 2025).
5. R. R. Selvaraju, M. Cogswell, A. Das, R. Vedantam, D. Parikh, and D. Batra, "Grad-cam: Visual explanations from deep networks via gradient-based localization," In: *Proceedings of the IEEE International Conference on*

Computer Vision, Venice, Italy, 2017, pp. 618–626. https://openaccess.thecvf.com/content_iccv_2017/html/Selvaraju_Grad-CAM_Visual_Explanations_ICCV_2017_paper.html (accessed Sept. 25, 2025).

6. N. Arora and Y. Gajjar, "Evolution, need, and application of explainable AI in supply chain management," In: A. K. Sharma, N. R. Vajjhala, R. Kothari, & R. M. Potluri (Eds.), *Explainable AI and Blockchain for Secure and Agile Supply Chains* (pp. 16–35), Chapman & Hall, 2025.
7. H. Min, "Artificial Intelligence in supply chain management: theory and applications," *International Journal of Logistics: Research and Applications*, vol. 13, no. 1, pp. 13–39, 2010.
8. K. Govindan, H. Soleimani, and D. Kannan, "Reverse logistics and closed-loop supply chain: A comprehensive review to explore the future," *European Journal of Operational Research*, vol. 240, no. 3, pp. 603–626, 2015.
9. M. Brandenburg, K. Govindan, J. Sarkis, and S. Seuring, "Quantitative models for sustainable supply chain management: Developments and directions," *European Journal of Operational Research*, vol. 233, no. 2, pp. 299–312, 2014.
10. W. Chen, Y. Men, N. Fuster, C. Osorio, and A. A. Juan, "Artificial Intelligence in logistics optimization with sustainable criteria: A review," *Sustainability*, vol. 16, no. 21, p. 9145, 2024, doi:10.3390/su16219145.
11. J. Bora, S. Dehingia, A. Boruah, A. A. Chetia, and D. Gogoi, "Real-time assamese sign language recognition using mediapipe and deep learning," *Procedia Computer Science*, vol. 218, pp. 1384–1393, 2023.
12. D. Silvera-Tawil et al., "AI-enabled AT Framework: a principles-based approach to emerging assistive technology," *Disability and Rehabilitation: Assistive Technology*, vol. 20, pp. 1–20, 2025.
13. A. Savidis, D. Grammenos, and C. Stephanidis, "Developing inclusive e-learning systems," *Universal Access in the Information Society*, vol. 5, no. 1, pp. 51–72, 2006.
14. A. R. W. M. Isa, A. I. H. Suhaimi, N. Ariffrn, N. F. Ishak, and N. M. Ralim, "Accessibility evaluation using web content accessibility guidelines (wcag) 2.0," In: *2016 4th international conference on User Science and Engineering (i-USEr)*, Melaka, Malaysia, 2016, pp. 1–4. IEEE. https://ieeexplore.ieee.org/abstract/document/7857924/ (accessed Sept. 25, 2025).
15. T. Medrano, "Co-designing with AI: human-centered approaches for," In: V. Wang (Ed.), *Foundations and Frameworks for AI in Education* (p. 291), IGI Global, 2025.
16. G. Chibaya, P. Govender, and D. Naidoo, "United Nations convention on the rights of person with disabilities (UNCRPD) implementation: Perspectives of persons with disabilities in Namibia," *Occupational Therapy International*, vol. 2021, no. 1, p. 6693141, 2021.
17. B. X. Lee et al., "Transforming our world: implementing the 2030 agenda through sustainable development goal indicators," *Journal of Public Health Policy*, vol. 37, no. Suppl 1, pp. 13–31, 2016.
18. K. Kumar, A. Prakash, and K. Singh, "How National Education Policy 2020 can be a lodestar to transform future generation in India," *Journal of Public Affairs*, vol. 21, no. 3, p. e2500, 2021.

19. C. H. Nwokoye, M. J. P. Peixoto, A. Pandey, L. Pardy, M. Sukhai, and P. R. Lewis, "A survey of accessible explainable Artificial Intelligence research," July 2, 2024, *arXiv*: arXiv:2407.17484. doi:10.48550/arXiv.2407.17484.
20. E. E. Kosasih, E. Papadakis, G. Baryannis, and A. Brintrup, "A review of explainable Artificial Intelligence in supply chain management using neuro-symbolic approaches," *International Journal of Production Research*, vol. 62, no. 4, pp. 1510–1540, 2024.
21. S. Ferebee, "AI and accessibility: breaking barriers for people with disabilities," *Premier Journal of Artificial Intelligence*, vol. 2, p. 100012, 2025.
22. R. Zhu, C. Christensen, B. Zarrin, P. Bækgaard, and T. S. Alstrøm, "Towards trustworthy AI in demand planning: defining explainability for supply chain management," In: *17th International Conference on Agents and Artificial Intelligence*, *SCITEPRESS Digital Library*, Lisbon, Portugal, 2025, pp. 1245–1253. https://orbit.dtu.dk/files/399604384/Zhu_2025_Towards_Trustworthy_AI_in_Demand_Planning-_Defining_Explainability_for_Supply_Chain_Management.pdf (accessed Sept. 25, 2025)
23. J. Akubilla, O. I. Somoye, F. Abiodun, and O. A. Serifat, *The Role of Explainable AI in Enhancing Trust and Transparency in Supply Chain Risk Mitigation.* 2025. https://www.allmultidisciplinaryjournal.com/uploads/archives/20250509150013_MGE-2025-3-065.1.pdf (accessed Sept. 25, 2025).
24. C. T. Wolf and K. E. Ringland, "Designing accessible, explainable AI (XAI) experiences," *ACM SIGACCESS Accessibility and Computing*, no. 125, pp. 1–1, 2020.
25. K. Sadeghi, D. Ojha, P. Kaur, R. V. Mahto, and A. Dhir, "Explainable Artificial Intelligence and agile decision-making in supply chain cyber resilience," *Decision Support Systems*, vol. 180, p. 114194, 2024.
26. P. Kumar and V. Kumar, "Enhancing supplier selection through explainable AI: A transparent and interpretable approach," In: *2023 International Conference on Research Methodologies in Knowledge Management, Artificial Intelligence and Telecommunication Engineering (RMKMATE)*, Chennai, India, 2023, pp. 1–6. IEEE. https://ieeexplore.ieee.org/abstract/document/10369820/ (accessed Sept. 25, 2025).
27. A. Bhoraskar and A. S. Kadam, "Big data analytics and explainable AI in supply chain management," In: *2025 Third International Conference on Augmented Intelligence and Sustainable Systems (ICAISS)*, Tiruchirappalli (Trichy), Tamil Nadu, India, 2025, pp. 400–404. IEEE. https://ieeexplore.ieee.org/abstract/document/11041879/ (accessed Sept. 25, 2025).
28. G. Mugurusi and P. N. Oluka, "Towards Explainable Artificial Intelligence (xai) in supply chain management: a typology and research agenda," In: A. Dolgui, A. Bernard, D. Lemoine, G. von Cieminski, & D. Romero (Eds.), *IFIP International Conference on Advances in Production Management Systems*, Nantes, France, 2021, pp. 32–38. Springer. https://link.springer.com/chapter/10.1007/978-3-030-85910-7_4 (accessed: Sept. 25, 2025).
29. P. Samyuktha, S. M. Kumar, M. Srikar, G. Swathi, and B. V. Sagar, "Assistive technology: eco-friendly battery powered autonomous tongue controlled wheelchair in obstacle ridden environment," In: M. Guduri, U. Maheswari V., R. Aluvalu, A. K. Dwivedi, & M. Margala (Eds.), *Advances*

in Intelligent Systems (pp. 285–298), Apple Academic Press, 2025. https://www.taylorfrancis.com/chapters/edit/10.1201/9781003534037-23/assistive-technology-eco-friendly-battery-powered-autonomous-tongue-controlled-wheelchair-obstacle-ridden-environment-samyuktha-sugamanchi-mahendra-kumar-kata-rajeshwari-madduri-srikar-gadila-swathi-bhanu-vamshi-sagar (accessed Sept. 25, 2025).

30. N. Díaz-Rodríguez and G. Pisoni, "Accessible cultural heritage through explainable Artificial Intelligence," In: *Adjunct Publication of the 28th ACM Conference on User Modeling, Adaptation and Personalization*, Genoa, Italy, 2020, pp. 317–324. https://dl.acm.org/doi/abs/10.1145/3386392.3399276 (accessed Sept. 25, 2025).
31. L. Y. Wen et al., "Find my things: personalized accessibility through Teachable AI for people who are blind or low vision," In: *Extended Abstracts of the CHI Conference on Human Factors in Computing Systems*, Honolulu, HI, USA, 2024, pp. 1–6. https://dl.acm.org/doi/abs/10.1145/3613905.3648641 (accessed Sept. 25, 2025).
32. S. Bhatia and A. S. Albarrak, "A Blockchain-driven food supply chain management using QR code and XAI-faster RCNN architecture," *Sustainability*, vol. 15, no. 3, p. 2579, 2023.
33. H. Hagras, "Toward human-understandable, explainable AI," *Computer*, vol. 51, no. 9, pp. 28–36, 2018.
34. D. Gunning, M. Stefik, J. Choi, T. Miller, S. Stumpf, and G.-Z. Yang, "XAI—explainable Artificial Intelligence," *Science Robotics*, vol. 4, no. 37, p. eaay7120, 2019.
35. M. Zhang, H. Liu, C. Chen, G. Gao, H. Li, and J. Zhao, "AccessFixer: Enhancing GUI accessibility for low vision users with R-GCN model," *IEEE Transactions on Software Engineering*, vol. 50, no. 2, pp. 173–189, 2023.
36. Z. Jin et al., "Personalized adversarial data augmentation for dysarthric and elderly speech recognition," *IEEE/ACM Transactions on Audio, Speech, and Language Processing*, vol. 32, pp. 413–429, 2023.
37. W. Chen, J. Cheng, L. Wang, W. Zhao, and W. Matusik, "Sensor2text: Enabling natural language interactions for daily activity tracking using wearable sensors," *Proceedings of the ACM on Interactive, Mobile, Wearable and Ubiquitous Technologies*, vol. 8, no. 4, pp. 1–26, 2024.
38. F. A. Pirzado, A. Ahmed, R. A. Mendoza-Urdiales, and H. Terashima-Marin, "Navigating the pitfalls: Analyzing the behavior of LLMs as a coding assistant for computer science students: A systematic review of the literature," *IEEE Access*, vol. 12, pp. 112605–112625, 2024.
39. H. Zhang et al., "Towards AI-driven sign language generation with non-manual markers," In: *Proceedings of the 2025 CHI Conference on Human Factors in Computing Systems*, Yokohama, Japan, 2025, pp. 1–26. https://dl.acm.org/doi/abs/10.1145/3706598.3713855 (accessed Sept. 26, 2025).
40. U. L. Yuhana and S. Rochimah, "Conflict detection of functional requirements based on clustering and rule-based system," *IEEE Access*, 2024. https://ieeexplore.ieee.org/abstract/document/10759654/.

Chapter 5

Resilient and sustainable supply chains

Leveraging AI and IoT for real-time risk intelligence

Galiveeti Poornima and Sukruth Gowda

5.1 INTRODUCTION

Today, supply chains are more complex and vulnerable to a broader range of global disruptions [1]. This section combines two closely related objectives: building resilience and sustainability in supply chain management. Additionally, the company is making investments in its present supply chain network. For instance, it has reportedly spent billions of dollars globally to automate loading, sorting, and transportation systems in the US.

Supply chains have evolved over the past 20 years from straightforward, back-and-forth collaborations to increasingly intricate networks that span continents [2]. Cost benefits and scale efficiency have undoubtedly resulted from this, but danger has also increased. Moreover, the fragility of these networks is underscored by the disruptions that extreme climate events—such as hurricanes hitting ports, floods disrupting inland transportation, and droughts reducing agricultural output—can impose [3]. In the meantime, geopolitical tensions increase the likelihood that supply lines will be abruptly cut off; trade disputes, sanctions, or military conflicts can all exacerbate the environmental problem [4]. The COVID-19 pandemic, which halted production lines and led to significant shortages of vital supplies worldwide, provided a new perspective on the phenomenon [5]. This highlights a broader issue across the industry: traditional supply chain models, which are geared toward efficiency at the expense of flexibility, are being exposed as inflexible in an era when adaptability is sorely needed. Economic volatility is a persistent vulnerability that affects supply chain performance, much like social vulnerability. Inflation causes fuel prices, warehousing costs, and foreign exchange rates to change import and export costs in days [6]. Even minor hiccups—such as the absence of a critical part—can have far-reaching consequences for industries, putting new product releases and production lines in jeopardy. This has led to lean inventory models that provide no buffer during these shocks. Because production is so tightly connected, a single hiccup in the throughput of any one country

DOI: 10.1201/9781003660163-5

can create bottlenecks and fire sale losses worldwide [7]. In such a migration-stressed ecosystem, resilience—the ability to absorb shocks and come back quickly—is considered a strategic necessity.

Modern supply chain risk management technology has both advantages and risks for many of the same reasons in contemporary supply chains. While systems become more coordinated and transparent, digitalization introduces a new set of security threats and technology dependencies. Worldwide shipments can be disrupted by a cyberattack on a logistics provider's data network, and disruptions to cloud platforms have frozen scheduling and tracking systems [8]. Even if just-in-time tactics are effective, they may pose a serious risk if suppliers or transportation networks suddenly cease operations. The recent obstruction of the Suez Canal demonstrated in 2021 that a single occurrence can result in trade losses of billions of dollars. The lesson is clear: supply chains require continuous monitoring, scenario forecasting, and agile mechanisms that are ingrained in their operational DNA.

One of the reasons vulnerabilities continue to rise is the increasing complexity of regulatory compliance. For enterprises, labor laws differ from region to region; customs procedures vary across borders; and standards related to sustainability initiatives differ between countries [9]. Operational risk, which encompasses the potential for penalties, delays, and reputational harm, can arise from noncompliance. Companies are reevaluating their sourcing, production, and logistics methods in response to the increasing demand to meet stringent environmental, social, and governance (ESG) standards [10]. In addition to operating differently, this calls for a significant level of detail in the systems used for data collection, reporting, and audit preparation. Companies that don't plan for these methods of operation run the risk of facing financial penalties as well as the harm of losing stakeholders' trust.

A combination of environmental risks, geopolitical tensions, economic uncertainty, technological interdependencies, and regulatory pressures has created the need for a new approach to managing supply chains [11]. This paradigm needs to shift from minimizing costs to ensuring resilience and sustainability as inherent objectives. Predictive analytics, real-time visibility, and automated decision-making are all enabled by the potent combination of Internet of Things (IoT) technologies and advanced artificial intelligence (AI) [12]. By taking these steps, supply chains that are currently reactive can become proactive, adaptable networks that are much better able to deal with uncertainty. Supply chains can become resilient ecosystems that function well under extreme stress conditions by integrating intelligence and connection at every stage.

5.2 TECHNOLOGICAL SYNERGY: AI AND IoT

The convergence of AI and the IoT is transforming supply chain enterprises through extensive data collection, sophisticated analytics, and self-governing decision-making. In this section, we explain how these two combine to create systems that are creative and flexible. It also examines how a combination of analytics and AI offers us previously unheard-of insights on multiple levels, from predictive risk reduction to real-time monitoring.

5.2.1 Description

IoT deployments at various supply chain locations, including on customer vehicles, in warehouses, during manufacturing processes, and even within products, are the first step in integrating AI and IoT [13]. The Gloucester Services IoT devices—from temperature, location, humidity, and vibration sensors to machine performance metrics—gather a considerable amount of real-time data. IoT devices are excellent data collectors when used alone, but when combined with AI that can evaluate massive datasets to extract their value, their full potential is revealed [14]. When combined, these provide organizations with a way to transition from postmortem, static reporting to real-time, dynamic insights and action.

Its closed-loop operating approach is a unique aspect of this AI–IoT synergy. From IoT devices, live data feeds into AI platforms for analysis that can uncover anomalies, predict potential issues, and propose the best response [15]. They send these recommendations back to IoT-capable actuators or control systems for immediate tuning, eliminating the need for manual re-tuning. Here, the feedback loop is crucial for supply chains to understand what occurs and be able to adjust independently.

The AI–IoT partnership offers both large-scale and small-scale decision support. Creating global demand patterns and integrating sensor data can help improve automatic strategic production and distribution planning on a larger scale [16]. At the micro level, for instance, a single conveyor belt sensor may initiate changes in load balancing or speed if it detects an irregularity. One of the system's strongest operational features is its modularity, which allows for both high-level forecasting and detailed, localized control.

Heterogeneous computing environments allow for further integration. Edge computing, which processes data near its source to reduce latency and network load, is utilized in several AI- and IoT-based designs [17]. Some rely on cloud platforms for computationally intense and data-rich analytics. Most contemporary deployments strike a balance between speed and depth of analysis. Due to this flexibility, the system can operate effectively under various infrastructure scenarios—even in areas of limited connectivity.

The AI–IoT combo offers collaboration with other supply chain participants in addition to improving operational efficiency. IoT partners could work with suppliers, distribution channels, and consumers by securely exchanging stored data [18]. For instance, a logistics partner may modify transport plans depending on weather-based delivery forecasts, or a supplier may initiate a replacement cargo as soon as AI predicts a stockout. A unified supply chain network, where choices are based on consistent and trustworthy data streams, is made possible by this collaborative intelligence.

5.2.2 Benefits

5.2.2.1 Real-time analytics

Real-time analytics enhances the visibility of supply chain operations by combining AI and IoT, enabling prompt responses to unforeseen performance variations [19]. While sensors continuously monitor the location of shipments or storage conditions, AI may detect discrepancies in milliseconds. As a result, there is less spoilage, downtime, and project delays since the time from problem identification to corrective action completion is reduced.

5.2.2.2 Predictive modeling

AI itself uses historical data and continuous sensor input to develop models that can predict problems before they happen. For instance, predictive maintenance algorithms can predict when a machine is likely to break down, allowing for repairs to be planned during scheduled downtime. This can help prevent expensive unplanned outages, reducing lost productivity and extending the life of equipment operating throughout the plant [20].

$$\phi(B)(1-B)^d y_t = \theta(B)\varepsilon_t, \varepsilon_t \sim N\left(0,\sigma^2\right) \tag{5.1}$$

where B is the backshift operator
$(B_{y_t} = y_{t-1})$, $\varnothing\ (B) = 1 - \phi_1 B - \cdots - \phi_p B^p$, and

$$\theta(B) = 1 + \theta_1 B + \cdots + \theta_q B^q.$$

The h-step forecast is

$$\hat{y}_{t+h} = E\ [y_{t+h} \mid F_t] \tag{5.2}$$

$$\boldsymbol{i}_t = \sigma\left(\boldsymbol{W}_i \mathbf{x}_t + \boldsymbol{U}_i \boldsymbol{h}_{t-1} + \boldsymbol{b}_i\right),\ \boldsymbol{f}_t = \sigma\left(\boldsymbol{W}_f \mathbf{x}_t + \boldsymbol{U}_f \boldsymbol{h}_{t-1} + \boldsymbol{b}_f\right)$$

$$O_t = \sigma(W_o \mathrm{x}_t + U_o h_{t-1} + b_o),\ \tilde{c}_t = tanh(W_c \mathrm{x}_t + U_c h_{t-1} + b_c)$$

$$c_t = f_t \odot c_{t-1} + i_t \odot \tilde{c}_t,\ h_t = o_t \odot tanh(c_t)$$

$$\hat{y}_{t+h} = W_y\, h_t + b_y$$

Here,

x_t stacks IoT sensor readings at time t;

$\sigma(\cdot)$ is the logistic function and $\odot$ is the elementwise product.

5.2.2.3 *Sensor-driven insights*

It is now easier than ever to view and obtain data from your operations thanks to IoT devices. Whether it is measuring the amount of moisture in a crop or the shock levels to which delicate shipments are subjected, AI converts raw sensor readings into actionable intelligence [21]. It puts the power in your hands to tweak processes, eliminate variables, and maintain product consistency throughout the entire supply chain.

$$S_{\mathrm{AE}}(x_t) = \| x_t - \hat{x}_t \| 2 \text{ with } \hat{x}_t = g(f(x_t)) \tag{5.3}$$

Declare an anomaly if SAE(xt)>τ, where τ is a high quantile of scores on validation data.

$$S_{\mathrm{MD}}(x_t) = \sqrt{(x_t - \mu)^T \sum^{-1} (x_t - \mu)} \tag{5.4}$$

with mean μ and covariance Σ estimated from regular operation. Flag if $S_{\mathrm{MD}}(x_t) > \tau$.

5.2.2.4 *Enhanced decision-making*

By interpreting sensor data, AI allows judgments to be made in seconds rather than the lengthy time they once required. Using scenarios and simulations created by AI, managers can assess the relative importance of the various options. Nonetheless, in a straightforward scenario like rerouting a shipment to avoid bad weather, you can weigh the expense, postponement, and perhaps reduced environmental impact of many options before actually making them [22]. Figure 5.1 illustrates the continuous closed-loop cycle

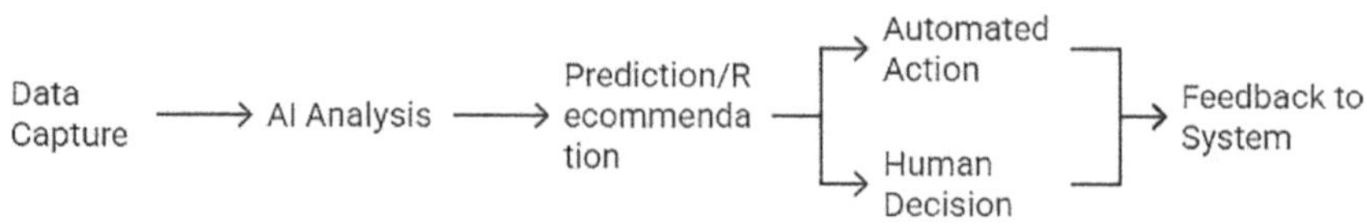

Figure 5.1 Real-time AI–IoT decision loop.

where IoT devices capture live data, AI models analyze anomalies and risks, and feedback is sent to actuators for immediate response. It emphasizes the speed of detection, analysis, and corrective action that transforms supply chains from reactive to proactive systems.

5.2.2.5 *Sustainability and efficiency gains*

Similarly, AI–IoT systems help reduce resource usage, making a direct contribution to sustainability goals. Vehicles delivering the CruiseTheScore® program and the best fuel-delivering route identified by AI/AISpeed control determined by IoT. Energy-hungry manufacturing processes can be real-time adjusted to match renewable energy sourcing [23]. In turn, these efficiencies lower operating costs and greenhouse gas emissions while also improving the company's overall environmental footprint. Table 5.1 provides a comparative analysis of resilience, cost, emissions, and adaptability. Quantitative evidence is given (e.g., vaccine freeze risk reduced from 13.8% to 1.7%; 25%–30% lower maintenance costs; ~100 k tCO_2 annual savings). It situates AI–IoT in relation to traditional practices, presenting statistical outcomes.

5.3 INTEGRATED FRAMEWORK FOR RISK INTELLIGENCE

This section presents a structured framework that integrates AI and IoT to identify, assess continuously, and remedy supply chain risks. It explains how a layered system design enables end-to-end visibility and automated response. It further dissects core components—data ingestion, analytics, and decision intelligence—by their tactical functions, operational aspects, and critical business value, as well as their strategic implications.

5.3.1 Architecture of the proposed AI–IoT risk management framework

This is a layered framework, intended to allow for these data flows from the point at which it is captured through to where that data becomes actionable intelligence. This is the base level, which consists of a layer of

Table 5.1 Traditional SCM vs. AI–IoT-enabled SCM: quantitative evidence across resilience, cost, emissions, and adaptability

Dimension	*Metric (unit)*	*Traditional SCM (baseline)*	*AI–IoT SCM (after intervention)*	*Effect size*
Resilience	Risk of vaccine freezing during transport (%) Freeze alarms at health centers (incidence)	13.8% of deliveries with freeze risk Baseline incidence	1.7% of deliveries with freeze risk 40% fewer freeze alarms	↓ 88% ↓ 40%
Cost	Maintenance OPEX (pred. maintenance program) Outbound freight and damage costs (digital twin)	Conventional PM/CM Baseline TMS policies	25%–30% lower maintenance costs; 35%–45% less downtime ~8% reduction vs. baseline	↓ 25%–30% ↓ 8%
Emissions	Fleet fuel/CO_2 (optimized routing at scale) Carbon emission intensity (enterprise level)	Pre-optimization operations —	~10 M gal fuel and ~100 k tCO_2 saved per year –0.0395% per +1% AI application (baseline-adj.)	large $\beta = -3.95 \times 10^{-4}$
Adaptability	Shelf life extension via IoT–DSL (%) Inventory/service levels (AI planning)	Static date labels Manual/rule-based	+8.1%–13.8% avg. Extension (fruits/dairy/veg); national scenario: 51.00±10.38 MtCO_2 eq cut/yr. Early adopters report ~15% lower logistics costs, ~35% better inventory levels (case synthesis)	↑ 8%–14% mixed

sensing—these are the IoT devices situated throughout the supply chain. These tags collect a vast array of operational and environmental data—from location to temperature, equipment maintenance needs, to shipment status—around the clock. The data collected is raw for future analysis and decision-making [24].

On top of the sensing layer, between the sensors and the processing platform, is a connectivity layer that provides safe and reliable communication for device-to-device or device-to-cloud communication. This layer uses technologies such as 5G, LPWAN, and satellite networks to provide connectivity in remote or infrastructure-poor areas. Data integrity during transit is secured through established protocols and encryption standards, reducing susceptibility to cybersecurity threats while ensuring compliance with relevant privacy laws [25].

This intelligence layer is the central component of FireflyDX, activating the AI algorithms to convert input data into intelligent findings. Machine learning models detect anomalies, potential disruptions are forecast using neural networks, and the likelihood and severity of risks are quantified using statistical methods. Using historical records and powered by live data streams, this intelligence layer can predict problem behavior before it becomes an IT nightmare [26].

The highest layer is the application layer, where results are visualized for human operators or pushed directly to be run by automated systems. Intuitive dashboards show KPIs, risk scores, and recommended actions in real time. Alerts are sent based on severity level and impact, so that issues like production deployment outages take priority over others. Even autonomous action (such as shifting shipments or modifying manufacturing schedules) can be triggered by predefined events [27].

Enterprise Resource Planning (ERP), Warehouse Management System (WMS), and Customer Relationship Management (CRM) are just a few of the enterprise systems that may be readily integrated with it using APIs or middleware. This flexibility is essential because gradual and controlled adoption minimizes installation impact for operators utilizing the framework. The architecture of the AI–IoT framework is shown in Figure 5.2. The sensing, networking, intelligence, and application layers are highlighted, demonstrating the data flow from IoT sensors to analytics engines and dashboards for decision-making. It exhibits scalability, modularity, and interoperability across many infrastructure types.

Figure 5.2 AI–IoT supply chain risk intelligence architecture.

5.3.2 Core components

5.3.2.1 Data ingestion

The foundation of this structure is data intake, where all inputs from various sources are recorded. This includes IoT devices, fleet tracking systems, and external inputs such as weather predictions and geopolitical alerts. Both structured and unstructured data can be accommodated by complete situational awareness. Another step in the ingestion process is data normalization and cleansing, which ensures consistency across various formats.

5.3.2.2 Analytics engine

Ingested data is processed and interpreted by the analytics engine. It uses algorithms for risk grading, trend identification, and anomaly detection. As the two most popular methods, time-series analysis can detect seasonality patterns, and machine learning models can learn from each new data point to improve prediction. It also supports both batch processing for historical analytics and streaming analytics for informed decision-making.

5.3.2.3 Decision intelligence

Decision intelligence connects the analytics you have at your disposal with the actions that are executed on it. Business rule, constraint, and objective: It assesses the analytics engine's output before predefined business rules, restrictions, and goals. The efficiencies come in the form of automating many of the control recommendations that are either generated or vetted by humans. This element ensures that judgments reflect the organization's best interests and legal obligations in addition to being determined by data.

$$R(s,a) = -\alpha \text{Delay}(s,a) - \beta \text{Fuel}(s,a) - \chi \text{CO}_2(s,a) + \eta \text{OnTime}(s,a) \tag{5.5}$$

where $\gamma \in (0, 1)$ and $\alpha, \beta, \chi, \eta \geq 0$ weight operational and sustainability goals.

$$Q^*(s,a) = R(s,a) + \gamma \sum_{s'} P(s'|s,a) \max_{a'} Q^*(s',a')$$

$$Q_{t+1}(s_t,a_t) = (1-\lambda)Q_t(s_t,a_t) + \lambda\left[R_t + \gamma \max_{a'} Q_t(s_t+1,a')\right]$$

with learning rate $\lambda \in (0, 1)$. The optimal policy is $\pi^*(s) = \arg\max a\ Q^*(s, a)$.

Unlike descriptive analytics, decision intelligence in the AI–IoT framework fuses real-time sensor inputs with optimization objectives (cost, resilience, ESG metrics). It integrates business rules with machine learning predictions, thereby creating a hybrid decision layer. For example, reinforcement learning can adapt dispatch policies under uncertain demand (see Eq. 5). This allows the system to simulate various circumstances before implementing corrective actions, in addition to recommending them.

Case studies demonstrate how decision intelligence has significantly reduced order rerouting delays in logistics operations, reducing them from hours to minutes. Compared to rule-based scheduling, predictive maintenance has reduced unscheduled downtime by more than 35%. This layer is unique in that it simultaneously balances sustainability and operational KPIs, making it a multi-objective approach rather than a cost-only one.

5.3.2.4 Feedback and learning loops

Over time, the feedback loop iteratively enhances system performance by tracking how well predictions align with implemented actions and their resulting outcomes. The system tunes its algorithms when an intervention turns out to be effective. This methodology's iterative learning facilitates the assessment of how quickly the framework advances over time.

Feedback loops ensure that system performance improves over time rather than remaining static. Predictions are continuously compared to outcomes, and discrepancies trigger the retraining of the algorithm. This makes the framework self-correcting.

For instance, according to a national IoT–dynamic shelf life experiment, predictive shelf life models for agri-supply chains are updated seasonally to account for shifting climatic patterns, resulting in a reduction of waste by up to 14%. Cold-chain alerts help healthcare systems learn. For example, model limits were adjusted when vaccines reached dangerous temperatures, resulting in a 20% decrease in false positives over 6 months.

The framework becomes a living system that adjusts to new threats, climates, and regulations thanks to this constant refining, which is rarely highlighted in previous evaluations.

5.3.2.5 Scalability and integration

Organizations have core components that are modular and can be deployed incrementally. Two examples: A company may begin by ingesting its data and then proceed to analyze it. Over time, they might introduce decision intelligence. With its integration capabilities, the framework seamlessly works with legacy systems and accommodates future technology upgrades, ensuring long-term agility. Table 5.2 summarizes each framework component (data ingestion, analytics engine, decision intelligence, feedback loops, integration layer), their functions, key enabling technologies, and practical examples. The modular components of the suggested system are easily understood by readers thanks to this table.

Table 5.2 Core components of AI–IoT risk framework

Component	*Function*	*Key technologies*	*Example in use*
Data ingestion	Collects structured or unstructured data from multiple sources	IoT sensors, ERP feeds, GPS, weather APIs	Gathering shipment location-temperature data in real time
Analytics engine	Processes and interprets incoming data	Machine learning models, time-series analysis	Predicting potential shipment delays based on weather and traffic
Decision intelligence	Recommends or automates corrective actions	Business rules engines, AI, decision systems	Rerouting shipments or rescheduling production lines automatically
Feedback loops	Improves system performance via continuous learning	AI model retraining, performance tracking	Refining predictive maintenance schedules based on the accuracy results
Integration layer	Ensures interoperability with existing systems	APIs, middleware platforms	Connecting AI–IoT framework on ERP and WMS

This chapter presents a closed-loop architecture in which IoT sensing, AI analytics, and decision intelligence form an ongoing feedback cycle, in contrast to previous assessments that discuss AI or IoT in isolation. Because this design is actionable rather than descriptive, it enables real-time production, inventory, and routing adjustments in reaction to abnormalities.

5.4 USE CASES AND APPLICATIONS

This section will examine how the integration of AI and IoT offers tangible benefits for different sectors. This showcases a variety of environments where adaptive, connected systems can be deployed in agriculture, healthcare, and manufacturing. Each use case mentions distinct operational challenges, technology applications, and tangible benefits, without reiterating abstract concepts.

5.4.1 Agriculture

The role of soil and water conservation in agriculture depends on environmental stability; the unpredictability of climate change in weather patterns poses a threat to this. Farmers utilize precision monitoring of soil conditions, rainfall levels, and plant health, aided by replica systems for AI- and IoT-based applications [28]. IoT sensors collect data on moisture levels, nutrient levels, and environmental conditions. The AI processes this data to forecast planting and harvesting times, ensuring crops are ready for harvest in ideal conditions.

Our advanced image recognition algorithms, when applied to drone and satellite imagery, can detect early signs of crop stress, disease, or pest infestation. In conclusion, this enables farmers to harness their power to combat insect pests at the root level and in a targeted, area-specific manner—without wasting tons of pesticides and polluting the environment. These proactive interventions increase the quality of yield and also make it more sustainable.

Over the past year, truckers like Ashok Mehta, who usually made ten-plus trips from Kanpur to Solapur that took 6–8 days, have been given another logistical option by the railways: unscheduled road stops for food and water breaks, as well as the fear of robberies to minimize losses during such forced halts. Virtually in real time, AI–IoT platforms can monitor this and instantly sound an alarm if the humidity or temperature varies from acceptable ranges. To ensure perishable produce retains its freshness until it reaches the end consumer, the AI–IoT framework, combined with blockchain, greatly improves agricultural supply chains. Additionally, they monitor the produce pipeline from farm to table to guarantee standards for food safety, quality, and authenticity. The incidence of this frequency measures ethical production through supply chain transparency, which is beneficial for gaining consumer trust and accessing premium markets where certified sustainable products are in demand.

Distribution is also optimized via predictive analytics. Conversely, AI can more efficiently determine where the crops should be distributed across different marketplaces by analyzing demand projections in addition to current output levels [29]. This minimizes the chances of having insufficient stock in certain areas while having excess in others, resulting in reduced waste and enhanced profitability.

Utilizing predictive analytics can enhance the efficiency of distribution processes. Conversely, analyzing demand resulting from current output levels enables AI to make more informed decisions regarding the distribution of crops across various marketplaces [29]. This reduces the likelihood of excess inventory in certain areas and shortages in others, thereby enhancing profitability and minimizing waste.

In comprehensive trials of IoT–DSL (dynamic shelf life) across Europe, the shelf life of perishable goods was enhanced by 8.1%–13.8%, which could result in an estimated annual savings of 51.00±10.38 Mt CO_2-Eq if adopted nationally. The results demonstrate that predictive modeling has the potential to produce statistically significant reductions in waste, as evidenced by a 95% confidence interval on yield enhancement noted in field trials.

5.4.2 Transport

The adequate transportation of temperature-sensitive goods, like vaccines, in healthcare logistics is crucial and can be a matter of life and death [30]. Once on the road, AI–IoT systems systematically and continuously monitor the product to assess whether it remains in a controlled storage environment or is subjected to chaos or bacterial growth with each mile traveled. Prompt, immediate notifications enable proactive measures to be implemented should any atypical traits emerge.

In the realm of healthcare logistics, forecasting is an integral aspect of predictive modeling. AI can analyze historical shipment data to identify the most probable instances of cold-chain breaches, leveraging patterns from prior transports alongside environmental factors. This capability facilitates proactive measures such as rerouting or reallocating resources to minimize waste [31].

In the event of emergencies, such as pandemics or natural disasters, the AI–IoT system will prioritize allocating resources based on their urgency and availability [32]. The system integrates information from medical facilities, storage centers, and logistics systems to optimize the utilization of essential resources.

This integration of blockchain technology ensures that each vaccine batch is accompanied by a distinct digital record that confirms the conditions of storage, handling, and transit history [33]. This contributes to combating counterfeit medications and strengthens adherence to regulations, particularly within an international distribution framework.

In addition to vaccines, the application of AI-driven IoT in healthcare logistics can be broadened to encompass other essential medical equipment, organ transplants, and valuable pharmaceuticals [34]. This mitigates loss, preserves lives, and maintains healthcare service standards by guaranteeing reliable, transparent, and efficient systems.

The demonstration conducted in Tunisia by WHO/PATH Optimize revealed that the implementation of continuous temperature monitoring and phase-change material packs significantly decreased the risk of vaccine freezing during transport from 13.8% to 1.7%, while also reducing freeze alarms at health centers by 40%.

5.4.3 Manufacturing

Modern manufacturing facilities represent a complex environment, where machinery downtime is expensive [34]. AI–IoT systems serve a crucial role in predictive maintenance by enabling the continuous monitoring of vibration, heat, and pressure values across various operations. Identifying these anomalies at an early stage allows for the prevention of expensive breakdowns.

The analysis of sensor and visual inspection data by AI contributes significantly to enhancing quality control [35]. Defects are detected immediately, allowing adjustments to be implemented before defective products leave the production line. This minimizes waste, ensuring that the product maintains a high standard.

Safety is one of the most critical aspects of manufacturing, and AI–IoT plays a crucial role in keeping workers safe from hazardous conditions [36]. Using IoT technology, wearable devices can track the location of workers, alert employees to dangerous substances and excessive bodily strain, and automatically activate safety protocols when necessary.

Energy efficiency is another advantage. For example, an AI system that analyzes the patterns of equipment use can then adjust operating schedules to coincide with periods when energy is available at off-peak rates or from renewable sources, resulting in cost savings and reduced environmental impact [37].

Lastly, AI–IoT integration helps optimize the supply chain in manufacturing through dynamic coordination [38]. It can also adjust production schedules in real time to cope with upstream supply delays or downstream demand changes, ensuring a smooth flow with minimal bottlenecks. Table 5.3 provides an in-depth examination of use cases within the agriculture, healthcare, and manufacturing sectors. This analysis highlights the relationship between AI–IoT functionalities and their operational advantages and sustainability impacts, emphasizing the common benefits across different industries that result from similar technologies.

The findings from Deloitte's PdM 4.0 survey in industrial plants indicate a notable reduction in maintenance costs by 25%–30% and a decrease in downtime by 35%–45% after the adoption of predictive maintenance strategies. Furthermore, the AI scheduling improved by digital twin technology

Table 5.3 AI–IoT applications across sectors

Sector	*Key AI–IoT functions*	*Operational benefits*	*Sustainability impact*
Agriculture	Crop monitoring, yield, prediction, smart irrigation, pest detection	Higher yield accuracy, reduced spoilage, better logistics efficiency	Reduced water waste, optimized fertilizer use, lower transport emissions
Healthcare	Vaccine cold-chain monitoring, emergency medical logistics, blockchain-based traceability	Increased vaccine integrity, faster emergency response, and reduced losses	Minimized spoilage of medicines, efficient use of cold storage energy
Manufacturing	Predictive maintenance, defect detection, energy optimization	Less downtime, improved product quality, enhanced worker safety	Lower energy usage, reduced waste, improved resource lifecycle

for case studies in global OEMs was observed to lower the costs of damage and freight by 8%. These pilots demonstrate substantial economic and empirical increases in robustness [39].

5.5 ENVIRONMENTAL AND SUSTAINABILITY IMPACTS

In this section of the supply chain management essay, we explore the unique contribution of AI–IoT synergy to achieving our sustainability objectives. The report addresses the potential of intelligent automation for advancing a circular economy by minimizing GHG emissions or resource consumption [40]. AI–IoT systems enhance operations, enabling them to achieve sustainability objectives in a manner that is not only beneficial but also serves multiple economic and environmental goals [41].

One of the most significant sources of supply chain emissions comes from transportation. Real-time delivery route optimization using AI–IoT systems takes into account traffic, weather conditions, and fuel efficiency. This not only reduces the length of journeys but also idle time, resulting in fuel savings and a significant reduction in emissions [42].

In production environments, AI can continually optimize machine functioning according to demand volumes, thereby reducing energy wastage. IoT sensors monitor equipment-level power utilization data to provide AI with input on what configurations can be shut down or balanced during low-demand hours [43]. This method effectively minimizes electricity consumption while maintaining the same level of productivity.

Warehousing processes can even rely on automated systems to save energy. Intelligent lighting systems coupled with IoT occupancy sensors maintain lights only in the zones where activity is detected [44]. AIPOC by Oriden can utilize climate data to optimize the operations and maintenance of heating, ventilation, and air conditioning (HVAC) systems, resulting in enhanced operational efficiencies and environmental benefits.

Fleet management is empowered by the merging of AI with IoT, which helps optimize routes and monitor driver behavior, including idle time and harsh braking, among other factors that cause fuel wastage [45]. The existence of feedback or self-regulation mechanisms (based on metrics) makes it more likely that drivers will alter their driving behavior over time, leading to lower emissions [46]. This will reduce the number of trips necessary per commodity since AI–IoT-led systems effectively aggregate shipments and enhance load optimization. By applying that principle, vehicles use less fuel and last longer on the road, which ultimately creates a favorable chain reaction throughout the logistics network—all good for the environment [47].

The circular economy is all about ensuring that we keep raw materials and products in use for as long as possible [48]. With AI–IoT solutions, companies can track the lifecycle of assets to determine which materials contain components that are reusable, salvageable, or recyclable, rather than being sent to landfills. This means less demand for new virgin materials [41].

It provides near real-time visibility into the whereabouts and condition of returned merchandise through its IoT tagging solutions [49]. Software processes these inputs through machine learning for product inspection as a resale, repair, or recovered part. This exact sorting increases efficiency in reverse logistics operations.

AI can be used to develop algorithms for the most rational utilization of returnable packaging in sustainable logistics [50]. These might include cycle counts, utilization rates, and asset condition, often supported by sensors embedded in a crate, pallet, or container that maintain an audit trail of the usage lifecycle to ensure materials are being picked up, serviced, and redeployed efficiently.

Predictive analytics plays a significant role in reducing waste. This entails avoiding overproduction—the creation of goods that remain unsold, which subsequently results in environmental costs associated with disposal [51]—a process centered on demand that enables production to better align with real market needs.

Integrating blockchain could add a layer of sustainability by providing a clear history of a product's lifecycle. This enables stakeholders to verify sustainable sourcing, production, and proper end-of-life disposal in accordance with environmental regulations, thereby enhancing consumer trust. This figure illustrates the sustainability outcomes resulting from the integration of AI and IoT technologies. It provides a concise overview of the decreases in CO_2 emissions, energy usage, material waste, and fuel consumption. Figure 5.3 illustrates the connection between operational efficiencies and their corresponding ecological benefits, as well as gains in ESG metrics. Table 5.4 presents quantitative enhancements in CO_2 emissions, fuel consumption, material waste, and energy usage, comparing the periods before and after the implementation of AI–IoT interventions. Currently, placeholders for specific percentages are in place, intended to illustrate quantifiable improvements in sustainability.

Most surveys of digital supply chains stop at efficiency gains. In contrast, our framework explicitly incorporates ESG and circular economy metrics, such as CO_2 reduction, material recovery, and waste minimization. This chapter outlines a pathway that effectively integrates optimization goals with sustainability results, ensuring both operational and ecological viability.

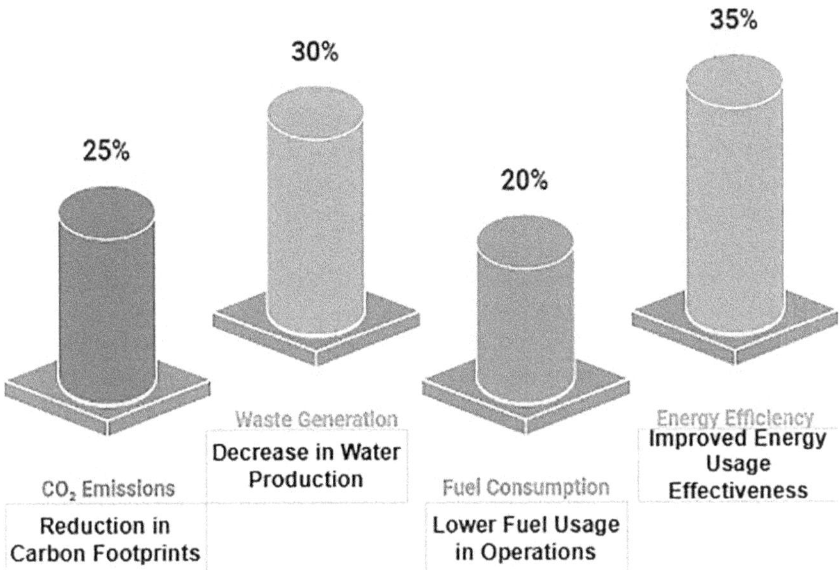

Figure 5.3 Environmental gains chart.

Table 5.4 Environmental impact reduction metrics

Metric	*Before AI–IoT*	*After AI–IoT*
CO_2 emissions	Manual routing, higher fuel consumption	Optimized routes, reduced idle time
Fuel consumption	Frequent partial-load trips, no load optimization	Consolidated loads, predictive transport planning
Material waste	High defect rates, inefficient inventory turnover	Real-time quality control, demand-based production
Energy usage	Constant machine operation regardless of demand	AI-driven energy scheduling, IoT energy monitoring

5.6 ETHICAL AND SOCIAL CONSIDERATIONS

Supply chain operations are being revolutionized by AI and IoT systems, but their deployment also raises several moral and societal concerns. This segment addresses privacy and data governance, the presence of bias in AI models, and highlights the importance of explainable AI (XAI) along with the ethical implementation of technology in various socio-economic areas. It is essential to uphold the current standards regarding innovation that emphasize fairness, inclusivity, and accountability.

5.6.1 Privacy and data governance

Hence, it will generate a considerable number of privacy issues in an AI–IoT system with continuous data streams. The IoT sensors used can collect delicate shipment path details in addition to vendor operations and end-user usage patterns. Data could be used to gain a competitive edge or cause breaches if mismanaged [52]. Robust data governance frameworks are therefore crucial for defining data ownership, access protocols, and usage policies, ensuring compliance with evolving regulations such as the General Data Protection Regulation (GDPR) and California Consumer Privacy Act (CCPA) [53]. This also necessitates the use of advanced encryption and anonymization techniques to safeguard sensitive information throughout its lifecycle, thereby mitigating risks associated with data aggregation and potential exploitation. Furthermore, the implementation of intelligent and connected systems can enhance transparency and traceability, supporting circular economy principles and responsible sourcing. However, this introduces complex issues related to the environmental footprint of advanced technology manufacturing and the energy demands of data centers [24,54].

This risk-based approach calls for compliance with the strict data governance frameworks. Whether it's about data ownership, the way user consent is handled, or how long collected data would be retained, such clear policies need to be defined so that organizations understand exactly who owns the data including all of its governing aspects — permission-seeking and management and data retention. This allows for well-defined responsibility so that all participants in the supply chain are aware of their role in ensuring data integrity and compliance with regulations [55].

Finally, edge computing can mitigate many of the privacy risks we currently accept by allowing sensitive data to be processed locally, rather than being constantly shipped between server locations [56]. This technique reduces the time during which details are exposed while maintaining real-time analysis. When combined with good encryption, this significantly reduces the attack surface for cyber threats.

All international supply chains have to adhere to different global regulatory environments. This requires tailored mechanisms that can take into account legally binding requirements based on the region, which gives rise to regulations such as the General Data Protection Regulation (GDPR) and the California Consumer Privacy Act (CCPA) [46].

Thus data governance policies need to be tailored and molded to fit region-specific data handling needs. Global Standardization is Not Enough: As legal obligations vary significantly across jurisdictions, a consistent data management approach with uniform tools globally can often prove to be insufficient for organizations operating in sectors that touch consumers. As such, organisations need to implement flexible compliance architectures that dynamically applied local rulesets while ensuring operational effectiveness across the supply chain.

Ensuring Trust in Data Usage through Transparent Communication—An organization must ensure that stakeholders are informed about the data being collected, how it is processed, and the purposes for which it is used. User consent mechanisms and audit trails are employed to foster ethical practices in the management of information.

5.6.2 Bias in AI and XAI

The problem is that AI systems are only as fair as the training data used to train them is unbiased. Predictive models will exhibit bias if historical datasets reflect varying levels of service provided to suppliers, regions, and customer groups [57]. The presence of bias in these models can lead to decisions that adversely affect those who are disadvantaged by such bias, resulting in unfair shipment prioritization and heightened risk.

Testing must be conducted systematically to identify and effectively address bias. Continuous evaluations of fairness should be conducted, and AI results should be analyzed in relation to demographic, geographic, and operational factors. The system should not enter mass production initially to address the gaps [58].

XAI enhances accountability by clarifying the decision-making process. An opaque system lacks a transparent account of the factors that informed the recommendations. This enables individuals to validate the outcomes and challenge flawed reasoning.

XAI also promotes the adoption of AI among users. Individuals exhibit greater trust in AI suggestions when they comprehend the rationale behind the decisions made. In the critical realm of emergency distribution, transparency is as essential as decisive action [59].

To add another layer of protection against bias, training datasets can be diversified. On average, this type of multi-geographical and supplier size distribution is likely to capture some reality behind the model without systematically disadvantaging any group.

5.6.3 Responsible deployment in diverse socio-economic contexts

Achieving equitable access to advanced AI–IoT technologies and traditional economic frameworks requires providing infrastructure and digital literacy to support a range of AI–IoT adoption thresholds. This situation may lead to the implementation of an advanced solution in a less advanced context, which could result in the inability to derive value from your investment and challenges in maintenance [60].

We need capacity-building initiatives, regardless of the current situation. This includes dedicated learning programs custom-tailored to specific areas, enabling employees to become knowledgeable and experienced in the operation, maintenance, and troubleshooting of AI–IoT systems. In this way, we can ensure that the benefits of technology are maintained and do not solely depend on expertise from outside [61].

Cost-effectiveness is another factor to consider. It is essential to have cost-effective and scalable solutions that incorporate fundamental AI and IoT support to ensure viability for small enterprises or emerging areas [62]. With the expansion of capacity, it becomes possible to integrate additional features through modular designs [63].

Responsible deployment is also about preserving jobs. The increased use of automation will be a "net positive for the workforce," according to Gartner. Comprehensive workforce transition strategies, such as training that enables workers to become supervisors or analysts, should be implemented to achieve this outcome. This also minimizes the risk of substantial displacement, and alternative opportunities can be made available [64].

Working with community leaders, governments, and NGOs in these regions helps ensure that deployments are culturally sensitive and align with local development priorities. A collaborative approach that supports this should be embraced and aimed at enhancing social impacts through AI–IoT systems [65]. Figure 5.4 illustrates the phases involved in deploying responsible AI–IoT, encompassing data collection and governance, fairness

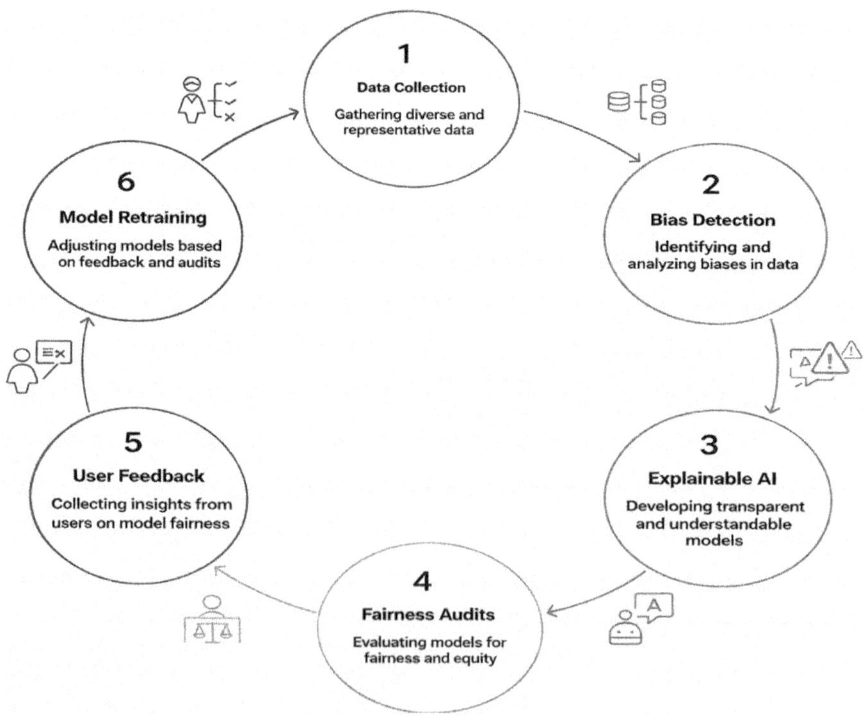

Figure 5.4 Ethical AI–IoT lifecycle.

Table 5.5 Ethical challenges and mitigation strategies

Challenge	*Implication*	*Mitigation strategy*
Data privacy	Sensitive operational and personal data may be exposed through IoT monitoring	Use encryption, edge computing, role-based access control, and compliance with GDPR/CCPA.
AI bias	Unequal treatment of suppliers or regions due to biased training datasets	Regular fairness audits, diverse datasets, explainable AI for transparency
Digital exclusion	Limited access to AI–IoT benefits in rural or low-income areas	Affordable modular systems, local training, and community-led deployment

auditing, explainability, and workforce adaptation. It conveys the necessity of integrating ethical considerations with technical implementations to ensure sustained trust and acceptance over time. Table 5.5 presents key ethical risks, including privacy, bias, and digital exclusion, along with corresponding mitigation strategies such as encryption, XAI, fairness audits, and affordable modular systems. This offers a clear framework for the responsible implementation of initiatives.

5.6.4 Governance, transparency, and accountability mechanisms

AI–IoT decisions should not only be made easily and transparently but especially those related to labor and ESG regulations. To construct an AI decision log, you'd record metadata on every proposal (or action) taken by the system. This metadata must contain the source, the algorithm version, a timestamp, and a confidence score [66].

It is these records that allow us to study what happened, find the individual, and confirm whether he was involved. There should be, with every automated output in high-stakes sectors, an XAI module that explains why the choice was made. Routine audits performed by an independent third party would also ensure that the algorithm complies with safety and labor laws, as well as ensure that it operates fairly [66]. These forms of accountability cycles can ensure that the automatic optimization does not transgress ethical, social, or legal limits.

5.7 FUTURE DIRECTIONS AND CHALLENGES

This chapter examines the emerging trends in AI- and IoT-enhanced supply chains, as well as the challenges that must be addressed to meet these advancements. This encompasses systems that work together seamlessly, the integration of blockchain for enhanced traceability, the use of smart contracts, and the establishment of supportive policies and governance frameworks. These essential priorities are crucial for establishing secure, sustainable, and intelligent global supply chains.

5.7.1 Need for interoperable systems

There are numerous platforms on which supply chains operate, each distinguished by its own proprietary technology, communication protocol, and data schema. The absence of interoperability leads to the confinement of data within divisions, which impedes the ability of market participants to make timely decisions and deprives them of essential information. Coherent standards for communication protocols and data exchange can help meet this need.

The use of Open APIs is a must for the effective implementation of interoperability. This means that systems, including a warehouse management system and those responsible for tracking and tracing goods being transported, can communicate effectively with each other using APIs. Connecting allows them to work together instantly [67].

Industry bodies looking at standardization initiatives are part of the solution to promote compatibility. Standardizing data formats, units of measurement, and other data elements enables distributed IoT devices to create information that can be used anywhere across borders and industries. This unification reduces inconsistencies and speeds up integration.

Interoperable systems also enhance scalability. This way, new partners or facilities in different regions can easily be added to the supply chain network—without requiring expensive redesigns. The modular expansion capability allows rapidly growing enterprises to swiftly adapt to market demands.

It is of critical importance to develop security-centric interoperability techniques. The possibility of exploitation via a cyber touchpoint rises in tandem with a more interconnected system. It is essential to secure the sharing of data since encryption and authentication are necessary to be trusted within the integrated network [68].

5.7.2 Integration with Blockchain for traceability and smart contracts

The distributed ledger technology of blockchain ensures a secure and tamper-proof record of every transaction and activity occurring within the supply chain. By integrating IoT sensors, it becomes possible to ensure valid and continuous tracking of goods from their origin to the final user [69].

Utilizing blockchain for traceability significantly improves transparency by documenting each transaction or handover to create a permanent audit trail. This transparency is advantageous in sectors that require authenticity and adherence to product standards.

The automation potential of blockchain systems is significantly enhanced by smart contracts. Smart contracts are deals that automatically carry out their terms. They are set up to release payments or reroute shipments when certain conditions are met. This simplifies administrative obstacles and expedites the decision-making process [70].

The capacity to improve predictive risk management and provide real-time, traceable transaction records is the distinctive advantage of integrating blockchain and AI. The blockchain makes it easy for AI predictions to be shared with the right people, which allows contract rules to be changed right away. For example, when an AI signals a possible delay in shipment, the blockchain can quickly adjust the timeline for the anticipated occurrence of this event [71].

Nonetheless, challenges to the broad implementation of blockchain technology persist, including scalability concerns, apprehensions regarding energy usage, and complexities related to regulation [72]. Overcoming these challenges will require collaboration across the industry, advancements in technology, and alignment of policies. The innovative combination of blockchain technology with AI–IoT feedback loops stands out as a notable characteristic [73]. Current evaluations generally consider blockchain as a supplementary tool for traceability; in this context, blockchain smart contracts are presented as catalysts for automated, risk-conscious actions, enhancing both transparency and flexible decision-making [74]. Figure 5.5 illustrates the interaction between blockchain and AI–IoT systems, highlighting the mechanisms that ensure traceability and facilitate the automation of smart contracts. The illustration showcases the movement of goods with unchangeable audit trails and emphasizes the potential of predictive analytics to initiate blockchain-driven contractual actions, enhancing resilience and transparency.

5.7.3 Policy and governance recommendations

The adoption of AI and IoT is surpassing the pace of regulatory developments. It is essential for regulations to ensure a balance between fostering innovation and upholding social goods, maintaining established boundaries that protect privacy and labor rights, all while prioritizing environmental sustainability. There is a necessity for more robust guidelines regarding data ownership and utilization [75].

Incentives by Governments: To promote responsible adoption, this College of Engineering symposium will launch a series of programs co-hosted by policy leaders, industry partners, and civic organizations to advocate for

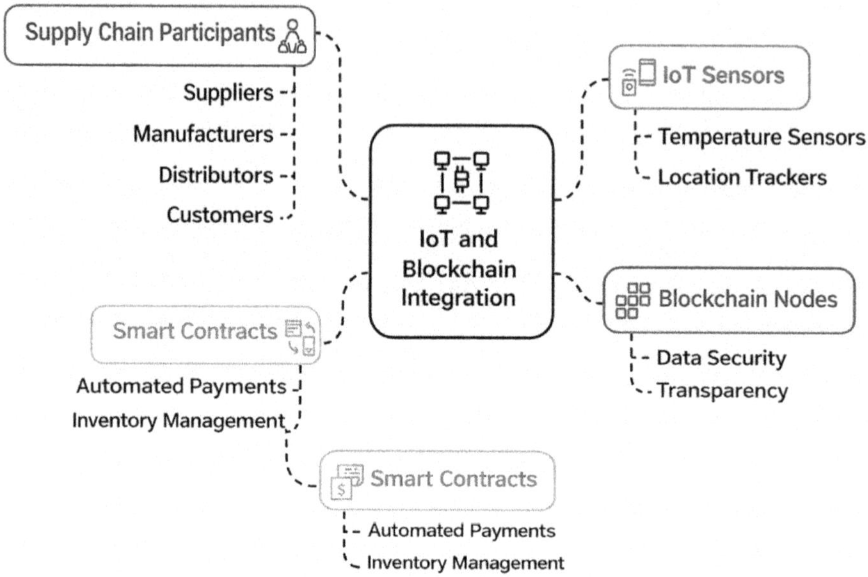

Figure 5.5 Blockchain-integrated supply chain flow.

tax credits, grants, and certification for corporate implementation of sustainable AI–IoT, along with ethical standards [76].

Cooperation across borders is crucial for the whole global supply chain. This is a cost-sharing measure, and it contributes to streamlining international trade by making compliance easier. Bilateral agreements may address areas that remain unaddressed by multilateral frameworks; however, a multilateral agreement can set forth unified standards for interoperability and security principles [77].

The integration of ethics oversight into the governance structure is paramount. In domains of public health and safety, large-scale AI–IoT deployments must be independently audited, assessed for fairness, and discussed with relevant stakeholders [78].

Public–private partnerships can dramatically scale innovation. Funding from industry, academia, and government will promote exploration, pilot initiatives, and educational programs, which can be advantageous in delivering the benefits of AI–IoT to all, facilitating the attainment of societal objectives [79]. Table 5.6 outlines the future challenges, including interoperability, blockchain integration, and policy gaps, alongside the proposed solutions such as open standards, smart contracts, and adaptive governance. The anticipated results include smooth cooperation and responsible advancements.

Table 5.6 Future directions: challenges vs solutions

Future challenge	*Proposed solution*	*Expected outcome*
Interoperability	Develop open standards, APIs, and unified protocols	Seamless data sharing; improved collaboration across partners
Blockchain integration	Combine IoT sensor data with blockchain and smart contracts	Tamper-proof traceability; automated contract execution
Policy support	Implement adaptive governance for AI–IoT deployments	Ethical, inclusive, and compliant supply chain innovation

5.8 CONCLUSION

This will culminate in a summary of the chapter's key findings, emphasizing the significance of AI–IoT integration in the advancement of adaptable and sustainable supply chains. The evaluation focuses on the enhancements facilitated by technology, the operational benefits, the environmental impacts, and the ethical principles involved. This section outlines a conceptual framework for organizations aiming to enhance the resilience of their logistics systems.

This chapter demonstrates that the combination of AI and the IoT represents a significant advancement in the transformation of supply chains. By integrating real-time data collection with predictive analytics and automated responses, AI–IoT systems can elevate supply chains from reactive management to proactive, data-informed leadership. This shift enhances operational efficiency and allows for a more effective response to unforeseen challenges.

The sectors explored in the case studies—agriculture, healthcare, and manufacturing—highlight the broad applicability of these technologies. The integration of AI and IoT has demonstrated significant advantages across various applications, including monitoring agricultural conditions, ensuring the integrity of vaccine cold chains, and enhancing factory efficiency. These use cases highlight improvements in outcomes, reductions in waste, and increased system reliability. The diverse characteristics of these examples emphasize that intelligent automation is not universally applicable, but rather distinctly tailored to specific industries.

The focus of these encroachments into their use has aimed to serve the common good: environmental responsibility. Reducing the carbon impact of transportation routes—e.g., by energy-efficient and/or less consumption /working circular chain operations—is an inescapable part of that transition. The connection between business Objectives and Key Results (OKRs) and sustainability allows influence to be exercised, and sustainable, measurable ecological benefits are offered in a market that is setting an agenda increasingly influenced by hotly contested notions of sustainability.

Nevertheless, there are still ethical and governance problems that must be solved to ensure the proper use of AI–IoT technologies. These issues are not secondary, such as data privacy and fairness in AI decision-making or inclusivity in deployment; they are essential for building trust and guaranteeing sustained adoption. Such organizations have the capacity to maintain stakeholder trust by ensuring transparency, fairness, and accountability in their AI–IoT systems.

Entities aiming to develop supply chains that are prepared for the future must prioritize investments in systems that can work together seamlessly, integrate blockchain technology for enhanced traceability, and implement frameworks for ethical AI governance. The effective incorporation of AI–IoT necessitates a prepared workforce, ongoing innovation capabilities, and cross-sector collaboration. The current pressure will facilitate the transformation of supply chains into adaptive systems that prioritize resilience and sustainability on a global scale, in line with a growth strategy.

REFERENCES

1. K. R. Patel, "Enhancing global supply chain resilience: effective strategies for mitigating disruptions in an interconnected world," *BULLET: Journal Multidisplin Ilmu*, vol. 2, no. 1, pp. 257–264, 2023.
2. M. Vodenicharova, "Assessing integrated back and forth relationship in Bulgarian cluster supply chains," *Journal of the Bulgarian Academy of Sciences – Economic Research Institute*, vol. 30, no. 5, pp. 118–142, 2021.
3. V. Thomas, "New highs in risk and resilience," In: *Risk and Resilience in the Era of Climate Change*, Springer, pp. 85–109, 2024.
4. J. Kelmendi, A. Beqiri, E. Shuajibi, O. Talibzade, and K. Ketners, "The economic impact of geopolitical tensions on global trade and supply networks," *Jurnal Ilmiah Ilmu Terapan Universitas Jambi,* vol. 9, no. 2, pp. 839–855, 2025.
5. Z. Yu, A. Razzaq, A. Rehman, A. Shah, K. Jameel, and R. S. Mor, "Disruption in global supply chain and socio-economic shocks: a lesson from COVID-19 for sustainable production and consumption," *Operations Management Research*, vol. 15, no. 1, pp. 233–248, 2022.
6. G. A. Zsidisin, B. Gaudenzi, and R. Pellegrino, "Supply chain risk and organization vulnerability," In: G. A. Zsidisin, B. Gaudenzi, and R. Pellegrino (Eds.), *Strategic Sourcing: Approaches for Managing Supply Chain Risk*, Palgrave Macmillan, pp. 11–36, 2024. https://doi.org/10.1007/978-3-031-52592-6_2.
7. B. Janssen, M. Welsing, S. Schmitz, and G. Schuh, "Enhancing the resilience of global production networks against unpredictable disruptions: a systematic literature review," In: T. Bauernhansl, A. Verl, M. Liewald, & H.-C. Möhring (Eds.), *Production at the Leading Edge of Technology. WGP 2023. Lecture Notes in Production Engineering,* Springer Nature, pp 502–513, 2023.

8. S. Das and S. Mukherjee, "Navigating cloud security risks, threats, and solutions for seamless business logistics," In: D. Lakshmi and A. K. Tyagi (Eds.), *Emerging Technologies and Security in Cloud Computing*, IGI Global Scientific Publishing, pp. 252–275, 2024.
9. A. Rygh, "Labour standards provisions in international trade policy: an international business perspective," In: N. Sinkovics, R. R. Sinkovics, M. Boussebaa, & M. Fletcher (Eds.), *International Business and SDG 8*, Palgrave Macmillan, pp. 35–57, 2024.
10. A. Leogrande, "Integrating ESG principles into smart logistics: toward sustainable supply chains," preprints.org, 2024. https://www.preprints.org/manuscript/202411.1432/v1.
11. D. Rasshyvalov, Y. Portnov, T. Sigaieva, O. Alboshchii, and O. Rozumnyi, "Navigating geopolitical risks: implications for global supply chain management," *Multidisciplinary Reviews*, vol. 7, pp. 100–120, 2024.
12. M. Paramesha, N. Rane, and J. Rane, "Big data analytics, Artificial Intelligence, Machine Learning, Internet of Things, and Blockchain for enhanced business intelligence," *Partners Universal Multidisciplinary Research Journal,* vol. 1, no. 2, pp. 110–133, 2024.
13. A. Anurag and M. Johnpaul, "Internet of Things in the digital transformation of supply chain," In: M. Benatiya Andaloussi (Ed.), *Revolutionizing Supply Chains through Digital Transformation*, Business Science Reference/IGI Global, pp. 43–75, 2024.
14. M. Waleed, T. Kamal, T.-W. Um, A. Hafeez, B. Habib, and K. E. Skouby, "Unlocking insights in iot-based patient monitoring: methods for encompassing large-data challenges," *Sensors*, vol. 23, no. 15, p. 6760, 2023.
15. W. Villegas-Ch, J. Garcıa-Ortiz, and S. Sa´nchez-Viteri, "Toward intelligent monitoring in IoT: AI applications for real-time analysis and prediction," *IEEE Access*, vol. 12, pp. 40368–40386, 2024.
16. K. Kumar, V. Kumar, and Seema, "Integration of Artificial Intelligence and Machine Learning for Internet of Things," In: *Proceedings of the International Conference on Sustainable Communication Networks and Application (ICSCNA 2023)*, Bharath Niketan Engineering College in Theni, Tamil Nadu, India, pp. 491–497, 2023.
17. G. Yao and L. Gupta, "A survey on the use of partitioning in IoT-edge-AI applications," *arXiv* preprint arXiv:2406.00301, 2024.
18. H. T. T. Truong, M. Almeida, G. Karame, and C. Soriente, "Towards secure and decentralized sharing of iot data," In: *2019 IEEE International Conference on Blockchain (Blockchain)*, Atlanta, GA, United States, pp. 176–183, IEEE, 2019.
19. E. O. Alonge, N. L. Eyo-Udo, B. C. Ubanadu, A. I. Daraojimba, E. D. Balogun, and K. O. Ogunsola, "Real-time data analytics for enhancing supply chain efficiency," *Journal of Susspply Chain Management and Analytics*, vol. 10, no. 1, pp. 49–60, 2023.
20. M. Shaikh, P. Patil, P. V. Thokal, and D. B. Pardeshi, "Implementing Machine Learning for predictive maintenance in industrial machinery," In: *Proceedings of 2024 15th International Conference on Computing, Communication and Networking Technologies (ICCCNT),* Kamand, Himachal Pradesh, India, pp. 1–6, 2024.

21. G. K. Vani, R. Naveenkumar, R. Singha, R. Sharkar, and N. Kumar, "Advancing predictive data analytics in IoT and AI leveraging real time data for proactive operations and system resilience," *Nanotechnology Perceptions*, vol. 20, pp. 568–582, 2024.
22. B. Vashisht, "The role of AI in optimizing dispatching and route planning," *International Journal For Multidisciplinary Research*, vol. 5, pp. 1–13, 2023.
23. C. Sharanya, P. Radhakrishnan, N. Nirmalsingh, D. R. A. Kumar, S. Cloudin, and M. R. Joel, "Iot-driven innovations for sustainable resource use and waste reduction," In: E. Ozen, A. Singh, S. Taneja, R. Rajaram, and J. P. Davim (Eds.), *Sustainability, Innovation, and Consumer Preference (Advances in E-Business Research Series),* IGI Global Scientific Publishing, pp. 307–328, 2024.
24. A. H. Ikevuje, D. C. Anaba and U. T. Iheanyichukwu, "Optimizing supply chain operations using IoT devices and data analytics for improved efficiency," *Magna Scientia Advanced Research and Reviews*, 11(2), 70, (2024). https://doi.org/10.30574/msarr.2024.11.2.0107.
25. S. Rubbani, *Harmonizing Data Streams for Edge AI Protocols to Quantify Shannon's Bounds*, *Research Square*, 2023. https://doi.org/10.21203/rs.3.rs-3584273/v1.
26. S. Ahmad, "Next-gen IT operations with AI and ML: from reactive to proactive cloud management," 2025. https://doi.org/10.31219/osf.io/x4nb8_v1.
27. K. K. Kandakumar, "Reinventing operations with predictive dashboards and real-time intelligence," *Journal of Information Systems Engineering & Management*, vol. 10, p. 542, 2025. https://doi.org/10.52783/jisem.v10i58s.12632.
28. S. Ahmad, R. Kaushik, R. Ghatuary, A. Kotiyal, S. Jarial, and R. Kumar, "Utilizing iot and ai for soil health monitoring and enhancement in sustainable agriculture," In: M. Azrour, J. Mabrouki, and S. Ahmad (Eds.), *IoT and Advanced Intelligence Computation for Smart Agriculture*, CRC Press, pp. 110–125, 2025.
29. O. A. Elufioye, C. U. Ike, O. Odeyemi, F. O. Usman, and N. Z. Mhlongo, "AI-driven predictive analytics in agricultural supply chains: a review: assessing the benefits and challenges of AI in forecasting demand and optimizing supply in agriculture," *Computer Science & IT Research Journal*, vol. 5, no. 2, pp. 473–497, 2024.
30. B. S. Kumbhar, "Smart IoT and ml based vaccine storage system," *Journal of Advanced Research in Embedded System,* vol. 10, pp. 14–17, 2023.
31. V. Yadav, "AI-driven predictive models for healthcare supply chains: developing AI models to predict and optimize healthcare supply chains, especially during global health emergencies," *Progress in Medical Sciences*, vol. 6, no. 1, pp. 1–8, 2022.
32. H. Allioui, A. Allioui, and Y. Mourdi, "Maintaining effective logistics management during and after covid-19 pandemic: survey on the importance of Artificial Intelligence to enhance recovery strategies," *Opsearch*, vol. 61, no. 2, pp. 918–962, 2024.

33. P. Idayakumar, S. Mahadik, I. Paul, and S. Chaure, "Vaccine supply management using Blockchain," In: *Proceedings of 2022 International Conference on Computing, Communication, Security and Intelligent Systems (IC3SIS),* Kochi, Karukutty, Ernakulam District, pp. 1–8, 2022.
34. D. Rangarajan, A. Rangarajan, C. K. K. Reddy, and S. Doss, "Exploring the next-gen transformations in healthcare through the impact of AI and IoT," In: P. Paramasivan, S. S. Suman Rajest, K. Chinnusamy, R. Regin, and Ferdin Joe (Eds.), *Advances in Medical Technologies and Clinical Practice Book Series,* IGI Global, pp. 73–98, 2024.
35. S. Sundaram and A. Zeid, "Artificial Intelligence-based smart quality inspection for manufacturing," Micromachines, vol. 14, p. 570, 2023.
36. L. N. Mohamed, Y. Fernando, and A. B. D. Gani, "A conceptualization of industrial Internet of Things towards manufacturing safety performance," *Journal of Advanced Research in Business and Management Studies*, vol. 29, pp. 1–8, 2023.
37. T. M. Olatunde, A. C. Okwandu, D. O. Akande, and Z. Q. Sikhakhane, "Reviewing the role of Artificial Intelligence in energy efficiency optimization," *Engineering Science & Technology Journal*, vol. 5, no. 4, pp. 1243–56, 2024.
38. K. Ramesh, P. N. Renjith, S. Balasubramani, M. A. Bennet, S. Saritha, D. Pandey, and U. K. Kanike, "Leveraging cutting-edge technologies and innovative strategies to optimize the IoT and AI integration in supply chain management," In: D. Jose and P. Nanjundan (Eds.), *AI-Driven IoT Systems for Industry 4.0,* Routledge, pp. 159–173, 2024.
39. S. S. M. Prabu, R. Senthilraja, A. M. Ali, S. Jayapoorani, and M. Arun, "AI-driven predictive maintenance for smart manufacturing systems using digital twin technology," *International Journal of Computational and Experimental Science and Engineering*, 11(1), (2025). https://doi.org/10.22399/ijcesen.1099.
40. M. Mrad, M. A. Frikha, and Y. Boujelbène, "State-of-the-art review: Artificial Intelligence and robotics for reducing carbon emissions in supply chain management," *Logistics*, vol. 9, no. 3, Art. 104, 2025. https://doi.org/10.2139/ssrn.5274912.
41. V. Harish, R. Sharma, G. Rana, and B. Parashar, "Leveraging digital technologies for circular economy: exploring the role of AI, IoT, and Blockchain in overcoming future challenges," In: R. Kumar and S. N. Mohanty (Eds.), *Sustainable Economy Models in the Age of Industry 5.0*, pp. 53–72, Springer, 2025. https://doi.org/10.1007/978-981-96-4104-8_4.
42. Y. Leshchenko, I. Moroz, M. Yukhymchuk, and V. Lesko, "Optimization of decentralized last-mile delivery using Artificial Intelligence," In: *Proceedings of the 17th International Conference on Measurement and Control in Complex Systems (MCCS),* Vinnytsia National Technical University, 2024.
43. D. Jeevitha, D. V. Jose, S. N. Mohanty, and A. Latha, "Energy management in industry 4.0 using AI," In: D. Jose and P. Nanjundan (Eds.), *AI-Driven IoT Systems for Industry 4.0*, Routledge, pp. 349–363, 2024.

44. S. Rajesh and R. Raman, "IoT-driven energy savings in warehouses cloud-integrated lighting management," In: 2024 *International Conference on Advances in Data Engineering and Intelligent Computing Systems (ADICS)*, Chennai, India, pp. 1–6, 2024.
45. P. R. Potdar and S. Parikh, "Internet of Things (IoT) and Artificial Intelligence (AI) enabled framework for smart fleet management," *OPSEARCH,* (2025). https://doi.org/10.1007/s12597-025-00961-7.
46. F. Goedertier, B. Weijters, and P. Vanpaemel, "The longitudinal effect of digitally administered feedback on the eco-driving behavior of company car drivers," *Sustainability*, vol. 15, no, 24, p. 16571, 2023. https://doi.org/10.3390/su152416571.
47. C. Yaiprasert and A. N. Hidayanto, "AI-powered ensemble Machine Learning to optimize cost strategies in logistics business," *International Journal of Information Management Data Insights*, vol. 4, no. 1, p. 100209, 2023. https://doi.org/10.1016/j.jjimei.2023.100209.
48. S. S. Haghshenas, V. Astarita, S. S. Haghshenas, and G. Guido, "Artificial Intelligence-powered driver behavior analysis for fuel consumption optimization: a pathway to greener roads," In: 2024 *10th International Conference on Control, Decision and Information Technologies (CoDIT),* Valletta, Malta, pp. 116–122, 2024.
49. P. Joshi, V. Singh, N. Thapliyal, K. Joshi, A. Bhatt, and M. Mahur, "Intelligent IoT-enabled real-time monitoring system for logistics management," In: 2024 *International Conference on Computing, Sciences and Engineering (ICCSE),* Kuwait City, Kuwait, pp. 1–5, 2024.
50. W. Chen, Y. Men, N. Fuster, C. Osorio, and A. Juan, "Artificial Intelligence in logistics optimization with sustainable criteria: a review," *Sustainability*, vol. 16, p. 9145, 2024.
51. M. Gafarov, "Predictive analytics for sales forecasting and inventory management," *Multidisciplinary Reviews*, vol. 8, p. 109, 2024.
52. M. A. Amin, S. Kim, M. A. S. A. Rishat, Z. Tang, and H. Ahn, "A systematic literature review of privacy information disclosure in AI-Integrated Internet of Things (IoT) technologies," *Sustainability*, vol. 17, no. 1, p. 8, 2024. https://doi.org/10.3390/su17010008.
53. I. Zejjari and I. Benhayoun, "The use of Artificial Intelligence to advance sustainable supply chain: retrospective and future avenues explored through bibliometric analysis," *Discover Sustainability*, vol. 5, no. 1, 2024. https://doi.org/10.1007/s43621-024-00364-6.
54. N. Dacre, J. Yan, R. Frei, M. K. S. Al-Mhdawi, and H. Dong, "Advancing sustainable manufacturing: a systematic exploration of Industry 5.0 supply chains for sustainability, human-centricity, and resilience," *Production Planning & Control*, 1, 2024. https://doi.org/10.1080/09537287.2024.2380361.
55. D. S. P. Gokulakrishnan and S. Venkataraman, *Ensuring Data Integrity: Best Practices and Strategies in Pharmaceutical Industry*, Intelligent Pharmacy, 2024. https://doi.org/10.1016/j.ipha.2024.09.010.

56. S. Munusamy and K. R. Jothi, *Blockchain-Enabled Federated Learning with Edge Analytics for Secure and Efficient Electronic Health Records Management*, *Research Square*, 2025. https://doi.org/10.21203/rs.3.rs-6678464/v1.
57. K. S. Chadha, "Bias and fairness in Artificial Intelligence: methods and mitigation strategies," *International Journal for Research Publication and Seminars*, vol. 15, no. 3, p. 36, 2024. https://doi.org/10.36676/jrps.v15.i3.1425.
58. T. Stoecker, S. Bayer, and I. Weber, "Bias mitigation for AI-feedback loops in recommender systems: a systematic literature review and taxonomy," *arXiv* preprint, 2025. https://doi.org/10.48550/ARXIV.2509.00109.
59. J. Visave, "Transparency in AI for emergency management: building trust and accountability," *AI and Ethics*, vol. 5, no. 4, 3967, 2025. https://doi.org/10.1007/s43681-025-00692-x.
60. J. Xiao, Z. Xu, A. Xiao, X. Wang, and M. Škare, "Overcoming barriers and seizing opportunities in the innovative adoption of next-generation digital technologies," *Journal of Innovation & Knowledge*, vol. 9, no. 4, 100622, 2024. https://doi.org/10.1016/j.jik.2024.100622.
61. M. Rahaman, C. Lin, P. Pappachan, B. B. Gupta, and C. Hsu, "Privacy-centric AI and IoT solutions for smart rural farm monitoring and control," *Sensors*, 24(13), 4157, (2024). https://doi.org/10.3390/s24134157.
62. N. Li, T. D. Palaoag, H. Du, and T. Guo, "Design and optimization of smart campus framework based on artificial intelligence," *Journal of Information Systems Engineering & Management*, vol. 8, no. 3, p. 23086, 2023. https://www.jisem-journal.com/download/design-and-optimization-of-smart-campus-framework-based-on-artificial-intelligence22-13853.pdf.
63. B. M. Bazrov, "Effectiveness of modular technology," *Vestnik Mashinostroeniya*, vol. 103, p. 1042, 2024. https://doi.org/10.36652/0042-4633-2024-103-12-1042-1046.
64. L. Benos, V. Marinoudi, P. Busato, D. Kateris, S. Pearson, and D. Bochtis, "Benchmarking large language models in evaluating workforce risk of robotization: insights from agriculture," *AgriEngineering*, vol. 7, no. 4, p. 102, 2025. https://doi.org/10.3390/agriengineering7040102.
65. D. V. Dennison, M. Jain, T. Ganu, and A. Vashistha, "Designing culturally aligned AI systems for social good in non-western contexts", *arXiv* preprint, 2025. https://doi.org/10.48550/ARXIV.2509.16158.
66. R. Fulton, D. Fulton, N. Hayes, and S. Kaplan, "The transformation risk-benefit model of Artificial Intelligence: balancing risks and benefits through practical solutions and use cases," *International Journal of Artificial Intelligence and Applications* vol. 15, no. 2, pp. 1–22, 2024. https://arxiv.org/abs/2406.11863.
67. G. B. Yusupkhodjaeva, "Development of a unified digital transport and logistics intelligent platform based on the national operator," *E3S Web of Conferences*, vol. 461, p. 1055, 2023. https://doi.org/10.1051/e3sconf/202346101055.

68. M. Homaei, Ó. Mogollón-Gutiérrez, J. C. S. Núñez, M. Ávila, and A. Caro, "A review of digital twins and their application in cybersecurity based on Artificial Intelligence," *Artificial Intelligence Review*, vol. 57, no. 8, 2024. https://doi.org/10.1007/s10462-024-10805-3.
69. N. Subramanian, A. Joshi and D. Bagga, "Transparent and traceable food supply chain management," *arXiv*:2305.12188, 2023. https://doi.org/10.48550/ARXIV.2305.12188.
70. M. Ghaly, E. Elbeltagi, A. Elsmadony, and M. A. Tantawy, "Integration of Blockchain-enabled smart contracts in construction: SWOT framework and social network analysis," *Civil Engineering Journal*, vol. 10, no. 5, p. 1662, 2024. https://doi.org/10.28991/cej-2024-010-05-020.
71. Z. Zhang, Y. Rao, H. Xiao, X. Xiao and Y. Yang, "Proof of quality: a costless paradigm for trustless generative AI model inference on Blockchains," *arXiv* (Cornell University), 2024. https://doi.org/10.48550/arxiv.2405.17934.
72. A. A. Sharabati and E. R. Jreisat, "Blockchain technology implementation in supply chain management: a literature review of Blockchain technology implementation in supply chain management," *Sustainability*, vol. 16, no. 7, p. 2823, 2024. https://doi.org/10.3390/su16072823.
73. I. Abrar and J. A. Sheikh, "Current trends of Blockchain technology: architecture, applications, challenges, and opportunities," *Discover Internet of Things*, vol. 4, no. 1, 2024. https://doi.org/10.1007/s43926-024-00058-5.
74. B. C. Cheong, "Leveraging Blockchain for enhanced transparency and traceability in sustainable supply chains," *Discover Analytics*, vol. 3, no. 1, 2025. https://doi.org/10.1007/s44257-025-00032-7.
75. F. O. Olatoye, K. F. Awonuga, N. Z. Mhlongo, C. V. Ibeh, O. A. Elufioye, and N. L. Ndubuisi, "Review of AI and ethics in business: a comprehensive review of responsible AI practices and corporate responsibility," *International Journal of Science and Research Archive*, 11(1), 1433, (2024). https://doi.org/10.30574/ijsra.2024.11.1.0235.
76. G. Shirvani and S. Ghasemshirazi, "Towards sustainable IoT: challenges, solutions, and future directions for device longevity," 2024. *arXiv* (Cornell University), https://doi.org/10.48550/arxiv.2405.16421.
77. J. C. S. Gummadi, "Cybersecurity in international trade agreements: a new paradigm for economic diplomacy," *American Journal of Trade and Policy*, 11(1), 39, (2024). https://hal.science/hal-04909160v1/file/32.4.pdf.
78. A. M. Ruzbahani, "AI-protected Blockchain-based IoT environments: harnessing the future of network security and privacy," *arXiv* (Cornell University), 2024. https://doi.org/10.48550/arxiv.2405.13847.
79. A. M. Al-Ansi, A. Garad, M. Jaboob, and A. Al-Ansi, "Review of elevating elevating E-government: unleashing the power of AI and IOT for enhanced public services," *Heliyon*, vol. 10, no. 23, 2024. https://doi.org/10.1016/j.heliyon.2024.e40591.

Chapter 6

Ethical and social implications of AI in sustainable supply chain management

Shreya, Anushree, Ravi Prakash, and Rama Moorthy Hejamadi

6.1 INTRODUCTION

In the 21st century, artificial intelligence (AI) increasingly plays a major role in international firms' operations and a variety of fields, in particular, supply chain management (SCM). AI improves demand forecasting, inventory optimization, and logistics tracking, reducing costs and improving productivity, facilitating timely market responses.

AI provides speed and efficiency also in sustainable SCM (SSCM) while promoting social justice, environmental stewardship, and ethical leadership practices by requiring firms to be fair in their labour practices and transparent in their goods and services. On the other hand, AI prompts issues of responsibility and equity and whether or not, and to what degree, automation leads to job loss and inequality in society.

The complexity of supply systems means there is no effective way to ignore AI. Whether or not firms are recognized as socially responsible, the success of different supply systems is dependent upon responsible AI use that prioritizes inclusion, sustainability, and ethics.

6.1.1 Emergence of AI in supply chains

AI has advanced supply chains by enhancing information: forecasting, logistics, and real-time supervision, which helps to reduce costs and increase productivity. Still, the more pressing concern is AI's social and sustainability impacts, especially as entrusting such outcomes to machines may mean yielding human control.

6.1.2 Objectives for sustainability in contemporary business

Companies are putting a greater focus on sustainability, with supply chains helping to reduce greenhouse gases and ensure workers are being treated ethically. AI can optimize this process in a meaningful way by improving business processes and tracking supplier behaviour. Yet as organizations

DOI: 10.1201/9781003660163-6

embed these improvements into their sustainability strategy, leaders must be mindful of how they utilize AI and its potential to increase inequality or externalize environmental costs.

6.1.3 Moral conflicts in AI implementation

AI in supply chains has potential advantages but raises ethical questions about accountability, fairness, and transparency. Biased data and opaque decision-making can disadvantage marginalized communities and small suppliers. The ethical incorporation of AI will be important to balance considerations of inclusion, efficiency, and ethics.

6.2 METHODOLOGY

This research investigates the emerging association between AI and SSCM. By employing innovative exploratory and qualitative approaches, we aim to uncover how AI can revolutionize practices within the realm of sustainability in supply chains. Join us as we explore the potential of technology to create more efficient and environmentally friendly supply chain solutions. This methodological approach aims to capture not just the technological applications of AI, but also the ethical, environmental, and governance factors that come with its adoption. Rather than depending just on numerical statistics, this method emphasizes interpretation, synthesis of literature, and critical study of real-world activities. The technique is divided into four independent but interrelated sections: literature research, policy and governance framework analysis, case study evaluation, and an integrative analytical framework.

6.2.1 Literature review

The research procedure began with a thorough literature evaluation, which served as the theoretical basis for the investigation. The literature research was carried out in a methodical manner to ensure that varied academic perspectives and empirical findings were included. Scopus, IEEE Xplore, SpringerLink, ScienceDirect, and Google Scholar databases were searched for materials published between 2015 and 2024, the most active time of scholarly discussion on AI in SCM.

The literature search included terms like "AI in supply chains," "sustainable logistics," "green supply chains," "AI ethics," and "AI governance." Priority was given to peer-reviewed journal papers, scholarly publications, and conference proceedings. At the same time, industry studies and white papers from consulting companies and international organizations (such as the OECD and UNCTAD) were studied for practical insights.

The purpose of the literature review is threefold:

- To map the present status of research on AI applications in supply chains.
- To identify gaps in current research, particularly on social and environmental impacts.
- To highlight areas of consensus and debate concerning the ethical implications of AI.

This stage helped in situating the present study within the broader academic and professional discourse, ensuring that subsequent analysis is both contextually grounded and theoretically informed.

6.2.2 Policy and governance framework analysis

The second methodological phase focuses on analysing the policy and governance frameworks. Given that AI technologies traverse borders, it is critical to understand how various governance systems strive to regulate and standardize their use in supply chain scenarios. This step included a review of international, regional, and national frameworks. At the international level, papers such as the OECD AI Principles and ISO Standards were examined. These frameworks provide broad criteria for openness, accountability, and justice. At the regional level, the European Union's AI Act was thoroughly examined since it is one of the first comprehensive legislative efforts to control AI deployment, including high-risk applications pertinent to supply chains. At the national level, AI plans and policies from prominent economies, including India, the United States, and China, were analysed to identify differences in governance methods. To summarize the similarities and differences between these governance models, we employed a comparative tabulation approach. Risk management, ethical standards, sustainability emphasis, and enforcement measures were evaluated. This strategy not only identifies similarities but also reveals gaps in policy coverage, particularly when it comes to environmental sustainability and labour rights [1,2].

6.2.3 Case study approach

To supplement the theoretical and policy analysis, the study used a case study technique. This technique was selected because case studies provide a thorough evaluation of how AI is used in practice inside complicated supply chain systems. Three global firms were chosen as key case studies: Walmart, Maersk, and Unilever. These firms were chosen because they cover a wide range of industries (retail, shipping and logistics, and consumer products) and are known for pioneering AI-driven supply chain projects.

- To map the present status of research on AI applications in supply chains.
- To identify gaps in current research, particularly on social and environmental impacts.
- To highlight areas of consensus and debate concerning the ethical implications of AI.

Secondary sources for these case studies included sustainability reports, annual reports, media releases, and independent reviews conducted by NGOs and academic institutes. The inclusion of many case studies enabled cross-case comparisons, emphasizing both effective practices and recurring difficulties [3,4].

6.2.4 Analytical framework

Finally, the data from the preceding steps was combined into an analytical framework. This approach divides the function of AI in SSCM into three main dimensions:

a. Social implications include worker well-being, employment change, inclusion, equity, and the hazards of algorithmic prejudice. This dimension assesses how AI affects job patterns, labour rights, and social equality in supply chains.
b. Environmental implications emphasizing environmental advantages such as carbon footprint reduction, effective resource utilisation, and the creation of "green AI." It also includes issues like AI systems' energy use and electrical waste.
c. Governance frameworks investigate the role of legislative restrictions, ethical norms, and corporate governance principles in promoting responsible AI deployment. This dimension fills the gap between technological innovation and regulatory monitoring.

By organizing the investigation across these three aspects, the framework guarantees that the study is not just focused on technology, but rather gives a comprehensive assessment of AI in sustainable supply chains. It aids in balancing excitement about efficiency improvements with caution about ethical, social, and environmental concerns [5].

6.3 BACKGROUND

6.3.1 AI in supply chain administration

AI has emerged as a fundamental element of contemporary supply chains by enhancing forecasting precision, automating storage facilities, and allowing real-time monitoring of shipments [6]. These instruments enable businesses to decrease delays, cut expenses, and swiftly respond to unexpected disruptions in worldwide markets. Machine learning models forecast demand more accurately, generative AI mimics risks such as natural disasters to boost readiness, and computer vision improves quality assurance by identifying defects in manufacturing processes. Together, these applications enhance the efficiency, competitiveness, and resilience of supply chains [6] (Figure 6.1).

Yet, the impact of AI goes further than just operational efficiency. Its implementation also alters the dynamics among businesses, employees, providers, and the ecosystem. AI-driven decisions can influence job opportunities, choice of suppliers, and impact on the environment. This positions AI as not only a technological innovation but also a societal influence that must be judiciously aligned with ethical and sustainable principles to guarantee it serves all stakeholders instead of introducing new risks.

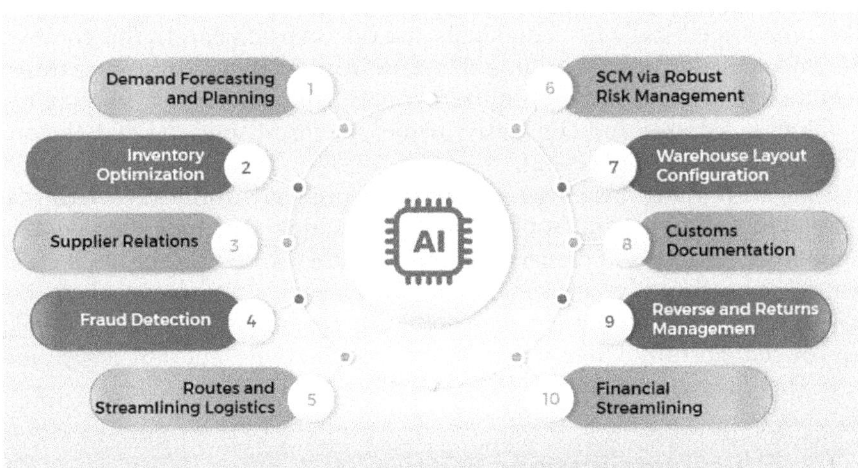

Figure 6.1 Artificial intelligence in supply chains administration.

Figure 6.2 Sustainable practices in supply chain.

6.3.2 Sustainable practices in supply chains

Supply chain sustainability is influenced by the ESG framework—environmental, social, and governance—focusing on minimizing emissions, guaranteeing equitable labour practices, and upholding transparent governance. An effective supply chain should not just provide products efficiently but also function ethically for individuals and the environment. In this context, AI enhances outcomes by minimizing overproduction, decreasing transportation emissions with optimized routes, and improving traceability, enabling businesses and consumers to verify ethical sourcing and responsible practices [7] (Figure 6.2).

Even with these benefits, AI can also pose environmental hazards. Developing and running sophisticated AI models demands considerable computing resources and energy, amplifying the carbon footprint. If companies prioritize only performance improvements, AI might weaken sustainability objectives rather than enhance them. Organizations should use AI thoughtfully while maintaining accountability and efficiency, and achieving their governance and sustainability objectives.

6.3.3 Ethical implications

AI in SSCM involves a number of ethical considerations that must be properly addressed to guarantee that technology is utilized appropriately. These issues may be grouped into three categories: openness and accountability, fairness and prejudice, and decision-making responsibility. Each of these criteria is critical in determining how AI effects the ethical underpinning of supply chains [8,9].

6.3.4 Transparency and accountability

One of the most important ethical issues in AI-powered supply chains is a lack of transparency, sometimes known as the "black box" dilemma. Many AI models, particularly those based on machine learning and deep learning, provide results without providing explicit explanations for how decisions were made. This is especially important in supply chains, as AI systems must select which suppliers to prioritize, how to distribute resources, and when to alter hazards. If these judgements are not explainable, stakeholders' faith in the system suffers, and it is unclear if the outcomes are fair or accurate [10].

This problem is intimately tied to accountability. When AI-driven judgements create harm, such as barring smaller suppliers or producing unfair trade results, it is sometimes unclear who should be held accountable: the AI developers, the managers who implement the systems, or the businesses that rely on them. Errors may go uncorrected in the absence of clear accountability procedures, and victims may not obtain justice. To address this, organizations should embrace explainable AI (XAI), create decision-making audit trails, and develop rules that explicitly allocate responsibilities at each level of AI use [10].

6.3.5 Fairness and bias

Another key ethical worry is the possibility of prejudice in AI systems. AI, which depends largely on historical and training data, has the potential to unintentionally exacerbate existing disparities. For example, if databases are dominated by information about large global suppliers, AI models may favour them over smaller or local enterprises. Poor-quality or under-represented data can exclude suppliers from poorer countries, worsening economic imbalances. To address pervasive bias, organizations should utilize diverse data sets, regularly test AI models for bias, and establish inclusive governance frameworks. AI can promote equity for all parties rather than deepen existing inequities [11].

6.3.6 Responsibilities for decision-making

AI-driven decision-making frequently puts business and efficiency ahead of morality. For example, it might advocate for cheaper suppliers who engage in abusive labour practices or prefer quicker but more environmentally

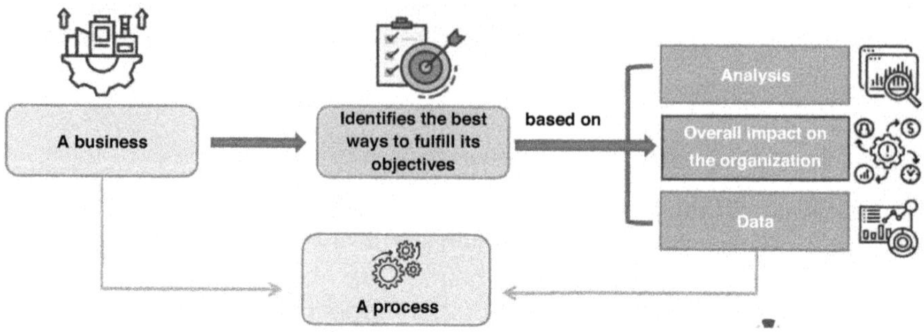

Figure 6.3 Decision-making flowchart.

damaging transit options. These situations show the dangers of giving AI too much freedom without human supervision [12] (Figure 6.3).

It is essential for organizations to ensure that decision-makers are competent managers with strong ethical judgement. This human touch not only encourages accountability but also preserves the integrity of decisions. Putting ethical standards into AI systems, such as removing suppliers that fail labour compliance checks, helps to align decision-making with sustainability goals. Furthermore, personnel must be educated to critically analyse AI results rather than just accepting them. In this sense, AI might help human decision-makers rather than take away the moral responsibility that comes with human judgement.

6.4 SOCIAL IMPLICATIONS

AI's inclusion in supply chains has important societal implications. These effects may be classified into three categories: employment and reskilling, worker well-being, and social equality and inclusion. Each of these domains influences how society views and reacts to AI adoption in SSCM [13].

6.4.1 Employment and skill development

One of the most debated societal effects of AI is its impact on jobs. As AI automates warehouses, shipping, and administrative operations, many conventional occupations will be displaced. Workers who formerly performed physical work may be displaced, resulting in unemployment in vulnerable populations. However, AI does not just eliminate jobs; it also create new ones. The growth of digital supply chains raises the demand for data analysts, AI supervisors, and sustainability managers. The difficulty is that displaced workers frequently lack the necessary skills to take on these new tasks. This leads to a mismatch between the disappearance of old occupations and the creation of new opportunities that need sophisticated training. To address this issue, organizations must engage in reskilling and upskilling initiatives. Workers should be trained on digital tools, data analysis, and AI systems. Governments and educational institutions also play an important role in revising curriculums to train future workers for AI-driven sectors. Without such efforts, AI may result in widespread inequity, with only highly qualified professionals benefiting while low-skilled individuals are left behind. This could undermine sustainability and threaten social stability. To ensure AI benefits society, proactive reskilling should accompany job disruption and reduce burnout by monitoring workloads. This promotes a supportive workplace culture. On the downside, AI can lead to excessive supervision of workers, with some companies tracking every move and decision. While this may boost efficiency, it can also increase stress and reduce employees' sense of privacy. Balancing AI surveillance benefits with worker rights is crucial. Companies should use AI to assist employees, not control them. Monitoring systems must prioritize safety and task balance over micromanagement [14][15][16].

6.4.2 Worker well-being

When used properly, AI can enhance job satisfaction by automating repetitive tasks, allowing workers to concentrate on meaningful and innovative work, which boosts both happiness and productivity. In conclusion, prioritizing worker welfare is essential when implementing AI. Responsible use of AI leads to safer, healthier, and more enjoyable workplaces (Figure 6.4).

Figure 6.4 Social impact of AI in supply chain.

6.4.3 Social equity and inclusion

Depending on how it is applied, AI has the potential to decrease or worsen societal inequality. Large firms frequently have the resources to implement advanced AI technology, but small enterprises may struggle to keep up. This establishes a distinction between "AI-rich" and "AI-poor" organizations. This mismatch in supply chains might hurt small suppliers and impede inclusion. On the other hand, AI has the ability to increase inclusivity. By enhancing traceability, AI can assist businesses in identifying and supporting ethical small suppliers. It can also provide more information to consumers about the social impact of their purchases, supporting more equitable practices across the supply chain [17].

6.5 ENVIRONMENTAL IMPLICATIONS

AI is strongly linked to environmental sustainability, but it has two consequences. On the one hand, training AI models requires a significant amount of energy, which contributes to the carbon impact. AI, on the other hand, helps to cut emissions and waste in supply chains by improving operational efficiency. The key environmental concerns are grouped into three categories: the carbon footprint of AI models, AI's beneficial contributions to sustainability, and cost–benefit analysis (Figure 6.5).

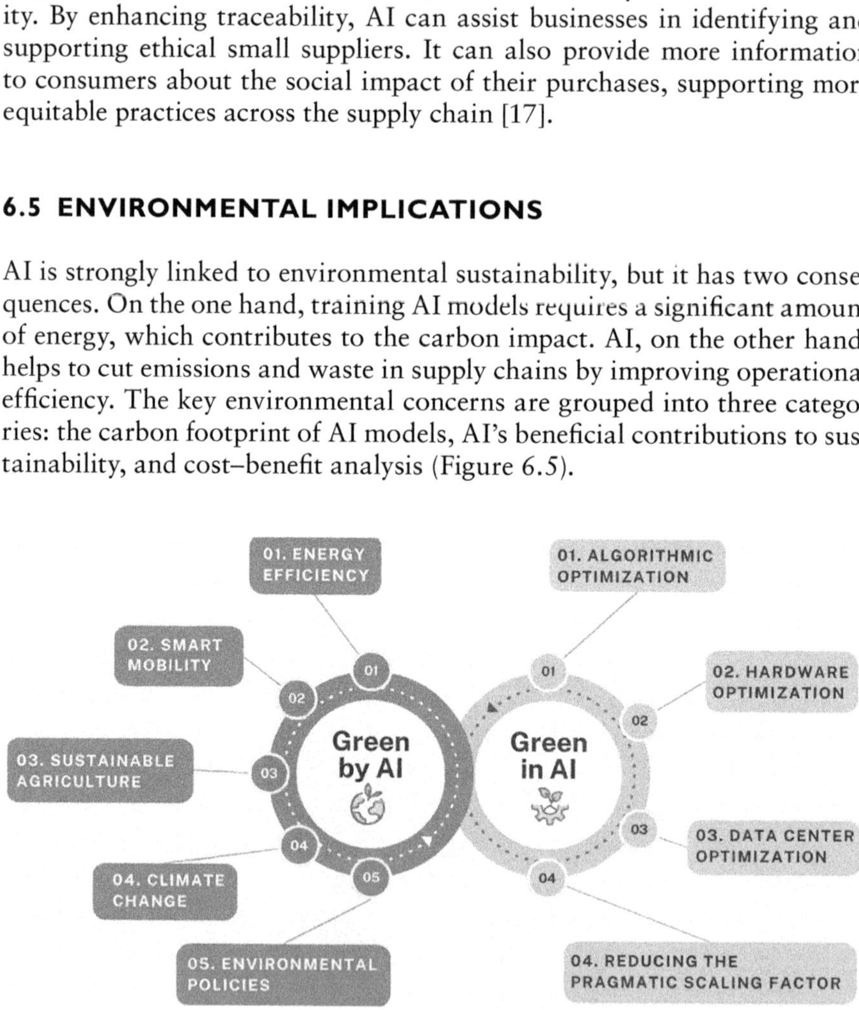

Figure 6.5 AI applications in green supply management.

6.5.1 The carbon footprint of AI models

Advanced AI models require a massive amount of processing resources to train. Data centres must process billions of data points, which consumes enormous quantities of power. Furthermore, these facilities consume a lot of water to cool, which puts pressure on the environment. For example, the energy required to train a large AI model can be comparable to the emissions created by multiple automobiles during their lifespan. This demonstrates that AI is not a neutral technology, but one with considerable environmental consequences. These environmental consequences become more important as AI grows internationally. As more businesses use AI in their supply chains, the overall footprint of training and operating models grows. AI's own sustainability aims may be jeopardized if energy-efficient procedures are not used. Adopting green AI principles is one possible approach. This includes leveraging renewable energy sources in data centres, optimizing algorithms to save energy, and reusing learnt models rather than creating new ones from start. As a result, while AI provides environmental solutions, its own footprint should not be overlooked. Responsible usage guarantees that AI does not become a source of the problem it was designed to address [18,19].

6.5.2 The positive environmental contributions of AI

Regardless of its footprint, AI delivers great tools for sustainability. In logistics, AI can determine the most fuel-efficient routes, lowering transportation emissions. AI in manufacturing can monitor energy use and provide waste reduction strategies [20].

AI also benefits the circular economy. Companies may build recycling and reuse systems by anticipating product life cycles. For example, AI can forecast when machinery parts will break, allowing them to be replaced and repurposed instead of destroyed [21].

Real-time emission tracking is another important addition. AI systems can track carbon emissions throughout supply chains and identify areas where reductions are achievable. This enables businesses to set realistic sustainability targets and track their progress. When applied effectively, AI may be both an efficiency tool and a sustainability facilitator, as demonstrated by these examples.

6.5.3 Balancing costs and benefits

The environmental issue of AI is to balance its costs and advantages. While AI uses energy, it has the potential to reduce emissions significantly through optimized logistics and waste reduction. The objective is to guarantee that the beneficial effects outweigh the negative. For example, while training an AI system may use energy, if that system saves millions of litres of petroleum each year by optimizing shipping routes, the net impact is beneficial. Companies must carefully evaluate these trade-offs. This balancing act needs continuous monitoring. Businesses must monitor both the direct expenses (AI footprint) and the indirect advantages (AI-enabled sustainability). Only then can they justify AI deployment in terms of environmental impact. Governments and regulators both have a role. They may assist in maximizing the environmental advantages of AI by promoting green AI standards and the usage of renewable energy in data centres. Thus, the environmental impact of AI is determined by how responsibly it is implemented. With the proper procedures, AI may transform from a polluter to a great ally in sustainability [22].

6.6 GOVERNANCE AND POLICY FRAMEWORKS

Strong governance frameworks are required for AI in SSCM. Without explicit restrictions, firms may use AI to prioritize short-term profitability above long-term social and environmental goals. Governance ensures that AI systems function fairly, openly, and in accordance with sustainability goals, therefore eliminating unforeseen consequences such as labour exploitation or environmental damage. Effective policy frameworks serve as guidelines for how AI is conceived, applied, and assessed in global supply chains [23,24].

6.6.1 The EU AI Act

The European Union's AI Act was one of the first comprehensive regulatory approaches to managing AI. It proposes a risk-based categorization system that categorizes AI applications according to the possible harm they might create. High-risk AI systems in critical industries like health and food delivery must meet strict requirements, including transparency, testing, and risk assessments before implementation. The EU AI Act sets standards to ensure that businesses use AI responsibly, protecting human rights, public health, and safety. It emphasizes that AI is a powerful tool that requires careful management to prevent negative consequences and encourages global supply chains to prioritize ethical AI and accountability [25] (Figure 6.6).

The AI Act, introduced within the European Union (2024), is the inaugural legal framework bringing a comprehensive risk-based classification system to AI. For high-risk AI applications, such as those associated with logistics, manufacturing, and healthcare, the law requires essential transparency checks, testing, and impact assessments prior to the receipt of the operation. One area where the mandatory compliance for the law will be felt is in supply chains. As an illustration, AI applications for companies

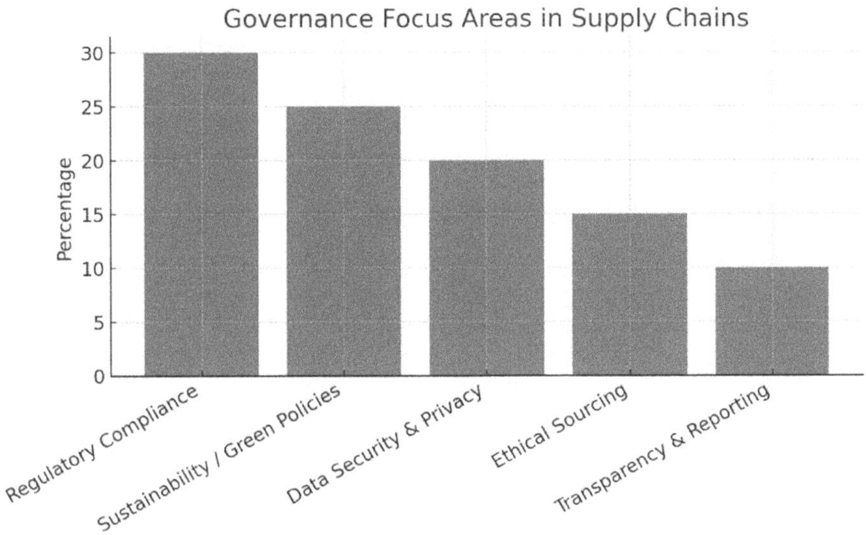

Figure 6.6 Governance focus area in supply chain.

like Maersk and Unilever rely on advanced AI for routing optimization and supplier monitoring. For example, Maersk's app relies on AI technology and would be classified as "high-risk" under the new EU AI Act, requiring environmental and safety assessments before being utilized. Such regulation ensures that AI does not repeat the mistakes of human-oriented systemic poor decisions in emissions or internationally prohibited types of trade. However, in practice, this would impose another burden in both public accountability and cost for smaller suppliers, representing another major deficiency in the new Act, as many small and medium-sized companies are relatively less able to pay the costs associated with such judgements. Nevertheless, the Act has already led to measurable enhancements in accountability and traceability in European supply chains, with studies reporting an approximate 20% improvement in the accuracy of carbon reporting (Engler & Renda, 2022). Overall, the EU AI Act broadens ethical oversight and environmental accountability, yet it still faces challenges in ensuring the involvement of smaller industry players [26].

6.6.2 OECD AI principles

The OECD has established AI principles focused on justice, transparency, and human-centred development. Unlike the legally binding EU AI Act, these principles are guidelines promoting voluntary compliance and international collaboration. They highlight the importance of protecting human rights, promoting inclusivity, and ensuring that AI benefits society. The OECD's recommendations emphasize the need for cross-border

Table 6.1 Governance and policy frameworks in supply chains

Framework	*Focus*	*Issues*
EU AI Act	Risk levels, safety, rights, green rules	Costly, tough for SMEs
OECD AI Principles	Fair, accountable, global rules	Voluntary, weak control
ISO Standards	Common rules, data, risk management	Slow, non-binding
National Policies	Promote AI, ethics, data laws	Different rules, uneven use

collaboration in ensuring sustainable supply chains. Supply chains are fundamentally global, encompassing many nations with distinct laws and cultural settings. The OECD principles promote harmonization, which reduces the possibility of contradictory legislation and promotes a common ethical framework. This global collaboration is critical because unethical activities might easily transfer to places without harmonization [27] (Table 6.1).

The OECD AI Principles prioritize fairness, accountability, and human-centred design with the goal of harmonizing ethical frameworks for AI through voluntary global convergence. These principles influence companies like Walmart in SCM to integrate ethical and transparent AI. For example, Walmart's desire to deploy automation in logistics aligns with the OECD fairness principle, which refers to the recommendation of maintaining human discretion in automated approaches. In accordance with these principles, Walmart publicly announced a reskilling plan for over 10,000 employees working in its warehouses in 2023, similarly responding to the ethically troublesome workforce displacement associated with automation. Although OECD principles are not legally binding, they do provide an ethical and operational framework that pushes large corporations to embrace socially responsible AI. Nevertheless, they are only voluntary guidelines, meaning their effectiveness largely hinges on a corporation's choice to comply with them, leading to inconsistent application across industries and geographic regions. Therefore, while the OECD framework is advantageous in contributing social responsibility and fairness, it cannot compel organizations to act accordingly, as it lacks regulatory force.

6.6.3 ISO standards

The International Organization for Standardization (ISO) develops technical recommendations and standards to promote ethical and sustainable AI inclusion in supply chains. These standards address topics such as traceability, accountability, auditability, and data governance, providing enterprises with a systematic approach to ensuring compliance and best practices. Businesses that use ISO standards may demonstrate their commitment to transparency and sustainability in a quantitative way. The ISO standards are extremely significant in supply chains because they provide a common language across businesses and nations. These standards contribute to uniformity and confidence in global trade, where suppliers and manufacturers operate under a variety of situations. They also prevent "greenwashing," in which businesses pretend to be sustainable but do not genuinely follow responsible practices [28].

The ISO creates technical specifications that encourage ethical and sustainable AI use within supply chains. Standards like ISO 31000 (risk management) and ISO/IEC 42001 (AI management systems) provide a structured path to transparency, accountability, and data governance. For example, Unilever has used auditing tools that adhere to the ISO framework to increase supplier transparency and ensure sustainability compliance. This has resulted in a 25% decrease in unverifiable sourcing claims, as well as dramatic improvements in data accuracy in its global supply chain. Overall, ISO standards allow organizations to provide quantitative evidence of responsible and sustainable practices. However, ISO standards are not written with legal enforcement and typically progress at a slower pace than the headwind of technological advancements such as AI. Ultimately, ISO standards can increase sustainability reporting consistency, trust, and credibility, but they need to be updated more pragmatically in order to be sufficient for today's technology-based supply chains.

6.6.4 National policies

National policies have a significant impact on the governance and regulation of supply chains. Governments throughout the world are enacting new rules, standards, and ethical frameworks to guarantee that supply chains are transparent, fair, and ecologically sustainable. To increase accountability and deter harmful activities, several governments have introduced data protection rules, labour rights safeguards, and green supply chain policies. These rules not only protect consumers and workers, but they also encourage corporations to pursue responsible innovation while adhering to local and international laws. At the same time, national policies present issues owing to regional variations. While industrialized economies frequently apply severe sustainability and compliance criteria, many developing countries may have lax enforcement owing to budgetary restrictions. This inconsistency can result in regulatory fragmentation, making it difficult for international firms to comply with varied national regulations. When national rules are properly harmonized, they become strong instruments for advancing responsible SCM, as they lay the groundwork for ethical governance, build public trust, and promote sustainable corporate practices [29,30].

6.7 APPLICATIONS AND CASE STUDIES

Case studies from top organizations demonstrate how AI is already influencing supply chains in both good and negative ways. By examining instances from Walmart, Maersk, and Unilever, it becomes obvious that AI may increase efficiency and sustainability while also introducing ethical quandaries such as labour displacement, inequality, and privacy issues. These real-world examples demonstrate the significance of combining technological advancement with social responsibility [31].

6.7.1 Walmart

Walmart has used AI to optimize logistics and increase transportation efficiency. The firm uses AI-driven route planning to minimize fuel usage, delivery times, and operating expenses. These enhancements not only assist Walmart's bottom line but also lower greenhouse gas emissions, therefore supporting the company's environmental goals. This example shows how AI may produce a "win–win" situation in which both efficiency and environmental goals are met. However, Walmart's use of automation has created ethical questions, notably about employment. As AI systems take over warehouse operations and logistical activities, many workers will be displaced. This demonstrates the social trade-offs associated with AI adoption: although customers and shareholders profit from increased efficiency, employees incur the cost of automation [32]. Walmart's instance demonstrates the need for firms to balance efficiency benefits with investments in worker reskilling and fair labour standards.

Walmart has utilized AI to improve its logistics operations and transportation efficiency. According to the Walmart Sustainability Report (2024), the company's AI-based route optimization system decreased fuel consumption by 15% and logistics costs by 10% between 2020 and 2023. These metrics are measurable steps towards the company's objective of zero emissions by 2040. Walmart's AI initiative, which has been able to leverage an operational and environmental balance, contrasts with Amazon, which self-reportedly lowered fuel consumption by 8% during the same timeframe. Instead of just questioning, Walmart's case suggests agency and accountability in the possibility of marrying AI efficiency and profitability with environmental sustainability. However, in part, the increase in automation has resulted in ethical doubts concerning employment, as warehouse automation and algorithmic scheduling have displaced some workers. Therefore, while Walmart's example is useful in illustrating AI-driven efficiency, it remains important for organizations to first drive efficiency and follow up with ethical transition planning for workers, including employee education, upskilling talents, and fair labour conditions.

6.7.2 Maersk

Maersk, one of the world's major shipping businesses, employs AI to optimize worldwide shipping routes. AI algorithms analyse massive quantities of data, including weather patterns, port congestion, and fuel prices, to determine the most effective routes. This enables Maersk to save gasoline, reduce operating expenses, and deliver goods faster. From an environmental standpoint, route optimization cuts emissions dramatically, making global shipping more sustainable. At the same time, AI adoption by industry heavyweights such as Maersk might put smaller shipping businesses at a competitive disadvantage since they lack the means to invest in new technology. This exacerbates industrial inequality, as major businesses become more powerful while smaller enterprises struggle to compete. Thus, although Maersk's AI-driven efficiency helps to ensure sustainability, it also emphasizes the significance of ensuring that innovation does not prohibit smaller stakeholders from participating in the global economy [33].

Maersk, one of the largest shipping companies in the world, utilizes AI to improve global shipping routes due to its ability to analyse large data sets on things like weather conditions, port congestion, and fuel prices. By utilizing AI-enabled route optimization, Maersk has achieved a 6% reduction in average fuel consumption per voyage, along with a reduction in CO emissions of almost 1 million tonnes per year (Maersk Annual Report, 2024). This performance puts it at the forefront of its peers in the industry, where other similar organizations saw only a 3%–4% reduction in emissions. While these developments provide a clear environmental sustainability benefit, they also present ethical and economic considerations. The high cost of creating and implementing AI systems, even if it is 100 million dollars, creates a digital divide between large organizations such as Maersk and smaller shipping companies that lack the necessary finances to deploy similar technology. Therefore, while Maersk's AI strategy exemplifies that technology adoption can produce measurable sustainability outcomes, it also underscores the moral imperative for inclusive innovation policy to ensure smaller participants in the industry are not negatively affected in the transition from traditional supply chains to intelligent supply chain systems.

6.7.3 Unilever

Unilever is using AI to improve sustainable sourcing and supply chain traceability. The firm uses AI technologies to monitor supplier practices, verify sustainability compliance, and offer customers credible information about where and how products are produced. This increases customer trust and helps Unilever's reputation as a responsible brand. AI also enables Unilever to discover and collaborate with ethical suppliers, reinforcing its commitment to sustainability. However, Unilever's actions pose privacy issues, particularly when it comes to collecting supplier data. AI systems collect extensive information from suppliers to check sustainability compliance, raising concerns about data ownership and security. Smaller vendors may be under pressure to give sensitive information without fully knowing how it will be utilized. This instance demonstrates that although AI improves transparency and accountability, it also necessitates strict governance to safeguard privacy rights along the supply chain.

Unilever is utilizing AI to improve its sustainable sourcing and supply chain traceability. According to the Unilever Impact Report (2023), Unilever's investments in AI-supported supplier monitoring and compliance tracking have resulted in a 40% increase in supply chain traceability and a 25% decrease in sustainability violations over the last 3 years. These metrics support Unilever's sustainability practices and allow for increased consumer confidence, as customer confidence scores increased by 15% because of increased transparency in the supply chain. Yet, when subjects reported being monitored by AI systems and the collection of supplier data, they expressed concerns regarding ethical issues of privacy and data ownership. The results of surveys showed that nearly 30% of small suppliers showed alarm regarding how their data was being used. These results underscore the importance of prioritizing data governance and data transparency protocols for suppliers. This example demonstrates that leveraging AI system transparency can increase sustainability and corporate reputation, but a responsible governance framework is necessary to protect the ethical use of data and increase supplier trust in AI systems.

6.8 FUTURE DIRECTIONS AND CONCLUSIONS

The future of AI in SSCM relies on balancing sustainability, equality, and efficiency. Sustainability protects the environment, fairness ensures equitable treatment for suppliers and employees, and efficiency keeps supply chains competitive. Without this balance, AI could do more harm than good [34].

6.8.1 Future pathways

Combining blockchain, IoT, and AI can enhance supply chains by ensuring transparent records, enabling real-time data collection, and providing valuable insights, creating traceable and responsible systems that resist greenwashing. Embracing ethical labour practices with AI is crucial. Companies should ensure automation doesn't increase inequality and promote reskilling and inclusivity. Investments in worker training and equitable supplier collaborations are key to socially responsible AI deployment. The development of green AI is a top goal. AI's environmental effect can be decreased by creating AI models that use less energy and powering data centres with renewable energy. By doing this, AI is certain to advance sustainability rather than exacerbate climate problems. Furthermore, inclusive AI policy should eliminate the digital gap by ensuring that both developed and poor nations may equally benefit from technological breakthroughs.

6.8.2 Conclusion

Through increased openness, fewer emissions, and improved efficiency, AI may significantly improve sustainable supply chains. However, it runs the risk of escalating inequality and damaging the environment in the absence of adequate ethical oversight. Adoption of AI must be done responsibly going forward, with a foundation in diversity, fairness, and accountability. When applied properly, AI may make supply chains faster, more effective, more equitable, greener, and more inclusive. The goal is to strike the ideal balance now so that AI can promote long-term advancement rather than cause harm down the road.

REFERENCES

1. M. Hufty, "Investigating policy processes: The governance analytical frame-work (GAF)," In: U. Wiesmann & H. Hurni (Eds.), *Research for Sustainable Development: Foundations, Experiences, and Perspectives* (Vol. 6, pp. 403–424). Geographica Bernensia in Cooperation with NCCR North-South, University of Bern, 2011.
2. G. Capano, M. Howlett, and M. Ramesh, "Bringing governments back in: Governance and governing in comparative policy analysis," *Journal of Comparative Policy Analysis: Research and Practice,* vol. 17, no. 4, pp. 311–321, 2015.
3. Sengupta, Krishnakali, "How can generative ai drive logistics transformation?" Maersk Insights, 2024. https://www.maersk.com/insights/digitalisation/2024/01/03/how-can-generative-ai-drive-logistics-transformation (Accessed August 24, 2025).
4. TSG Strategy, "Walmart—AI-enabled demand forecasting and inventory optimization in retail," Case study report, TSG Strategy, 2025. https://www.tsgstrategy.com/assets/casestudies/casestudy_walmart.pdf (Accessed August 24, 2025).
5. R. S. Gaonkar and N. Viswanadham, "Analytical framework for the management of risk in supply chains," *IEEE Transactions on Automation Science and Engineering*, vol. 4, no. 2, pp. 265–273, 2007.
6. H. Min, "Artificial Intelligence in supply chain management: theory and applications," *International Journal of Logistics: Research and Applications*, vol. 13, no. 1, pp. 13–39, 2010.
7. S. Seuring, J. Sarkis, M. Mu¨ller, and P. Rao, "Sustainability and supply chain management: An introduction to the special issue," *International Journal of Production Economics,* vol. 111, no. 2, pp. 154–159, 2008.

8. E. Simangunsong, L. C. Hendry, and M. Stevenson, "Managing supply chain uncertainty with emerging ethical issues," *International Journal of Operations & Production Management*, vol. 36, pp. 1272–1307, 2016. https://www.plattform-i40.de/sites/default/files/Report_Industrie%204.0_engl_1.pdf.
9. M. Pournader, H. Ghaderi, A. Hassanzadegan, and B. Fahimnia, "Artificial Intelligence applications in supply chain management," *International Journal of Production Economics*, vol. 241, p. 108250, 2021.
10. N. Diakopoulos, "Accountability, transparency," *The Oxford Handbook of Ethics of AI*, vol. 17, no. 4, p. 197, 2020.
11. GEP, "Responsible AI governance in supply chains: All you need to know," GEP Blog Post, 2023. https://www.gep.com/blog/technology/ responsible-ai-governance-in-supply-chains-all-you-need-to-know (Accessed August 24, 2025).
12. Z. Wu and M. Pagell, "Balancing priorities: Decision-making in sustainable supply chain management," *Journal of Operations Management*, vol. 29, no. 6, pp. 577–590, 2011.
13. O. Bak, C. Jordan, and J. Midgley, "The adoption of soft skills in supply chain and understanding their current role in supply chain management skills agenda: A UK perspective," *Benchmarking: An International Journal*, vol. 26, no. 3, pp. 1063–1079, 2019.
14. A. Foroughi, "Supply chain workforce training: addressing the digital skills gap," *Higher Education, Skills and Work-Based Learning*, vol. 11, no. 3, pp. 683–696, 2021.
15. H. Mayer, L. Yee, M. Chui, and R. Roberts, *Superagency in the Workplace: Empowering People to Unlock AI's Full Potential* (Corporate Report). McKinsey & Company. Retrieved August 24, 2025 from McKinsey & Company Website.
16. M. Quinlan and R. K. Sokas, "Community campaigns, supply chains, and protecting the health and well-being of workers," *American Journal of Public Health*, vol. 99, no. S3, pp. S538–S546, 2009.
17. J. Kroes, A. Land, A. S. Manikas, and F. Klein, "Gender diversity and injustice among supply chain executives: exploring outcomes that advance social justice," *International Journal of Operations & Production Management*, vol. 45, no. 3, pp. 677–699, 2025.
18. MLCO2, CodeCarbon: Track Carbon Emissions of AI Experiments (Version 2.x) [Computer software]. GitHub, 2023. Retrieved August 24, 2025, from https://github.com/mlco2/codecarbon.
19. G. Tamburrini, "The AI carbon footprint and responsibilities of ai scientists," Philosophies, vol. 7, no. 1, p. 4, 2022.
20. J. Patalas-Maliszewska, M. Szmol-da, and H. Losyk, "Integrating Artificial Intelligence into the supply chain in order to enhance sustainable production: A systematic literature review," *Sustainability*, vol. 16, no. 16, p. 7110, 2024.
21. Wikipedia Contributors, *Environmental Impact of Artificial Intelligence. Wikipedia,* 2025, April 11. Retrieved August 24, 2025, from https://en.wikipedia.org/wiki/Environmental_impact_of_artificial_intelligence.
22. M. Bj¨orklund, H. Gillstrom, and G. Kovacs, "Balancing costs and benefits in the food redistribution supply chain," *The International Journal of Logistics Management*, vol. 36, no. 7, pp. 178–199, 2025.

23. M. Brandenburg, T. Gruchmann, and N. Oelze, "Sustainable supply chain management: A conceptual framework and future research perspectives," *Sustainability*, vol. 11, no. 24, p. 7239, 2019.
24. J. Wang and B. Ran, "Sustainable collaborative governance in supply chain," *Sustainability*, vol. 10, no. 1, p. 171, 2018.
25. A. C. Engler and A. Renda, *Reconciling the AI Value Chain with the EU's Artificial Intelligence Act*. CEPS, 2022.
26. European Union, *The EU Artificial Intelligence Act. European Union*, 2024.
27. A. Green and L. Lamby, *The Supply, Demand and Characteristics of the AI Workforce across OECD Countries* (OECD Social, Employment and Migration Working Papers No. 287, pp. 1–62). OECD Publishing.
28. A. Chiarini, "Designing an environmental sustainable supply chain through ISO 14001 standard," *Management of Environmental Quality: An International Journal*, vol. 24, no. 1, pp. 16–33, 2012.
29. M. J. Ferrantino, "Policies to improve the supply chain: what needs to be done?" In D. K. Elms & P. Low (Eds.), *Global Value Chains in a Changing World* (pp. 263–277). World Trade Organization; Fung Global Institute; Temasek Foundation Centre for Trade & Negotiations, 2013.
30. R. Z. Farahani, N. Asgari, and H. Davarzani (Eds.). *Supply Chain and Logistics in National, International and Governmental Environment: Concepts and Models*. Springer, 2009.
31. B. L. Aylak, "Sustaiscm: Intelligent supply chain process automation with agentic AI for sustainability and cost efficiency," *Sustainability*, vol. 17, no. 6, p. 2453, 2025.
32. J. Frazer, "Walmart and the new supply chain reality: Ai, automation, and resilience," Logistics Viewpoints Article, 2025. https://logisticsviewpoints.com/2025/03/19/ walmart-and-the-new-supply-chain-reality-ai-automation-and-resilience/ (Accessed August 24, 2025).
33. J. Frazer et al., "Harnessing ai for sustainable shipping and green ports: Challenges and opportunities," *Applied Sciences*, vol. 14, no. 14, p. 5994, 2024.
34. S. Culey, "AI tools should free unilever supply chain professionals from the routine," *Supply Chain Movement Magazine*, 2022. https://www.supplychainmovement.com/ai-tools-should-free-supply-chain-professionals-from-the-routine/ (Accessed August 24, 2025).

Chapter 7

Responsible AI and Blockchain-Based Smart Contracts for Industries

Saravanan Chinnappan and
Sanjay Mallenahalli Basavaraj

7.1 INTRODUCTION TO RESPONSIBLE ARTIFICIAL INTELLIGENCE AND BLOCKCHAIN-BASED SMART CONTRACTS FOR INDUSTRIES

Industries are being transformed intensively across the globe by the empowered increasing pace of digitalization and the merging of intelligent automation. With progressively organized and advanced functional ecosystems, the necessity for reliable, responsive, and open systems has become more critical. Two innovative technologies are leading this responsibility: Blockchain and Artificial Intelligence (AI). Though each of them can separately add value to the industrial community, their union in the form of AI-powered smart contracts unleashes a standard shift in industrial process management, monitoring, and optimization [1].

Blockchain affords an unchangeable, distributed space and provides security. Smart contracts are contracts that are self-executing on a blockchain platform in coded form. Intermediaries are not required to interact and collaborate with smart contracts. This helps to reduce administrative cost, fraud, risks related to human error, and other disputes [2].

The contracts are transparent and traceable by nature, with high levels of trust in multistakeholder environments like manufacturing, logistics, supply chain, and finance. Their potential has, however, been constrained to this point in time by their deterministic nature. Traditional smart contracts execute pre-coded responses to pre-conditioned inputs with no provision for contextual understanding, learning, or real-time decision-making [3].

Conversely, contemporary industrial environments are governed by conditions that are seldom fixed. Machines wear out over time, market scenarios shift rapidly, geopolitical occurrences restructure value chains overnight, and user requirements evolve dynamically. Such a world calls for systems that possess the capability to learn, deduce from past trends, react to anomalies, and make forecasted decisions under uncertainty. AI is applicable in this case. When applied to intelligent contracts, AI enhances their capabilities beyond hard automation, making them work more as smart agents than as simple digital scripts.

 DOI: 10.1201/9781003660163-7

When you bring AI into contracts, it's not all smooth sailing. Sure, it promises speed and smarter decisions, but it also adds a fair bit of mess: legal, ethical, and even day-to-day operational issues. And this isn't just theory. Think about it: if an anomaly detection system gets it wrong, it might shut down a production line that didn't need to stop, or order spare parts no one actually asked for. Both cost money; both frustrate people. That's why the idea of responsible AI really matters here. It's about building AI that stays fair, explainable, and accountable. In other words, AI should support people, not steamroll over human judgment or fairness [4].

Now, throw blockchain smart contracts into the mix, and things get even more interesting. Normally, contracts on a blockchain are strict; they do exactly what's written, no matter what. But the real world doesn't work like that. With AI involved, contracts could adapt. Picture a factory filled with Internet of Things (IoT) sensors. An AI-driven contract could detects when a machine starts malfunctioning, predict a breakdown before it happens, and automatically adjust supplier terms or schedule maintenance. And here's the key part: because ethics are baked in, these decisions don't happen in secret. Stakeholders can see them, and they line up with both company policies and outside regulations.

The chapter sketches a model for how this kind of system could actually be put together. It combines familiar tools: blockchain platforms like Ethereum, AI libraries such as TensorFlow and scikit-learn, and real-time data streams through oracles like Chainlink. Put together, these pieces let contracts do more than just "execute"—they can actually interpret what's happening and respond in context. It's still theoretical for now, yes, but it lays a solid base for future industrial applications and research.

By outlining the importance of blockchain–AI convergence and the need for ethical safeguarding under responsible AI, this chapter lays the foundation for in-depth analysis in the rest of the book. These will include an end-to-end architectural approach, industry-specific use cases, implementation challenges, and governance, security, and directions forward considerations for industrial scalability.

7.2 SMART CONTRACTS IN INDUSTRY: BEFORE AND AFTER AI

Smart contracts really changed the game in how industries operate digitally. The whole idea was cutting out middlemen and making things faster and more trustworthy. We're talking less paperwork and fewer mistakes in things like payments or records. Companies in finance or shipping saw real perks early on: cheaper ops and quicker deals getting done without all the back-and-forth [5].

But those first versions weren't perfect. They worked on strict if–then rules with no wiggle room. If a situation didn't match exactly what was programmed, it was game over. Like if your supplier got hit by a typhoon or demand suddenly spiked? The contract would freeze up, needing human help to fix it. Not exactly helpful when you're dealing with real-world chaos [6].

Take supply chains, for example. You've got moving parts everywhere—inventory swings, shipping delays from weather or political messes. Those old-school contracts could handle basic stuff like paying when a package arrives but had zero ability to adapt when things went sideways. They were more like robots following orders than actual problem-solvers [7].

Smart contracts are capable of processing real-time data streams such as traffic patterns, weather forecasts, and news alerts. As a result, a shipping contract is no longer limited to only tracking deliveries; it can dynamically reroute trucks or adjust deadlines in response to disturbances.

The ongoing technological transformation, especially within Industry 4.0 and the emerging Industry 5.0, marks interconnected devices and data-driven systems as the foundation of operations [8]. Smart contracts, once regarded as specialized applications of blockchain technology, are now undergoing rapid evolution, exhibiting adaptive and self-updating functionalities akin to autonomous systems. This progression not only expands the scope of opportunities but also brings challenges.

As smart contracts grow increasingly complex, the necessity for attentive oversight correspondingly increases. Allowing critical decision-making to autonomous digital contracts without adequate safeguards could result in significant risks [9]. Therefore, the establishment of governance structures is essential.

The ethical consequences of this evolution cannot be ignored. Transparency and well-defined ethical frameworks should be embedded to avoid undesirable consequences. Though adaptive smart contracts embrace progressive goals like sustainability and client satisfaction, Decision of human beings play a crucial role in the development and deployment.

Currently, these technologies remain in their initial stages. But the possibility of AI-driven contracts disrupting and transforming entire industries is increasingly evident. As their capabilities continue to advance, the need for strong oversight and ethical considerations will become even more demanding. So, it is essential to approach this technological evolution with a balance of optimism and prudence (Table 7.1).

7.3 EVOLUTION OF AI-DRIVEN SMART CONTRACTS FOR INDUSTRY

Smart contracts are popular as they avoid all unnecessary middle agents. Before the existence of smart contracts, deals were happening for weeks with a lot of documentation between various people. The smart contract

Table 7.1 Evolution of smart contracts—from traditional models to AI-driven systems

Feature	*Traditional smart contracts*	*AI-driven smart contracts*
Logic execution	Static, pre-coded rules	Adaptive, data-driven decision-making
Environmental awareness	None	Real-time context analysis using AI
Learning capability	Not supported	Supports learning from past data
Flexibility	Rigid	Dynamic and self-adjusting
Ethical decision-making	Absent	Ensured via responsible AI principles
Human oversight	Limited	Integrated as part of responsible AI

agreements have automated the process of traditional contracts. All transactions are securely stored related to the operations carried out from initial to final in the blockchain.

The problem came when businesses got messier. Basic smart contracts work great when everything goes smoothly, but that's not how the real world operates [10]. Trucks break down, sensors malfunction, and wars disrupt shipping—and the original contracts had no clue how to handle any of that. They'd just sit there stuck because nobody programmed them for these scenarios.

That's when people started integrating AI. Machine learning and prediction tools turned these rigid contracts into something that could actually adapt. Say you've got a factory with sensors all over the machinery. The old version would just record weird vibrations and call it a day. But throw in some AI, and now it's looking at those vibration patterns, figuring out that the machine is probably going to die next Tuesday, and automatically ordering parts while scheduling a repair crew [11].

Shipping got even more interesting. Basic contracts would tell you "Package delivered," and that's it. The AI versions are constantly juggling GPS coordinates, weather forecasts, and customs delays—rerouting drivers around thunderstorms, pushing back delivery windows when things get backed up, and even rewriting parts of the contract when it's clear the original timeline isn't happening [12]. All while keeping that bulletproof record of every decision.

Putting it all together, AI, blockchain, and ethical design are reshaping automation from rigid rule followers to proactive problem-solvers that learn as they go [13]. Organizations get agility plus trust, which matters when industries shift fast and unpredictably. It's about building systems that can think ahead while keeping everyone honest (Figure 7.1 and Table 7.2).

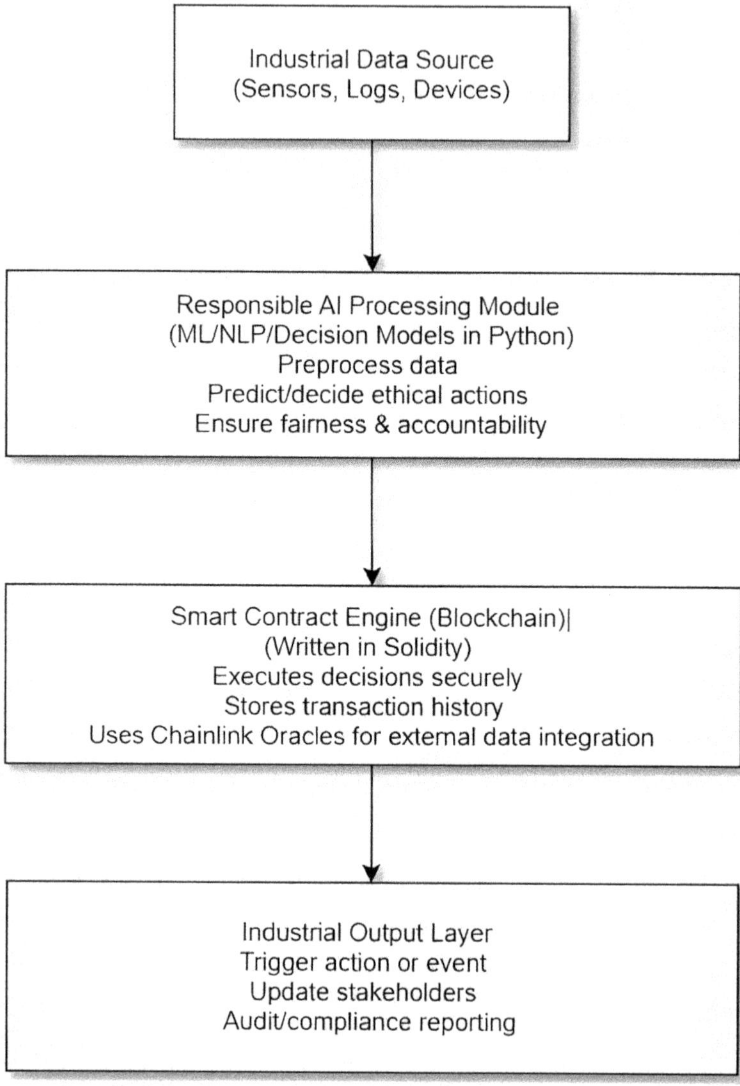

Figure 7.1 System architecture of responsible AI and blockchain-based smart contracts for industry.

7.4 APPLICATIONS OF AI-DRIVEN SMART CONTRACTS IN INDUSTRIAL OPERATIONS

AI and blockchain smart contracts together are seriously shaking up how businesses handle automation. Old setups worked fine for basic if–this–then–that execution, but real-world operations are messy. They need more

Table 7.2 Key tools and technologies in responsible AI+smart contract systems

Layer/component	*Technology/tool example*	*Description*
Data input layer	IoT sensors, industrial logs	Sources of operational and contextual data
AI processing module	Python, scikit-learn, NLP models	Performs learning, prediction, and ethical reasoning
Blockchain engine	Solidity, Hyperledger	Stores contracts and transaction logic
Oracles	Chainlink, Provable, Band Protocol	Bridges between off-chain data and contracts
Output layer	API endpoints, industrial actuators	Triggers actions and logs compliance events

than rigid rules—they need actual thinking. With AI baked in, smart contracts don't just run code; they analyze live data, learn from it, and adjust as they go [14].

Factory maintenance is a clear example. Traditionally, maintenance relied on scheduled checkups or waiting for sensors to signal danger. AI-powered contracts take it further. They monitor every data point: vibrations, heat, and energy consumption collected through IoT sensors. If bearings wear down or motors overheat, the system doesn't just send alerts. It acts. It can book repair teams, order replacement parts, reschedule production, and process payments through blockchain—all automatically. That means less downtime, fewer breakdowns, and longer machine life.

Inventory management also gets smarter. Instead of reacting when stock runs low, AI-enhanced contracts forecast needs by analyzing sales history, supplier performance, and even weather that could disrupt shipments. If a main supplier raises prices or fails to deliver, the system doesn't stall; it sources alternatives, negotiates deals, and places new orders itself. Production keeps moving smoothly without human delays.

Shipping logistics benefit in the same way. These contracts track trucks, ships, and other carriers in real time, using GPS, traffic feeds, and port data. If delays hit due to bad weather, traffic jams, or customs holdups, the system automatically reroutes shipments, adjusts deadlines, or switches carriers. Every step is logged on the blockchain, giving all stakeholders transparent visibility. That prevents disputes later and makes deliveries more reliable.

Highly regulated industries like pharma, agriculture, and chemicals see even greater value [15]. Compliance rules shift constantly, but AI-enhanced contracts can scan new regulations as they appear, interpret them, and update terms automatically. For instance, if new safety rules on chemical storage emerge, contracts can flag non-compliant facilities instantly, block shipments, alert managers, and log fixes. Companies stay audit-ready without the panic.

Dynamic pricing is becoming a big deal in industries where costs shift constantly [16]. Power companies and phone service providers now use smart contracts to watch how much customers consume, track what's happening with market rates, and study buying patterns as they happen. Picture a scorching summer day when everyone cranks up their AC—the energy system notices this surge, bumps up the rates, sends out new bills instantly, and alerts users about the changes. Every step gets locked into a record nobody can alter.

AI's main job here is turning dumb automation into something that actually thinks. These aren't your old-school contracts that just run through a checklist anymore. They watch for shifts in conditions, figure out what to do about them, and make calls without human input. Blockchain handles the trust side by keeping everything out in the open. The results speak for themselves: machines in factories stay running longer, supply chains bounce back faster when things go wrong, packages show up when they're supposed to, rules get followed without anyone having to babysit the process, and prices move with real-world conditions instead of staying stuck. Mix AI with blockchain, and you end up with systems that solve today's headaches while staying ready for whatever weird stuff tomorrow might throw at them.

7.5 INDUSTRIAL CASE STUDIES

To understand how responsible AI works with blockchain-based smart contracts in practice, we need to look at what companies are actually doing. The cases below show how these ideas have moved from theory into real applications that are making a difference in automation, tracking, and decision-making. When you examine these examples, you can see how organizations are changing the way they handle risk, follow regulations, and use data to make better choices.

7.5.1 Siemens: using AI to predict when machines need fixing

Siemens has been at the forefront of using AI and blockchain to change how factories maintain their equipment [17]. The old way of doing maintenance was to wait until something broke, which meant unexpected shutdowns and expensive emergency fixes. Siemens took a different approach by setting up AI systems that constantly watch the data coming from their machines—things like how hot they get, how much they vibrate, and how worn out parts are getting.

When the AI spots signs that a machine might fail soon, smart contracts built on blockchain automatically start a chain of actions. They book maintenance teams, order the right replacement parts from approved suppliers, change production schedules to make room for repairs, and let everyone who needs to know what's happening. Since everything gets recorded on blockchain, you can trace back through every step of what happened and why. This system has made their equipment more reliable, cut down on unexpected breakdowns, and helped them spend their maintenance budget more wisely.

7.5.2 IBM and Maersk: blockchain contracts for global trade

The TradeLens platform, developed by IBM and Maersk, demonstrates the power of combining AI-driven insights with blockchain contracts in international shipping. Global trade logistics face constant challenges, such as port delays, extreme weather, or customs holdups, where speed and accuracy are critical.

TradeLens leverages AI models to assess risks across the supply chain [18]. When risks such as port congestion or customs delays are detected, blockchain smart contracts step in to reroute shipments, generate updated documentation, and notify logistics partners and customs agencies automatically. All actions are logged on the blockchain, ensuring tamper-proof visibility. During the COVID-19 pandemic, the platform processed tens of millions of updates seamlessly, proving its resilience and scalability.

7.5.3 Walmart: food safety through intelligent traceability

Walmart provides another strong case for AI-enhanced blockchain contracts in food safety. Food supply chains require strict monitoring of freshness and safety, where lapses can become public health threats [19]. Walmart embedded blockchain smart contracts into its food network to track products from farm to shelf.

AI systems process data on shipping temperatures, expiration dates, supplier performance, and contamination alerts. If violations occur, such as perishable goods exposed to unsafe conditions, the smart contract halts distribution, issues recall alerts, and updates retail systems instantly. What once took days of manual tracing now takes seconds. This not only enhances consumer safety but also reduces waste, strengthens supplier accountability, and builds customer trust.

These case studies reveal how forward-thinking organizations are already harnessing intelligent and ethically aligned smart contracts. By embedding real-time analytics and trust mechanisms into contractual workflows, Siemens, IBM-Maersk, and Walmart are setting new benchmarks for efficiency, compliance, and agility in Industry 4.0.

7.6 CHALLENGES, RISKS, AND ETHICAL CONSIDERATIONS

As industries embrace blockchain, smart contracts, and AI, there is a need to evaluate the revolutionary scale of these technologies apart from challenges and ethics. With the ability of automation, intelligence, and trust within the systems, they can provide data privacy, justice for algorithms, uncertainty, stability of systems, and job disruption.

7.6.1 Data privacy and security

The availability of data received from external data sources regularly, IoT sensors, and enterprise systems has made blockchain oracles rely on AI-driven smart contracts. The transparency of blockchain provides security benefits, yet metadata operations, payment data, or personal identifiers are placed on-chain [20]. API compromise, insecure deployment of AI models in the cloud, and manipulation of oracles are a few of the integrity attacks against the system.

7.6.2 Model bias and ethical decision-making

AI models are no more unbiased than the training data sets upon which they were constructed. Societal and organizational biases—geographically, demographically, or by vendor behavior—are inherent in past data sets. Since biased data sets will be the basis for autonomous smart contract decisions, the outcomes will unknowingly perpetuate inequality or discrimination. An example is a purchasing contract that will always favor veteran suppliers based on a past performance record, local or small suppliers in preference to being competitively positioned. Such a policy could be contrary to the interests of an organization aiming to be inclusive, sustainable, or to enable the local economy. To mitigate such ethical concerns, the use of responsible AI principles is warranted. They comprise fairness, explainability, accountability, and transparency of AI system outcomes. Explainable AI (XAI) technologies will facilitate auditors and programmers in understanding which specific contract clause was activated, leading to scrutiny by humans.

7.6.3 Legal ambiguity and accountability

Legal accountability plays an important role in AI-based smart contracts. The systems behave in an independent manner with firm decisions that lead to consequences in real-world scenarios like payment operations, fine improvements, and stopping supply chains. Digital contract law regimes, automation liability, and AI ethics are wildly variable across jurisdictions.

Organizations will need to secure legal protection in advance until there are global standards. Low-level contract execution audit trails, multi-party approval procedures to handle high-impact choices, and delineating clear lines of responsibility for developers, data providers, and end-users are typical strategies. Traditional regulatory regimes and blockchain-based regulatory models introduced by international institutions are increasingly providing regulations for global conversations and leading industries to implement increasingly regulatory norms.

7.6.4 System reliability and technical malfunction

Major significant differences between traditional AI and smart contracts lie in finding out what to do. Smart contracts follow well-defined rules and regulations with a computation process. AI systems anticipate outcomes based on statistical logic and operate with a probabilistic approach. This inconsistency can produce mistakes where AI output starts automatic action without any sufficient context. For instance, when an AI tool incorrectly identifies a machine as failed on the basis of a single outlier data point, the smart contract needlessly halts production, thus incurring unnecessary economic losses. Edge cases happen with extremely frequent occurrence and undermine trust in automatic systems. To avoid this from occurring, industries are incorporating hybrid contract designs that combine human judgment with automation. Emergency stop mechanisms on critical system operations, multi-signature workflows based on feedback from various stakeholders, and backup logic to relay choices to human decision-makers in case of uncertainty are being increasingly implemented. Thorough testing, sandbox testing, and test environments also load these systems before releasing them into production.

7.6.5 Workforce impact and skill transition

As smart contracts automate processes in procurement, compliance, logistics, etc., no one would doubt that they can restructure the industrial workforce. Although automation will eliminate boring jobs, there is always fear of losing work, especially in boring decision-making, record-keeping, or monitoring jobs; but this transformation also provides new scenarios for human work in new domains like data quality management, AI model governance and auditing, blockchain system management, and smart contract lifecycle management. For this transition to be more efficient and effective, upskilling and reskilling activities need to be concentrated on, and certification programs along with industry–academia collaboration can equip workers with market-ready jobs.

It is not a question of obedience to the law to tackle these matters right from the beginning, but a critical step in designing responsible, dependable, and durable digital systems for generations to come.

7.7 RESPONSIBLE AI PRINCIPLES IN THE DEVELOPMENT OF BLOCKCHAIN-BASED SMART CONTRACTS FOR INDUSTRIES

Data security, XAI, and fairness are fundamental principles of responsible AI applied in the development of blockchain-based smart contracts for industries.

a. Data security ensures sensitive information handled by AI systems is protected from data leakage, data manipulation, and unauthorized access. The data security principle can be applied in blockchain-based smart contracts through immutable ledger, decentralized storage, and zero-knowledge proofs.

   Immutable Ledger: The immutable structure of blockchain ensures that once transactions are recorded, they cannot be tampered with, strengthening data integrity.

   Decentralized Storage: Sensitive training data can be handled by distributing it across nodes, reducing single points of failure and cyber-attack risks.

   Zero-Knowledge Proofs: These enable verification of contract conditions without revealing private data, thus maintaining the confidentiality of the data.

b. XAI principles ensure that AI decisions are understandable, transparent, and understandable by humans. XAI principles can be applied in blockchain-based smart contracts through traceable decision logic and model explainability modules. AI components integrated with smart contracts use interpretable models like SHapley Additive exPlanations (SHAP) to justify outcomes before execution.

c. Fairness: This principle ensures that decisions are bias-free and equitable across all user groups. It can be achieved through bias-aware AI models, transparent rule enforcement, and community governance and auditable fairness metrics.

   Bias-Aware AI Models: AI models embedded in smart contracts are trained with diverse, representative data sets to prevent systematic discrimination.

   Transparent Rule Enforcements: Smart contracts automatically enforce rules consistently for all parties, reducing human bias in contract execution.

   Community Governance: Decentralized Autonomous Organizations allow stakeholders to vote on fairness criteria, ensuring inclusive oversight of AI and contract rules.

7.8 FRAMEWORKS OR STANDARDS ADOPTION FOR RESPONSIBLE AI AND BLOCKCHAIN-BASED SMART CONTRACTS

The adoption of responsible AI and blockchain-based smart contracts across industries can be carried out by establishing a robust governance and compliance framework. This begins with implementing ISO/IEC 42001, which offers a structured approach to AI governance, ensuring accountability, transparency, and continuous improvement. It should be complemented by ISO/TR 23676, which defines blockchain governance principles to maintain trust, integrity, and stakeholder accountability within distributed ledger ecosystems.

Organizations should apply ISO/IEC 23894 to build a comprehensive AI risk management framework, enabling the identification, assessment, and mitigation of potential ethical, operational, and societal risks. Ensuring data privacy and protection through ISO/IEC 27701 is equally critical, particularly when handling sensitive or personal data within AI-enabled smart contracts. Alongside this, ISO/TR 23455 supports transparency and interoperability in smart contract design and execution, fostering trust and auditability.

To uphold fairness and ethical responsibility, it is essential to continuously monitor algorithmic bias using ISO/IEC TR 24027 and align system design and deployment with ethical guidelines outlined in ISO/IEC TR 24368. Collectively, these standards establish a unified, responsible, and transparent framework for deploying AI and blockchain technologies in industrial applications.

7.9 FUTURE PROSPECTS AND CONCLUSION

The convergence of blockchain-enabled smart contracts and responsible AI is less an innovation rooted in existing industrial infrastructure and more a paradigm shift in how industries cooperate, communicate, and decide. As technology change accelerates in industries, the two technologies are converging to redefine the very fabric of industrial ecosystems. Their complementary alignment is a powerful blueprint for achieving operational excellence and ethics, accountability, and sustainability in the long term.

Subsequent generations of intelligent contracts can anticipate escaping their current limitations. No longer restricted by rigid execution of pre-defined conditions, they are increasingly being programmed to perceive, infer, and respond to changing contexts. By integrating AI at the core of contract reasoning, the next generation of smart contracts will pick up on lessons learned, identify new patterns, and dynamically modify terms and conditions in real time. This nascent evolution is a move away from hard-wired automation to smart orchestration, where contracts no longer dictate rules but understand intent, manage risk, and dynamically respond to business and regulatory environments.

The potential for inter-industry cooperation will also expand exponentially. These systems will not operate in isolation but as part of interoperable ecosystems. On these shared platforms, manufacturers, suppliers, auditors, logistics providers, regulators, and even buyers will be able to exchange information securely through blockchain-validated smart contracts. These shared infrastructures could reshape international trade by streamlining processes, cutting down on bureaucratic delays, easing disputes, and making supply chains more transparent. Initiatives such as Hyperledger Fabric and other consortia-led blockchain platforms already provide the foundational frameworks for these federated systems.

As these technologies grow, the responsibility to ensure their ethical use grows alongside them. Large-scale intelligent automation is not only a technical development but also a social one, carrying implications that extend well beyond efficiency and cost. The call for responsible AI has therefore moved to the forefront—it is no longer a "nice to have" but a core design requirement. Future systems should be guided by clear rules, transparent decision-making, and mechanisms that allow for meaningful human oversight. Ethical responsibility must span the entire lifecycle, from the way algorithms are built and training data is selected to how decisions are logged, audited, and contested when necessary. Building and maintaining trust will also require that principles of fairness, inclusivity, and accountability be woven into every layer of the AI–blockchain ecosystem.

Regulatory systems will also need to adapt to these technological shifts. As autonomous smart contracts begin influencing decisions in finance and manufacturing, organizations must operate within clear regulatory guardrails. Developing these safeguards will require policymakers to work closely with industry experts and academic researchers to define standards around liability, consent, transparency, and mechanisms of appeal in contract enforcement.

Even with the technical and legal complexities involved, human considerations must remain central. As automation continues to advance, organizations will need concrete ways to address its socio-economic impact, particularly the challenges surrounding workforce changes. While intelligent automation will certainly replace certain roles, it will also create new opportunities in areas such as AI auditing, blockchain infrastructure management, ethics oversight, and data stewardship.

Finally, the union of AI and blockchain-based smart contracts represents more than just a technical milestone; it signals a broader shift in how industries operate, share information, and adapt within a data-driven economy. This chapter has highlighted how AI-enabled smart contracts can be designed not only to improve efficiency and autonomy but also to uphold values of transparency, resilience, and ethical responsibility. The challenges ahead, ranging from privacy and bias to questions of legality and labor, are complex, yet they can be addressed through collective foresight, thoughtful innovation, and a commitment to ensuring that technological progress benefits all stakeholders.

By embedding principles of ethical governance and responsible AI into technological innovation, industries can create ecosystems that go beyond the simple execution of code. These systems have the potential to build trust, enable collaboration, and support a digital future that is both unbiased and sustainable. The real challenge lies in designing technologies where progress aligns with purpose, ensuring that tomorrow's innovations serve not only business objectives but also the needs of society.

REFERENCES

1. Gaurav, A., Mishra, D., & Batra, R. Ethical AI in Industry 5.0: Frameworks and Applications. *International Journal of Information Management Data Insights*, 3(2), 100092 (2023).
2. Rai, P., Jain, A., & Tanwar, S. Blockchain and AI-Driven Framework for Industrial IoT. *IEEE Internet of Things Journal*, 9(9), 6613–6622 (2022).
3. Deloitte. *AI Ethics and Governance: Balancing Innovation and Responsibility*. Deloitte Insights Report. https://www2.deloitte.com/insights (2022).
4. Tapas, N., Tanwar, S., Patel, P., & Tyagi, S. Artificial Intelligence-Enabled Smart Contract for Industrial Automation: Architecture and Applications. *IEEE Access*, 9, 65699–65717 (2021).
5. Wang, J., Ye, X., & Liu, Y. Integration of Blockchain and Artificial Intelligence in Supply Chain: A Review. *Computers & Industrial Engineering*, 152, 107067 (2021).
6. Kaur, K., Rathee, G., & Iqbal, R. Blockchain Technology for Smart Contracts and Industrial IoT: Recent Advances and Future Directions. *Journal of Network and Computer Applications*, 176, 102918 (2021).
7. Dwivedi, A. D., Srivastava, G., Dhar, S., & Singh, R. S. A Decentralized Privacy-Preserving Healthcare Blockchain for IoT. *IEEE Transactions on Industrial Informatics*, 17(8), 5502–5510 (2021).
8. Hyperledger Foundation. *Hyperledger Fabric Whitepaper*. https://www.hyperledger.org (2021).
9. Aggarwal, R., & Kumar, N. Integration of AI and Blockchain for Sustainable Industrial Operations. Sustainable Computing: *Informatics and Systems*, 32, 100642 (2021).
10. IBM and Maersk. TradeLens: Digitizing the Global Supply Chain (*IBM* Blockchain Case Study) IBM and Maersk (2020).
11. World Economic Forum. *Redesigning Trust: Blockchain Deployment Toolkit*. World Economic Forum (2020).
12. Zhang, Y., White, J., Schmidt, D. C., & Lenz, G. Blockchain for IoT-based Smart Manufacturing: Challenges and Solutions. *Sensors*, 20(1), 142 (2020).

13. Sivathanu, B. Adoption of Blockchain Technology in Supply Chain and Logistics: An Integrated TAM-TOE Model. *Technology in Society*, 62, 101320 (2020).
14. Ferrag, M. A., Shu, L., Derhab, A., Maglaras, L., & Mukherjee, M. Blockchain and IoT-based Smart Contracts in Industry 4.0: A Review. *IEEE Access*, 7, 63989–64010 (2019).
15. European Commission. *Ethics Guidelines for Trustworthy AI*. European Commission (2019).
16. Xu, X., Weber, I., & Staples, M. *Architecture for Blockchain Applications*. Springer (2019).
17. Lacity, M., & Van Hoek, R. Leveraging Blockchain for Supply Chain Transparency. *MIT Sloan Management Review*, 60(4), 1–10 (2019).
18. De Filippi, P., & Wright, *A. Blockchain and the Law: The Rule of Code*. Harvard University Press (2018).
19. Casey, M. J., & Vigna, P. *The Truth Machine: The Blockchain and the Future of Everything*. St. Martin's Press (2018).
20. Alketbi, A., Nasir, Q., & Talib, M. A. A. Blockchain for Government Services—Use Cases, Security Benefits, and Challenges. In: *IEEE International Conference on Internet of Things and Intelligence System (IOTAIS)*, Jeddah, Saudi Arabia, pp. 110–115 (2018).

Chapter 8

The evolution of supply chains and the role of Blockchain in transforming transparency, trust, and efficiency

Chandrani Chakravorty, Mohammad Ayyan A Shaikh, and Divya Tarikere Lokeshwara

8.1 INTRODUCTION

Global trade has always relied on supply chains, which are complex networks that manage the production, movement, and delivery of goods and services to consumers. What began as small exchanges among local traders has gradually developed into extensive, digitally connected systems that link producers, suppliers, distributors, and retailers across the world. These systems have become the foundation of the global economy, ensuring that products reach markets efficiently through the smooth flow of materials, finances, and information. Over the years, supply chains have evolved from simple, company-centered structures into intricate global networks that demand real-time coordination, accuracy, and transparency at every stage.

As these networks have expanded in scale and complexity, their weaknesses have also become more visible. Events such as natural disasters, pandemics, political conflicts, and cyber-attacks have repeatedly exposed vulnerabilities in tracking, communication, and data management. Traditional tools like enterprise resource planning (ERP) and supply chain management (SCM) have improved documentation and efficiency but still operate within separate databases, leading to data duplication and slower information sharing. These challenges have created a need for a secure and shared system that allows multiple organizations to access accurate and verifiable data simultaneously.

Recent advancements in digital technologies such as blockchain, artificial intelligence (AI), the Internet of Things (IoT), and cloud computing have begun to reshape how supply chains function. Among these innovations, blockchain plays a particularly vital role because it enables transparent and tamper-resistant transaction recording through a decentralized system. By eliminating the reliance on a central authority and encouraging mutual trust among all participants, blockchain provides a collaborative framework that strengthens accountability and increases the overall resilience of global logistics systems.

DOI: 10.1201/9781003660163-8

8.1.1 Understanding supply chains: evolution and complexity

From the earliest trade routes to the present era of global production networks, the movement of goods from their origin to end consumers has remained a vital pillar of economic progress. Ancient civilizations such as Mesopotamia, Egypt, and Rome developed structured systems for transporting essential resources, including food, textiles, and metals, through both land and sea routes, establishing the foundations of coordination and accountability in trade. During the Middle Ages, merchant guilds and trade alliances such as the Hanseatic League introduced formal regulations and standards for long-distance commerce [1], reinforcing cooperation among different regions.

The Industrial Revolution of the 18th and 19th centuries brought a major transformation to supply chains through mass production, mechanized transportation systems like railways and steamships, and centralized manufacturing centers. These developments created the basis for today's global supply networks. In the 20th century, SCM emerged as a distinct discipline that evolved from logistics and inventory management practices [2]. It was further strengthened by post-World War II advancements in material handling, freight transport, and digital communication systems.

Concepts such as just-in-time, lean manufacturing, and the bullwhip effect became key principles in both industrial practice and academic research. The introduction of ERP systems in the 1990s enabled the integration of various supply chain functions into unified digital platforms, improving coordination and decision-making. The 21st century has witnessed accelerated globalization, where products are designed, manufactured, and assembled across multiple nations, adding both complexity and vulnerability to global networks. This progression has led to the adoption of strategies focused on resilience, adaptability, sustainability, and digital transformation. Emerging technologies such as AI, the IoT, and blockchain have become essential in ensuring secure data exchange, enhancing trust, and promoting transparency in modern supply chain operations (Figure 8.1).

8.1.1.1 Why supply chains are complex?

Modern supply chains operate across multiple time zones, legal jurisdictions, and regulatory frameworks, involving a vast number of autonomous partners. This distributed nature introduces difficulties in maintaining data consistency, real-time tracking, and accountability. Each participant often maintains separate records, creating fragmented data silos that hinder collaboration. Additionally, market volatility, demand fluctuations, and sustainability pressures require faster, data-driven decision-making [3].

Traditional information-sharing mechanisms—emails, spreadsheets, or centralized databases—struggle to provide trustworthy, tamper-proof records in such distributed environments. This lack of unified visibility

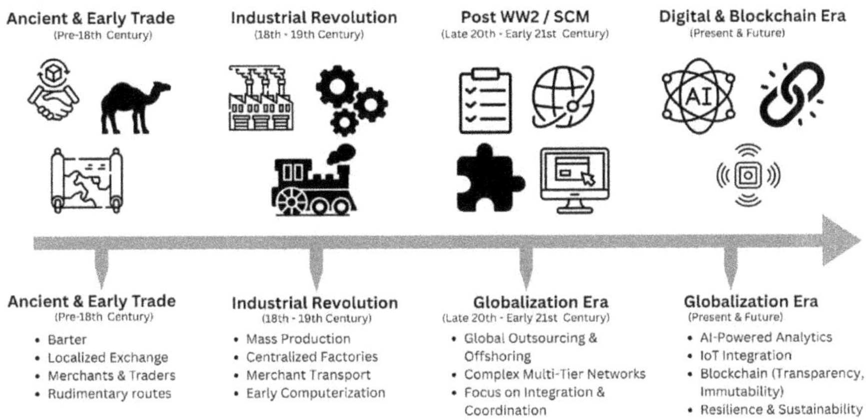

Figure 8.1 The evolution of supply chains: from barter to blockchain.

often results in inefficiencies, delays, counterfeit infiltration, and erosion of consumer confidence. The growing demand for traceability and ethical sourcing has made transparent, verifiable supply chain data an operational necessity rather than an optional enhancement.

8.1.1.2 Key stakeholders in a supply chain

A typical supply chain involves a diverse range of participants, each serving specific roles and pursuing distinct objectives as a product moves from raw materials to the end user.

i. **Raw Material Suppliers:** These are the foundational participants who provide the essential resources and basic inputs necessary for the initial stages of production.
ii. **Manufacturers or Producers:** These organizations transform raw materials into finished goods or intermediate components and function as the central units of production.
iii. **Distributors or Wholesalers:** These entities handle the bulk storage and distribution of goods, usually supplying products to retailers or other downstream partners.
iv. **Retailers:** These participants engage directly with consumers, selling products through physical stores or online platforms.
v. **Logistics Providers (3PLs or 4PLs):** These specialized companies manage crucial logistics activities such as transportation, warehousing, and the movement of goods across various points.
vi. **Customs and Regulatory Authorities:** These national and international agencies monitor the movement of goods across borders and ensure compliance with legal requirements, tariffs, and trade agreements.

vii. **Financial Institutions:** These key enablers offer payment solutions, credit facilities, and other financial services that support liquidity and transaction flow throughout the supply chain.
viii. **Consumers:** These are the final recipients and ultimate drivers of the supply chain, whose preferences, demands, and purchasing behavior influence production and distribution decisions.

8.1.2 Traditional supply chain challenges (pre-blockchain era)

Before the introduction of blockchain technology, supply chains encountered persistent difficulties that hindered efficiency, reliability, and profitability. Many of these challenges arose from fragmented data management, inadequate transparency, and a lack of trust among independent participants operating across various regions and organizations [4].

8.1.2.1 Counterfeiting: a major economic drain

The presence of counterfeit goods has been one of the most damaging issues in global supply chains. Sectors such as pharmaceuticals, electronics, automobiles, and luxury products have faced substantial losses due to the infiltration of fake goods into legitimate markets. By the year 2020, the estimated global value of counterfeit and pirated items had exceeded half a trillion US dollars annually [5]. Apart from financial damage, counterfeiting endangers consumer safety, weakens brand reputation, and supports unlawful trade networks. Conventional tracking and verification systems, which relied mainly on paper documentation or centralized databases, lacked strong mechanisms to confirm product authenticity throughout the supply chain.

8.1.2.2 Delays and operational inefficiencies

Traditional supply chains depended on disconnected databases, manual workflows, and fragmented communication systems. Such disjointed structures often resulted in delays in order processing, shipment tracking, and conflict resolution [4]. Every stage of a product's movement from manufacturing to final delivery introduced potential bottlenecks and miscommunication. The absence of real-time access to information regarding inventory, production, or transportation status further slowed down decision-making, increased operational costs, and reduced adaptability to market changes.

8.1.2.3 Consumer distrust and limited visibility

Modern consumers now expect transparency regarding product origin, ethical sourcing, and environmental compliance. However, conventional systems typically offered visibility only until the point of sale, making it difficult for customers to verify claims about authenticity, sustainability, or fair labor practices [6]. High-profile incidents involving unsafe or unethically produced goods have worsened consumer skepticism. Without dependable verification systems, public confidence in both products and the companies producing them remained weak.

8.1.2.4 *Quality control and waste management*

Ensuring consistent product quality across multi-layered global networks has long been a major concern. Inadequate monitoring and record-keeping often allowed defective or unsafe goods to reach consumers. Inefficient resource allocation and poor inventory control further contributed to unnecessary material and energy waste. Product recalls, when required, were often expensive and time-consuming because identifying affected batches across multiple suppliers and distributors was rarely straightforward [7]. The absence of traceable and trustworthy data also limited sustainability initiatives and made it difficult to maintain accountability throughout the product life cycle.

8.1.2.5 *Evolution of pre-blockchain supply chain technologies*

Before blockchain, several technological solutions attempted to address these challenges, though each had limitations. ERP systems focused on integrating internal business operations across departments, while SCM software aimed to improve the flow of goods and information across partner networks [8]. Electronic data interchange enabled standardized electronic communication between enterprises, streamlining certain transactions. Despite these advances, all of these solutions relied on centralized databases, which introduced single points of failure, vulnerabilities to tampering, and limited verifiability. As a result, disconnected systems and siloed data persisted, leaving many structural challenges unresolved.

8.1.3 Introducing blockchain: definition and core principles

Modern supply chains face increasing pressure to become more transparent, reliable, and secure. Traditional centralized systems, while efficient in data storage, often struggle with fragmented information, limited visibility, and vulnerability to manipulation. To address these limitations, blockchain technology has emerged as a powerful innovation—providing a decentralized framework for recording, sharing, and validating information across multiple participants. This development marks a crucial step in the technological shift toward greater trust, accountability, and collaboration within distributed supply chain networks.

8.1.3.1 Definition of blockchain

Blockchain can be defined as a distributed and tamper-resistant digital ledger that securely records transactions with precision and authenticity [9]. It functions as a continuously growing sequence of linked records known as blocks, each containing transaction details, a timestamp, and a cryptographic link to the preceding block. These blocks are arranged in chronological order, forming a chain that makes unauthorized alterations virtually impossible. Once a block is validated and added to the ledger, its data becomes permanent, ensuring long-term reliability and integrity of information.

Unlike traditional databases that depend on a central authority, blockchain operates through a decentralized network of nodes. Each node maintains an identical copy of the ledger and participates in verifying transactions. This distributed verification process removes the need for intermediaries, minimizing risks of manipulation, system failure, and unauthorized tampering.

8.1.3.2 Decentralization: collaborative verification in action

Decentralization represents the core principle of blockchain technology and serves as the foundation of its credibility and resilience. Instead of a single entity controlling the data, authority is shared equally among all network participants. It can be visualized as a shared document accessible to every member of a team, where everyone can view, validate, and confirm updates collectively rather than depending on one person to maintain records [10].

When a new transaction, similar to a proposed document change, is initiated, it is broadcast to all participants in the network. The transaction is added to the ledger only after a consensus is reached confirming that it meets predefined validation criteria. This collective verification mechanism prevents any single participant from altering data unilaterally and fosters a system based on mutual trust and authenticity. In the context of SCM, decentralization ensures that all stakeholders, including manufacturers, suppliers, and distributors, rely on the same verified and trusted record without depending on a central governing body.

8.1.3.3 Transparency: open yet secure information access

Transparency is another key characteristic that distinguishes blockchain from conventional systems. All authorized members of the network have access to identical transaction data, providing complete visibility across each stage of the supply chain [11]. This open access model promotes accountability, as every transaction is permanently recorded and can be independently verified. The combination of transparency and immutability builds trust among participants while maintaining the security and confidentiality of sensitive data. Although participant identities remain hidden behind cryptographic addresses for privacy, the actions themselves are visible and verifiable.

For global supply chains, such shared visibility ensures that manufacturers, regulators, and consumers can all trace the origin and movement of goods. This transparency helps prevent fraud, minimize inefficiencies, and promote ethical sourcing by maintaining a single, trustworthy version of the truth for all stakeholders involved.

8.1.3.4 Immutability: ensuring data cannot be altered

The immutability of blockchain gives it a unique edge over conventional systems. Once a transaction is validated and entered into a block, it becomes virtually impossible to alter or delete [12]. Each block contains a unique cryptographic hash based on the data it holds and the hash of the block before it, creating a continuous, secure link between all records (as illustrated in Figure 8.2).

If someone attempts to modify the data within an older block, that change alters its hash value, which would immediately invalidate all the blocks that follow. This built-in resistance to tampering provides an exceptionally secure audit trail, making blockchain one of the most trustworthy systems for recording and preserving supply chain transactions.

8.1.3.5 Connecting blockchain to the technological shift

By combining decentralization, transparency, and immutability, blockchain fundamentally redefines how organizations establish digital trust. Unlike traditional systems that depend on intermediaries or institutional oversight, blockchain relies on collective verification, creating a self-sustaining environment of trust among participants.

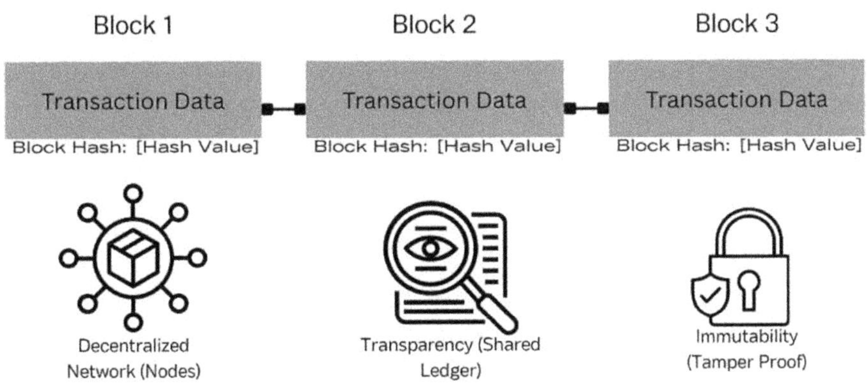

Figure 8.2 Core principles of blockchain technology.

This capability directly supports the ongoing transformation of supply chain management—from siloed, centralized platforms to interconnected, intelligent ecosystems. As companies increasingly prioritize verifiable authenticity and collaborative data sharing, blockchain stands as both a cornerstone and catalyst in building more transparent, efficient, and resilient global supply networks.

8.1.4 Blockchain's architecture and mechanism

Blockchain functions as a distributed ledger that securely records transactions across multiple stakeholders without relying on a central authority. Each transaction is grouped into a block, validated using consensus mechanisms, and cryptographically linked to previous entries, forming a continuous chain that cannot be altered without network-wide agreement. In supply chains, this ensures that data—such as product origin, batch numbers, or delivery confirmations—remains reliable and accessible to all authorized participants. Smart contracts further enhance efficiency by automating payments, quality checks, and compliance verification, reducing manual oversight and disputes. When combined with IoT and AI, blockchain enables near real-time monitoring, predictive analytics, and improved coordination across the supply network [13–19].

Structural Overview: A blockchain ledger is composed of sequential blocks, each containing a header with metadata like timestamps and cryptographic hashes, a body storing validated transaction data, and a reference linking it to the previous block. This interconnection preserves chronological order and makes tampering evident, as any modification would break the cryptographic chain, alerting all participating nodes [14].

The Role of Nodes: Nodes act as independent record keepers, storing and verifying transactions. Full nodes maintain complete copies of the ledger, while light nodes hold essential portions for validation. Nodes interact through a peer-to-peer network, broadcasting and synchronizing information without central oversight, which enhances transparency and reliability [15].

Transaction Flow: Data moves through a structured process: transaction initiation, broadcasting to nodes, validation using consensus rules, block formation, consensus approval, and final ledger update. This ensures all participants maintain identical, up-to-date records without intermediaries [16].

Consensus Mechanisms: Consensus algorithms determine which transactions are legitimate. Common approaches include Proof of Work (PoW), which uses computational puzzles; Proof of Stake (PoS), selecting validators based on staked tokens; and Proof of Authority (PoA), relying on trusted validators in private networks. Each method balances security, efficiency, and scalability according to network requirements [17].

Smart Contracts: These self-executing digital agreements automatically enforce predefined conditions, such as triggering payments once an IoT

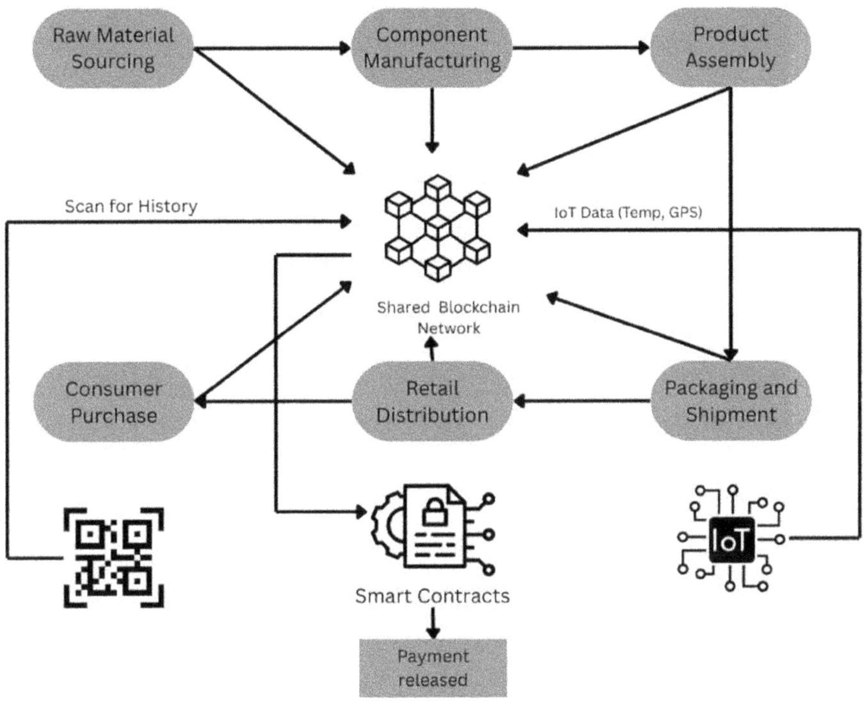

Figure 8.3 General architecture of blockchain networks and transaction flow.

device confirms delivery. By reducing manual intervention, they streamline processes and minimize disputes [18].

Blockchain Network Types: Networks may be public (open to anyone, e.g., Bitcoin and Ethereum), private (controlled by specific organizations for confidentiality and faster processing), or hybrid (combining elements of both for selective data sharing). Each model offers a different balance between transparency and privacy [19].

Significance in Supply Chain Integration: Blockchain's architecture strengthens supply chains by providing verifiable, tamper-resistant records from raw material sourcing to final delivery. Combined with smart contracts, it enhances operational accountability, mitigates fraud, prevents counterfeiting, and improves process reliability. Figure 8.3 illustrates how decentralization, consensus, and immutability work together to maintain network integrity.

8.1.5 Why blockchain matters for supply chains: initial impact

The core features of blockchain have directly addressed many of the persistent challenges that have hindered supply chain performance. Its ability to create a decentralized, tamper-proof, and transparent ledger of transactions has provided a transformative approach to managing complex networks [18]. The initial influence of blockchain on supply chains has been observed in several key areas:

a. **Enhanced Traceability:** Each product, component, and shipment has been tracked accurately from its origin to the final point of delivery, ensuring a permanent and verifiable audit trail [19].
b. **Increased Transparency:** All authorized stakeholders have gained equal access to synchronized, real-time information, thereby eliminating data asymmetry and strengthening trust among partners [20].
c. **Reduced Counterfeiting:** The immutable nature of blockchain has significantly restricted the entry of counterfeit or unauthorized goods, as the verified history of products cannot be falsified [21].
d. **Improved Efficiency:** The integration of smart contracts has automated numerous operational tasks, reduced manual errors, and accelerated transaction processing across the supply chain [22].
e. **Greater Accountability:** Recording every participant's action on an immutable ledger has enhanced responsibility, making inefficiencies, bottlenecks, or violations easier to identify and resolve [23].

8.1.5.1 The role of IoT in blockchain-powered supply chains

The integration of the IoT with blockchain has further advanced supply chain capabilities by linking physical product movement with secure digital records [19]. IoT devices such as Radio Frequency Identification (RFID) tags, sensors, and GPS trackers capture real-time data related to location, environmental conditions, and operational status [24]. These data streams are transmitted to blockchain platforms, where they are permanently stored and validated across the network.

For instance, in cold-chain logistics, temperature sensors continuously monitor perishable goods. When deviations from predefined thresholds occur, smart contracts can trigger automatic alerts or initiate insurance claims [25]. This capability reduces delays in risk management and ensures timely corrective actions.

By combining IoT with blockchain, supply chains achieve higher precision, reliability, and responsiveness. Continuous updates provide stakeholders with accurate information, enabling timely decisions and reducing reliance on manual reporting. The result is a system that supports autonomous operations, lowers risks of spoilage or loss, and strengthens overall resilience [26].

8.1.5.2 Consumer trust in blockchain-tracked products

A transparent and verifiable supply chain delivers significant advantages to end consumers, who emerge as primary beneficiaries of blockchain adoption. With blockchain, product information becomes directly accessible, allowing buyers to confirm authenticity, origin, and ethical sourcing practices [20].

For example, scanning a QR code on a product provides access to its complete blockchain record, including production details, certifications, and sustainability credentials [27]. This access empowers consumers to make informed purchasing decisions, strengthens confidence in brands, and motivates companies to uphold higher standards of safety, quality, and responsibility [28].

Independent verification of claims, rather than reliance on marketing statements, marks a shift in consumer–brand relationships. By offering transparency and accountability, blockchain fosters long-term trust and enhances brand reputation [29].

8.2 HOW DOES BLOCKCHAIN OPERATE WITHIN SUPPLY CHAINS?

Understanding the theoretical underpinnings of blockchain is one thing; grasping its practical application within the complex ecosystem of a supply chain requires a deeper dive into its operational mechanisms.

8.2.1 Incorporating data into blockchain

Understanding how data is added to a blockchain is essential for recognizing its importance in real-world supply chain operations. While theoretical concepts explain how trust is distributed, the practical process of incorporating data reveals how accuracy, security, and consensus are maintained among decentralized participants. Every new record follows a structured series of steps that ensure the information remains authentic, traceable, and tamper-resistant.

8.2.1.1 Entering and signing information

The process begins when a participant in the supply chain, such as a manufacturer recording product creation or a logistics partner confirming delivery, initiates a transaction [21]. This data is entered into the blockchain through an integrated software platform or an application programming interface (API) specifically designed for that purpose. Once entered, the information is digitally signed using the sender's private key, producing a unique cryptographic signature that verifies the source of the data and protects its integrity [26]. This signature functions like a secure seal, and if anyone attempts to modify the data afterward, the system can immediately detect the alteration.

8.2.1.2 Distributing and checking data

After a transaction is signed, it is instantly shared across the blockchain network so that every node receives the same copy [22]. Each node then independently validates the data to ensure it follows established rules and data structures. This process involves verifying the digital signature, checking data accuracy, and confirming that the sender has the necessary authorization to submit transactions—an especially vital step in permissioned or consortium blockchains [27]. Once validated, the transaction is placed in a temporary pool of pending records that await inclusion in the next block.

The network then activates its consensus mechanism, which may use models such as PoA, a system frequently applied in enterprise supply chains where trusted participants verify entries [28]. Through this collective consensus, the network determines the correct sequence of transactions, resolves conflicts, and ensures that all nodes agree on a unified version of the ledger. This collaborative validation maintains uniformity and reinforces the blockchain's reputation for data integrity and trust.

8.2.1.3 Forming and connecting blocks

Once a sufficient number of validated transactions have accumulated, the system creates a new block [23]. Each block contains the confirmed records, a timestamp, a unique hash value that serves as its digital fingerprint, and the hash of the previous block. The inclusion of the earlier block's hash securely links each block in chronological order, forming the structure known as the blockchain. Any attempt to modify information in a previous block would alter its hash, which would automatically invalidate every block that follows. This immediate detection of inconsistencies ensures that all recorded information remains permanent and verifiable.

8.2.1.4 Updating the distributed ledger

After the new block has been verified and accepted, it is permanently added to the blockchain and shared across all participating nodes [24]. Each node updates its local copy of the ledger, ensuring full synchronization throughout the network. This decentralized replication removes the need for a central authority or single point of control. Instead, every participant maintains an identical, verified version of the ledger, creating a system where trust is built into the design itself. The outcome is a transparent, secure, and resilient digital record that supports accountability and traceability across the supply chain.

8.2.1.5 Using digital tokens for asset representation

In certain blockchain-based supply chain systems, digital tokens are employed to represent goods, resources, or units of value [25]. These tokens may be fungible, meaning they are interchangeable and represent equal value, such as units of raw material, or non-fungible, representing unique assets with distinct identifiers like serialized products or certified components. Tokens play several important roles: they represent ownership as items move through various stages of production and logistics, allow real-time inventory tracking, facilitate automated payments through smart contracts linked to fiat or digital currencies, and support incentive systems that reward participants for meeting compliance or performance targets [29].

By tokenizing physical and digital assets, blockchain creates a bridge between the virtual record and the physical supply chain. This integration enhances traceability, promotes transparency, and enables faster and more secure financial settlements across all stages of the supply chain.

8.2.2 Complete transaction flow in supply chains

To understand how blockchain operates in a real supply chain environment, consider the movement of a high-value electronic product as it passes through several production and distribution stages, as illustrated in Figure 8.3. The process begins at the raw material stage, where a supplier extracts rare components and records essential details such as their source, quantity, and ethical compliance directly on the blockchain. This information is secured through the supplier's private key, ensuring authenticity and traceability right from the beginning [26].

Afterward, the components are delivered to the manufacturer, who documents their receipt, production specifications, and quality inspection outcomes. Each transaction creates new digital tokens that stay connected to the original material records, maintaining a seamless flow of information. When the assembly unit receives these components, it records details of product assembly, performance testing, and serial identification, thereby generating a separate token for every completed product [27].

Once the finished goods reach the packaging and logistics stage, shipment data such as seal identifiers, container numbers, and route confirmations are entered into the blockchain. The logistics partner then starts monitoring the product through IoT-based devices that automatically transmit live location and environmental information to the blockchain network [19]. Every handover, from distribution hubs to retail stores, is documented and digitally signed by all participants. When the final consumer purchases the product, ownership of the digital token is transferred to them. By scanning a QR or Near Field Communication (NFC) code, the buyer can immediately access the complete product history—from the extraction of raw materials to the final sale [20].

If the product needs servicing or return, the blockchain record serves as verified proof of its authenticity, warranty details, and complete transaction trail [27]. Throughout the entire process, every entry is digitally signed, validated, distributed, and permanently stored in the blockchain ledger.

Smart contracts can automatically perform predefined actions, such as releasing payment once delivery is confirmed or triggering inspections when compliance rules are violated. This transparent and interconnected framework highlights blockchain's capacity to promote trust, accountability, and automation across all levels of the supply chain.

8.2.3 Main security and working tools

The reliability of blockchain in supply chain operations relies on several essential cryptographic and computational mechanisms that provide security and support efficient performance. These mechanisms form the technical foundation that ensures immutability, authenticity, and trust among all participants.

8.2.3.1 Hash functions: safeguarding data integrity

A hash function is a mathematical algorithm that converts input data of any size into a fixed-length string, referred to as a hash or digital fingerprint [28]. Each hash is unique to its specific input, meaning that even a slight change in the original data results in a completely different output. This process is one-way, making it impossible to reconstruct the original data from the hash, thereby ensuring privacy while allowing verification. In blockchain, the hash of each block is incorporated into the next block, creating a sequential link. Any attempt to modify a previous record changes its hash, which breaks the chain and immediately alerts all participants to potential tampering [12]. This mechanism ensures that all data stored on the blockchain remains permanent, verifiable, and resistant to unauthorized alterations.

8.2.3.2 Digital signatures: verifying identity and authenticity

Digital signatures use pairs of cryptographic keys to authenticate the origin and integrity of a message or transaction [29]. They act as electronic equivalents of handwritten signatures—unique, verifiable, and non-repudiable. Among the most efficient cryptographic methods is elliptic curve cryptography (ECC), which provides strong security with smaller key sizes compared to older methods like RSA [30]. In practice, users sign the hash of their transaction using their private key, and others verify it using the corresponding public key. If both match, the system confirms that the sender is legitimate and that the information has not been altered.

For enhanced security, systems may employ multi-signature (multi-sig) verification, requiring multiple parties to authorize a single transaction [31]. For example, in a supply chain scenario, both a warehouse manager and an external auditor might sign off on high-value transfers, thereby reducing the risk of unilateral fraud or data manipulation.

8.2.3.3 Smart contracts: automating trust and compliance

Smart contracts are self-executing programs written into the blockchain that automatically enforce agreed-upon conditions when specific triggers occur [16]. In supply chain operations, they can perform functions such as releasing payments when goods pass inspection, notifying customs when shipments reach borders, processing insurance claims when sensors detect temperature violations, or reordering materials when stock levels fall below a threshold.

By automating these workflows, smart contracts reduce manual intervention, eliminate delays, and ensure accuracy across distributed systems. However, because smart contracts are immutable once deployed, rigorous pre-deployment audits and code testing are essential to prevent operational errors or vulnerabilities [32].

8.2.4 Privacy methods: weighing openness and secrets

While blockchain's openness strengthens accountability, organizations often require confidentiality for sensitive business information such as pricing models, proprietary formulas, or customer data. To balance these competing needs, modern blockchain systems integrate advanced cryptographic techniques that protect private data while preserving the benefits of transparency.

8.2.4.1 Zero-knowledge proofs

A zero-knowledge proof (ZKP) allows one party to prove the truth of a statement to another without revealing any additional details [33]. In supply chain applications, ZKPs can demonstrate that a product meets sustainability standards without disclosing supplier identities, verify shipment origins without revealing routes, or confirm fair labor compliance without exposing individual employee data. This approach builds verifiable trust while preserving business confidentiality and competitive advantage.

8.2.4.2 Homomorphic encryption

Homomorphic encryption (HE) enables computations to be performed on encrypted data without decrypting it first [34]. Although computationally intensive, this technology allows different organizations to collaborate on data analysis while keeping proprietary information hidden. For example, multiple companies can jointly analyze logistics performance or demand forecasts using encrypted datasets, ensuring collective insights without exposing individual data sources. Such techniques are becoming increasingly vital for secure data sharing and analytics in multi-stakeholder supply chain ecosystems.

8.2.5 New tech boosting blockchain

The potential of blockchain multiplies when combined with complementary digital systems.

8.2.5.1 AI for anomaly detection and optimization

AI analyzes blockchain-generated data to detect irregularities such as transaction delays, fraudulent entries, or deviations from normal operational patterns [35]. Beyond anomaly detection, AI can forecast supply disruptions caused by weather or market fluctuations and recommend adaptive strategies. When trained on verified blockchain records, AI systems produce more accurate predictions and optimize logistics routes and inventory levels, improving overall resilience.

8.2.5.2 Quantum-resistant cryptography

The advent of quantum computing presents future security challenges to existing encryption methods like ECC [36]. To safeguard data integrity, post-quantum cryptographic (PQC) schemes are being developed to resist quantum attacks. Incorporating PQC into blockchain protocols will be essential for ensuring long-term security in digital supply ecosystems.

8.2.5.3 Cross-platform interoperability

With the presence of multiple blockchain frameworks, such as Hyperledger Fabric and Ethereum, interoperability solutions have become increasingly important [37]. Protocols that enable cross-chain asset transfers, secure data sharing, and atomic swaps help integrate previously fragmented supply networks into unified and cohesive ecosystems. Such interoperable systems allow organizations to collaborate effectively without the need to migrate entirely to a single blockchain platform.

These integrations represent a movement toward intelligent, secure, and interconnected supply chain ecosystems, combining verified data integrity with adaptive analytics to enhance decision-making and operational efficiency.

8.3 STRATEGIC APPLICATIONS OF BLOCKCHAIN IN SCM

Moving beyond the theoretical, blockchain offers profound strategic advantages that can redefine how supply chains operate, fostering unprecedented levels of trust, efficiency, and resilience [38].

8.3.1 Defining blockchain's role and boundaries in supply chains

Blockchain's transformative power becomes even greater when combined with other advanced technologies that enhance its scalability, intelligence, and resilience within supply chain environments.

8.3.1.1 AI for predictive monitoring

When combined with blockchain, AI can analyze immutable transaction records to identify anomalies, inefficiencies, or signs of fraud [35]. By examining historical data, AI systems can detect unusual delays, atypical routing patterns, or potential disruptions in the supply chain before they escalate. Predictive analytics also supports more accurate demand forecasting, route optimization, and dynamic allocation of resources, enabling organizations to make proactive, data-driven decisions.

8.3.1.2 Quantum-safe cryptography

The emergence of quantum computing poses potential risks to current cryptographic algorithms, such as ECC, which may become vulnerable to decryption [36]. To mitigate these risks, researchers are developing PQC techniques designed to resist quantum-based attacks. Implementing these advanced algorithms in blockchain systems ensures that supply chain data remains secure over the long term, safeguarding sensitive records against future computational threats.

8.3.1.3 Interoperability and cross-chain integration

Organizations often operate on different blockchain platforms, including Hyperledger, Ethereum, or Corda, which can result in fragmented and isolated data silos. Interoperability frameworks, also known as cross-chain bridges or hubs, facilitate secure data exchange and asset transfers between these independent systems [37]. Such frameworks enable collaboration across industries and regions, providing a unified perspective on supply chain data while preserving the autonomy of each network. This seamless integration is crucial for establishing a globally connected, transparent, and efficient digital supply chain ecosystem.

8.3.2 Building a blockchain-based supply chain: a comprehensive approach

Implementing blockchain effectively within a supply chain necessitates a comprehensive and holistic strategic approach that extends far beyond mere technological integration.

8.3.2.1 The "single source of truth" paradigm

One of the most significant and transformative contributions of blockchain to SCM is its ability to establish a definitive "single source of truth" [39]. In conventional supply chains, each participant typically maintains separate and isolated records, which often results in inconsistencies, prolonged disputes, and a lack of shared, real-time visibility. Blockchain addresses this challenge by providing a shared, immutable ledger in which all essential supply chain events are recorded and collectively verified by authorized participants. This approach eliminates data silos, reduces the time and effort spent on reconciliation, and ensures that every participant works from the same verified information. As a result, disagreements are minimized, and overall coordination across the supply chain is significantly improved.

8.3.2.2 Integration with existing enterprise systems (ERP, SCM)

Effective blockchain adoption in supply chains generally does not require a full replacement of existing enterprise systems. Instead, successful implementation relies on seamless integration with current ERP and SCM systems [40]. In this model, blockchain acts as an additional foundational layer that enhances trust and transparency in operations. Key data from ERP systems, such as order details and production schedules, and from SCM systems, including inventory movements and shipment updates, can be securely recorded on the blockchain. Conversely, verified blockchain data can inform traditional systems. APIs play a crucial role in enabling secure and efficient bidirectional data exchange, allowing existing workflows to continue while benefiting from blockchain's capabilities in data verification and immutability.

8.3.3 Driving transparency and verifiability in supply chains

Blockchain fundamentally transforms the depth and breadth of transparency and verifiability available throughout the entire supply chain, delivering significant benefits to both businesses and, ultimately, their consumers.

8.3.3.1 The role of hashes and QR codes

For physical products, establishing a reliable link between the digital blockchain record and the tangible item itself is absolutely essential. This crucial connection is frequently achieved through the use of unique identifiers in conjunction with technologies like QR codes [41].

a. **Unique Identifiers:** Each individual product, batch, or component is systematically assigned a unique identifier (e.g., a serial number or a specific batch ID).
b. **Hashing:** A cryptographic hash of critical product information (such as manufacturing date, location of production, and results of quality checks) can be securely recorded on the blockchain.
c. **QR Codes:** A QR code, physically printed on the product or its packaging, can contain a direct link to its corresponding blockchain record or a hash of its underlying data. When scanned by a smartphone, this QR code can direct the user to a public-facing interface that displays the immutable history of that specific item as recorded on the blockchain. This empowers consumers with the direct ability to verify authenticity, origin, and other product claims independently.

8.3.3.2 Addressing privacy concerns with transparency

While transparency is undeniably a core benefit of blockchain, businesses frequently harbor legitimate concerns about openly sharing highly sensitive information—such as proprietary manufacturing processes, competitive pricing strategies, or confidential customer data—with all their supply chain partners. This is precisely where advanced privacy-enhancing technologies become indispensable.

a. **Permissioned Blockchains:** In private or consortium blockchains, access to specific information can be strictly limited to only authorized participants, based on their predefined roles and permissions.
b. **Zero-Knowledge Proofs (ZKPs):** As previously discussed, ZKPs enable the verification of data validity without actually revealing the underlying sensitive information [33]. For instance, a manufacturer could prove they sourced materials from an ethically certified supplier without disclosing the supplier's actual name or specific contractual details.
c. **Off-Chain Data with On-Chain Hashes:** Highly sensitive data can be stored securely off-chain in traditional, more controllable databases, while only its cryptographic hash is recorded on the immutable blockchain. This provides a verifiable link to the data's integrity without exposing the data itself. If the off-chain data is ever tampered with, the hash on the blockchain will no longer match, immediately indicating a breach.
d. **Homomorphic Encryption:** Although still nascent in its widespread application, this advanced cryptographic technique allows computations to be performed directly on encrypted data, thereby preserving privacy during collaborative data analysis. This could enable secure data sharing for analytics without exposing raw, sensitive information.

8.3.4 Enhancing supply chain flexibility and resilience

In an era characterized by increasing global volatility and unforeseen disruptions, supply chain resilience has become an absolute paramount concern. Blockchain significantly contributes to this resilience by enabling the creation of faster, more adaptive, and inherently robust operational frameworks.

8.3.4.1 Automation through smart contracts for adaptability

Smart contracts introduce a revolutionary level of automation and adaptability into supply chains [16]. They possess the unique capability to be programmed to automatically trigger actions based on real-time data inputs and predefined conditions. For example:

a. **Dynamic Re-Routing:** If a shipment experiences an unforeseen delay due to emergent circumstances, a smart contract could be programmed to automatically initiate alternative transportation arrangements based on pre-approved rules and available capacity.
b. **Contingency Payments:** In the event of a detected product defect or a failure to meet predefined quality standards (verifiable via IoT sensors or third-party audits), smart contracts can automatically trigger partial refunds or penalty payments, significantly streamlining the dispute resolution process.
c. **Automated Replenishment:** Smart contracts can continuously monitor inventory levels and automatically place replenishment orders with pre-approved suppliers when stock falls below a certain threshold, thereby ensuring a continuous and uninterrupted supply. This profound level of automation minimizes manual human intervention, accelerates critical decision-making processes, and crucially, enables supply chains to react with far greater swiftness and agility to both disruptions and rapidly changing market conditions.

8.3.4.2 Cross-chain interoperability for seamless data flow

The increasing adoption of diverse blockchain platforms across various industries and by different companies necessitates the development of robust interoperability solutions [37]. Without such interoperability, the profound benefits of a "single source of truth" would remain fragmented. Cross-chain interoperability specifically enables:

a. **End-to-End Visibility:** Connecting disparate segments of a highly complex supply chain where various partners might be utilizing different underlying blockchain networks for their internal processes.

b. **Seamless Data Exchange:** Facilitating the smooth and secure transfer of validated data, digital assets, or tokens between independent blockchain networks.
c. **Holistic Risk Management:** By aggregating and correlating data from diverse blockchain sources, businesses can gain a far more comprehensive and nuanced view of systemic risks across their extended supply chain network. This seamless flow of verified information profoundly enhances overall supply chain visibility, improves coordination among partners, and significantly boosts responsiveness to unforeseen disruptions.

8.3.4.3 Impact on supply chain resilience and risk management

By providing real-time, immutable, and transparent data, blockchain significantly bolsters supply chain resilience and elevates risk management capabilities [42]. This includes:

a. **Early Warning Systems:** AI-powered anomaly detection, applied to blockchain-recorded data, can identify potential disruptions at their earliest stages, allowing for proactive and timely mitigation strategies.
b. **Improved Traceability for Recalls:** In the unfortunate event of a product recall, blockchain's granular traceability allows for the precise identification of affected batches, thereby minimizing the scope, cost, and impact of the recall.
c. **Enhanced Compliance:** Immutable records of compliance checks and various certifications simplify regulatory audits and significantly reduce inherent regulatory risks.
d. **Diversified Sourcing:** Transparent and verifiable supplier performance data, available on the blockchain, can inform more strategic decisions about diversifying supplier bases, effectively reducing an over-reliance on single points of failure.
e. **Dispute Resolution:** The availability of clear, verifiable, and immutable records substantially reduces the likelihood and complexity of disputes that can arise between various supply chain partners.

8.3.5 Broader benefits for consumers, businesses, and sustainability

The strategic implementation of blockchain within supply chains yields multifaceted benefits that extend far beyond mere operational efficiency, profoundly impacting consumers, businesses across all sectors, and critically, global sustainability efforts.

8.3.5.1 Consumer trust and brand loyalty

For consumers, blockchain offers an unprecedented and unparalleled level of transparency and assurance [20]. By empowering them to directly verify the authenticity, precise origin, and ethical sourcing of the products they purchase, blockchain actively cultivates greater trust in brands. This newfound transparency can fundamentally transform the consumer–brand relationship from one primarily based on marketing claims to one solidly built on verifiable facts. Increased trust, in turn, translates into significantly enhanced brand loyalty, as consumers are demonstrably more inclined to support companies that openly demonstrate integrity and accountability within their supply chains. This also opens new market opportunities for brands genuinely committed to ethical and sustainable practices.

8.3.5.2 Cost savings and operational efficiencies

Businesses stand to gain substantial cost savings and remarkable operational efficiencies through judicious blockchain adoption [4]. These benefits include:

a. **Reduced Administrative Costs:** Automation facilitated by smart contracts effectively eliminates voluminous manual paperwork, reduces errors from data entry, and streamlines reconciliation processes.
b. **Lower Fraud and Counterfeiting Losses:** Immutable and transparent tracking capabilities significantly reduce the devastating impact of illicit trade and counterfeit goods.
c. **Optimized Inventory Management:** Real-time, shared visibility of stock levels allows for more precise control, effectively reducing both wasteful overstocking and costly stockouts.
d **Faster Dispute Resolution:** Readily available and verifiable records streamline the process of resolving conflicts and disagreements among supply chain partners.
e. **Streamlined Audits:** Immutable audit trails simplify compliance checks and significantly reduce the associated auditing costs and complexities.
f. **Reduced Lead Times:** Faster information flow and automated processes collectively accelerate the movement of products throughout the entire chain, from production to delivery.

8.3.5.3 Sustainability and ethical sourcing verification

Blockchain provides an exceptionally powerful tool for meticulously verifying and transparently showcasing sustainability and ethical sourcing claims within supply chains [7]. This enables:

a. **Transparent Sourcing:** The ability to track raw materials from their absolute origin to ensure they are demonstrably sustainably harvested or ethically produced (e.g., verifying conflict-free minerals or fair-trade coffee).
b. **Carbon Footprint Tracking:** The capacity to accurately record and verify carbon emissions data at each distinct stage of the supply chain, fostering environmental accountability.
c. **Waste Reduction:** Improved visibility into waste streams and material flows, actively facilitating the implementation of circular economy initiatives.
d. **Labor Practices:** The means to verify strict adherence to fair labor practices and effectively eliminate instances of child labor or exploitation within the supply chain by tracking relevant certifications and audit results. By making these crucial claims verifiably immutable, blockchain actively assists companies in meeting the escalating consumer and regulatory demands for genuinely responsible business practices and provides a robust defense against misleading "greenwashing" claims.

8.3.5.4 Impact on supply chain finance

Blockchain holds the potential to profoundly revolutionize supply chain finance by significantly increasing trust, accelerating transaction speeds, and providing more reliable data for comprehensive risk assessment [43].

a. **Faster Payments:** Smart contracts can be programmed to automatically release payments upon verified delivery or the completion of specific milestones, drastically accelerating cash flow for suppliers, particularly benefiting small and medium-sized enterprises.
b. **Improved Access to Financing:** Immutable and transparent transaction histories provide financial institutions with more reliable and auditable data to accurately assess creditworthiness. This can potentially lead to lower interest rates and increased access to working capital for various supply chain participants.
c. **Invoice Factoring:** Blockchain can facilitate a more secure and efficient process for invoice factoring by providing instantly verifiable invoices and real-time payment statuses.
d. **Trade Finance:** The technology can streamline notoriously complex trade finance processes by digitizing critical documents and automating conditional payments, thereby reducing instances of fraud and mitigating delays. This holistic impact ultimately leads to the development of a more efficient, transparent, and liquid financial ecosystem operating within the broader supply chain.

8.4 REAL-WORLD IMPACT: BLOCKCHAIN CASE STUDIES IN 2025

The theoretical benefits of blockchain in SCM are increasingly being realized through tangible implementations across diverse industries. The following case studies illustrate how pioneering companies are leveraging this transformative technology to solve real-world problems and achieve significant operational improvements.

8.4.1 AgriDigital: digitizing agriculture with blockchain

Problem Statement: The agricultural supply chain for grains has long struggled with inefficiencies, limited visibility, and frequent delays in payments to farmers, who are often the most affected participants in the process. Manual documentation and fragmented record-keeping have made it difficult to verify produce quality, quantity, and ownership, often leading to disputes and inconsistent data among farmers, traders, and intermediaries. To overcome these long-standing issues, AgriDigital developed a blockchain-based platform that digitizes the entire grain supply chain—from post-harvest handling to market delivery—by providing a shared, real-time record of transactions for all stakeholders. Through this system, each grain delivery is securely logged as a verifiable blockchain entry that records ownership, quality parameters, and payment status. The inclusion of smart contracts further enhances efficiency by automatically releasing payments once delivery conditions are confirmed, eliminating unnecessary intermediaries and ensuring farmers receive timely compensation. By integrating automation, traceable data, and trust-based digital interactions, AgriDigital's approach has improved transparency, reduced administrative burdens, and strengthened the financial reliability of agricultural trade networks [44].

8.4.1.1 Technologies employed

a. **Blockchain Platform:** AgriDigital primarily leverages a permissioned blockchain, such as Hyperledger Fabric or a similar consortium-grade blockchain, chosen specifically for its enterprise-level privacy features and high performance.
b. **IoT Sensors:** There is potential for integration with quality testing equipment located at silos, which could automatically record critical grain quality data directly onto the blockchain, ensuring data integrity at the source.
c. **Smart Contracts:** These are used to automate payment triggers, which are activated based on the verified delivery of grain and adherence to predefined quality parameters.

d. **Digital Tokens:** The platform utilizes digital tokens to represent units of grain, thereby facilitating transparent and auditable ownership transfers across the supply chain.

8.4.1.2 Tangible outcomes and before/after scenario

a. **Before:** In the traditional system, farmers often endured waiting periods of days or even weeks for payment, frequently contending with extensive paper invoices and manual reconciliation processes. Disputes over grain quality and quantity were a common occurrence, leading to significant financial strain and a pervasive sense of distrust among stakeholders.
b. **After:** AgriDigital's platform dramatically reduced payment times for farmers, transforming weeks of waiting into mere minutes or even seconds following verified delivery. It provided unprecedented real-time visibility into grain ownership and the entire transaction history for all participants. The immutable blockchain record significantly minimized disputes regarding quality and quantity, fostering enhanced trust across the agricultural supply chain. This digital transformation has notably improved cash flow for farmers and streamlined operational efficiencies for grain buyers [44].

8.4.2 Intel in fruit distribution: ensuring freshness

Problem Statement: The perishable goods supply chain, particularly for sensitive fresh produce like fruit, grapples with immense challenges related to spoilage, rapid quality degradation, and inefficient tracking mechanisms [45]. Maintaining optimal environmental conditions, such as precise temperature and humidity levels, throughout the entire journey from the farm to the consumer's hands is critically important; any deviation from these conditions can lead to substantial waste and significant financial losses. Traditional tracking methods often lack real-time data inputs and robust, tamper-proof records, making it exceedingly difficult to pinpoint the exact stage where spoilage occurred or to accurately assign accountability for quality breaches.

Blockchain Solution Overview: Intel, together with several industry collaborators, has been exploring blockchain-driven solutions to improve the traceability and quality assurance of high-value perishable goods. The initiative focuses on embedding IoT sensors within product crates and pallets to continuously track vital environmental factors—such as temperature, humidity, and handling conditions—throughout transportation. The data collected in real time is securely stored on a blockchain network, forming a tamper-proof record that traces each product's journey from origin to destination. Smart contracts add another layer of automation by instantly notifying stakeholders when conditions fall outside predefined limits or by initiating insurance processes for products compromised in transit. This integration of blockchain with IoT not only strengthens transparency and accountability but also minimizes loss, ensures quality preservation, and enhances overall supply chain resilience [45].

8.4.2.1 Technologies employed

a. **Blockchain Platform:** The solution would potentially utilize a permissioned blockchain, such as Hyperledger Sawtooth or a bespoke blockchain solution, offering the necessary control and privacy for a consortium of food distributors and logistics providers.
b. **IoT Sensors:** Critical components include temperature, humidity, and location sensors embedded directly with the produce to provide continuous monitoring and data collection.
c. **Smart Contracts:** These automated agreements are used to trigger immediate alerts, enforce predefined quality control rules based on sensor data, and initiate predefined actions, such as automated insurance claims, when conditions are violated.
d. **Data Analytics (AI):** AI algorithms can be integrated to analyze the vast streams of sensor data patterns, enabling the prediction of potential spoilage risks and optimizing logistics routes.

8.4.2.2 Tangible outcomes and before/after scenario

a. **Before:** Spoilage constituted a significant and recurring problem, leading to substantial financial losses and potential reputational damage for businesses. Identifying the precise point of failure within the cold chain was consistently difficult, and disputes over product quality were commonplace.
b. **After:** The blockchain-powered solution provides enhanced real-time visibility into environmental conditions across the entire supply chain. It offers the ability to precisely identify when and where temperature deviations occurred, enabling swift corrective actions or accurate accountability. This proactive monitoring and automated alerting capability leads to reduced spoilage and waste. Ultimately, consumer confidence in the freshness and quality of produce is increased, as its entire journey is verifiably tracked on the blockchain.

8.4.3 MediLedger: authenticating pharmaceuticals

Problem Statement: The pharmaceutical supply chain is particularly susceptible to the insidious threat of counterfeiting, which not only poses severe public health risks but also inflicts significant economic losses upon legitimate drug manufacturers [5]. The Drug Supply Chain Security Act (DSCSA) in the United States mandates stricter traceability requirements; yet, traditional serialization methods alone are often insufficient to comprehensively prevent counterfeit drugs from infiltrating the legitimate supply chain. Therefore, maintaining the absolute integrity of pharmaceutical products and preventing their diversion are paramount concerns for the industry.

Blockchain Solution Overview: The MediLedger Project, a collaborative initiative driven by a consortium of pharmaceutical industry stakeholders, has successfully built a blockchain network specifically designed to verify the authenticity of pharmaceutical products and ensure compliance with DSCSA requirements [5]. This network is centered on creating an immutable, shared record of drug transactions between all trading partners, including manufacturers, wholesalers, and pharmacies. When a drug changes hands, its unique serial number and the details of the transaction are meticulously recorded on the blockchain, thereby establishing a verifiable and unalterable audit trail. This robust system enables any authorized participant to quickly and reliably verify the legitimate origin and transaction history of a pharmaceutical product.

8.4.3.1 Technologies employed

a. **Blockchain Platform:** The MediLedger solution specifically utilizes a permissioned blockchain built on robust protocols like Hyperledger Fabric. This choice ensures the necessary privacy and regulatory compliance demanded within the highly regulated pharmaceutical industry, where participating companies require controlled access to sensitive data.
b. **Digital Signatures:** Each transaction recorded on the network is cryptographically signed by the respective trading partners. This ensures non-repudiation, meaning a party cannot later deny having made a transaction.
c. **Smart Contracts:** These automated agreements are utilized to automate verification processes and are capable of flagging transactions that appear suspicious or deviate from predefined rules, enhancing security protocols.
d. **Global Unique Identifiers (GTINs, SNIs):** These standardized identifiers are crucially employed to uniquely identify each individual drug package and link it directly to its corresponding blockchain record, ensuring precise traceability.

8.4.3.2 Tangible outcomes and before/after scenario

a. **Before:** The pharmaceutical supply chain faced a significant and dangerous risk of counterfeit drugs entering the market, directly endangering patients. Manual and fragmented record-keeping practices made it exceedingly difficult to verify product authenticity quickly and reliably. Furthermore, achieving full DSCSA compliance was often a challenging and costly endeavor.

b. **After:** MediLedger provides a robust, immutable system for verifying pharmaceutical product authenticity across the entire supply chain. It has led to a reduced risk of counterfeiting and diversion due to the real-time, verifiable transaction history. The platform has streamlined DSCSA compliance by providing a shared, auditable ledger of ownership transfers; and most importantly, it has enhanced patient safety by ensuring that only legitimate drugs reach consumers.

8.4.4 TradeLens: optimizing global logistics

Problem Statement: Global shipping and logistics operations are notoriously complex, characterized by the involvement of numerous intermediaries, an overwhelming volume of paperwork, and frequent delays stemming from fragmented information exchange [4]. The persistent lack of real-time visibility into cargo movements, reliance on manual processes for customs clearance, and high levels of mistrust among independent entities collectively contribute to significant inefficiencies, elevated operational costs, and prolonged transit times for goods.

Blockchain Solution Overview: TradeLens, a groundbreaking joint venture between shipping giant Maersk and IBM, is a blockchain-based platform explicitly designed to digitize and streamline global supply chains, with a particular focus on ocean freight [4]. It establishes a shared, immutable ledger accessible to all authorized participants within the complex shipping ecosystem—including shippers, port authorities, customs bodies, ocean carriers, and various logistics providers. Critical documents, events, and data (such as bills of lading, customs declarations, and real-time vessel movements) are meticulously recorded on this blockchain, creating a single, comprehensive, and real-time view of the entire shipping journey for all involved parties.

8.4.4.1 Technologies employed

a. **Blockchain Platform:** TradeLens is built upon Hyperledger Fabric, a robust, permissioned blockchain framework that is particularly well-suited for consortium-based enterprise solutions due to its scalability and privacy features.
b. **Smart Contracts:** These are utilized to automate various processes, including customs declarations, calculations for demurrage (charges for delayed loading/unloading), and other conditional agreements, thereby reducing manual effort and potential disputes.

c. **Application Programming Interfaces:** APIs are crucial for enabling seamless integration between the TradeLens blockchain platform and the existing legacy systems of participating organizations, ensuring smooth data flow.
d. **IoT Integration:** There is potential for integration with smart containers or port equipment, allowing real-time tracking data to be automatically fed onto the blockchain, enhancing granular visibility of cargo.

8.4.4.2 Tangible outcomes and before/after scenario

a. **Before:** The industry suffered from extensive paperwork, pervasive manual data entry, and fragmented communication, which led to significant delays and high administrative costs. A severe lack of real-time visibility into shipment statuses caused widespread uncertainty and inefficiencies. Disputes over demurrage and detention charges were a common and costly occurrence.
b. **After:** TradeLens has resulted in reduced transit times and administrative costs, primarily due to digitized documentation and automated processes. It has provided enhanced real-time visibility of cargo movements across the entire journey. Improved collaboration and fostered trust among diverse stakeholders are achieved through a shared, immutable record. The platform has led to fewer disputes and significantly faster customs clearance, ultimately contributing to a more efficient and predictable global logistics network.

8.4.5 Everledger: verifying ethical electronics

Problem Statement: The electronics industry faces considerable challenges concerning ethical sourcing, especially regarding **conflict minerals** (such as tin, tantalum, tungsten, and gold) frequently utilized in electronic components [7]. Consumers and regulatory bodies are increasingly demanding verifiable proof that products are not associated with conflict zones or unethical labor practices. Tracing these specific minerals through inherently complex, multi-tiered global supply chains proves exceptionally difficult using traditional, opaque methods.

Blockchain Solution Overview: Everledger, a company initially renowned for its groundbreaking work in tracking high-value assets like diamonds, has successfully expanded its expertise to encompass other valuable assets and ethical sourcing verification, including electronic components and their underlying minerals (Figure 8.4). They leverage blockchain technology to establish a permanent, immutable digital record of the provenance of materials [7]. Every sequential step of the supply chain, from raw mineral extraction (mining) through refinement and subsequent component manufacturing, is meticulously recorded on the blockchain. This record is further

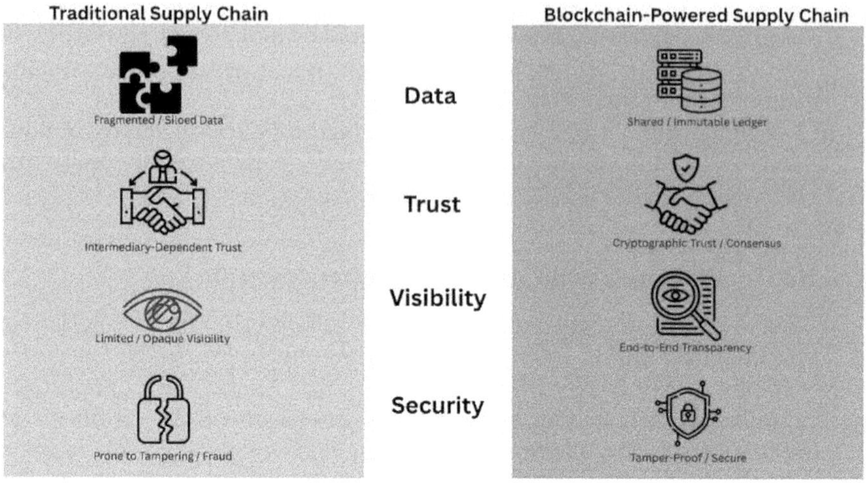

Figure 8.4 Traditional vs. blockchain-powered supply chain: a comparative view.

enriched with relevant certifications (e.g., fair-trade status, conflict-free designation). This comprehensive approach effectively creates a verifiable "digital passport" for each material or component, providing its complete and trustworthy history.

8.4.6 Technologies employed

a. **Blockchain Platform:** Everledger likely employs a permissioned blockchain platform specifically designed for tracking high-value assets, potentially incorporating private data channels to maintain confidentiality for sensitive information.
b. **Digital Identity/Tokens:** The system utilizes digital identities or tokens to represent the unique identity and specific attributes of the minerals or components being tracked.
c. **IoT Devices:** Potentially, IoT devices could be deployed at mining sites or processing facilities to automatically capture and upload key data, such as weight, location, and timestamps, directly to the blockchain.
d. **Smart Contracts:** These automated agreements are used to enforce ethical sourcing rules and can automatically trigger audits or alerts based on the verification of certifications, ensuring compliance.

8.4.7 Tangible outcomes and before/after scenario

a. **Before:** There was limited visibility into the true origin of conflict minerals, making it exceedingly difficult for electronics companies to definitively guarantee ethical sourcing. This created a high risk of reputational damage for businesses linked to unethical practices and posed significant challenges in meeting growing consumer and regulatory demands for transparency.

b. **After:** Everledger's solution has delivered enhanced transparency and verifiability of ethical sourcing for electronics components. It provides the ability to offer irrefutable proof that materials are genuinely conflict-free or meet other specified sustainability standards. This has resulted in increased consumer trust and bolstered brand reputation for companies demonstrably committed to ethical sourcing. Furthermore, it has improved compliance with critical regulations related to conflict minerals and human rights within supply chains.

8.5 CHALLENGES AND MITIGATION STRATEGIES IN BLOCKCHAIN ADOPTION

While blockchain offers immense potential for transforming supply chains, its widespread adoption is not without significant hurdles. Addressing these challenges through strategic planning and continuous technological advancements is crucial for fully realizing its profound promise.

8.5.1 Current hurdles to widespread blockchain adoption

Despite the compelling and widely recognized benefits, several practical and systemic obstacles continue to impede the rapid and pervasive integration of blockchain technology into complex global supply chains.

8.5.1.1 Scalability: addressing slow transaction speeds

Many existing blockchain networks, particularly public ones, can suffer from notably limited transaction processing speeds (measured in transactions per second, or Transactions Per Second (TPS) when compared to traditional centralized databases [46]. This poses a major concern for high-volume supply chains that routinely generate millions of transactions on a daily basis. The underlying consensus mechanisms, especially computationally intensive ones like PoW, require significant computational effort, which consequently leads to slower transaction finality. For a global supply chain that must handle a constant and rapid flow of goods and information, such slow transaction speeds can unfortunately negate many of the potential efficiency gains offered by blockchain.

8.5.1.2 Interoperability: bridging incompatible systems

The current blockchain landscape remains fragmented, with numerous distinct platforms (e.g., Hyperledger Fabric, Corda, Ethereum, alongside various proprietary private solutions) existing independently [37]. This diversity means that different companies, or even disparate departments within the same large organization, might adopt incompatible blockchain technologies. This pervasive lack of seamless interoperability makes it profoundly challenging to achieve true end-to-end visibility across a complex supply chain that spans multiple collaborating companies and potentially operates across several different blockchain networks. In essence, the problem of data silos can unfortunately migrate from traditional legacy systems to new, but still isolated, blockchain networks.

8.5.1.3 Regulatory compliance: navigating privacy laws and data management

Blockchain's foundational principles of immutability and transparency can, at times, directly conflict with existing stringent data privacy regulations, such as the General Data Protection Regulation (GDPR) in Europe or the California Consumer Privacy Act (CCPA) in the US [47]. For example, the "right to be forgotten" provision within GDPR directly contradicts the immutable nature of a blockchain ledger. Managing sensitive personal or proprietary business data within a transparent, distributed environment while simultaneously remaining fully compliant with diverse, evolving, and often inconsistent global regulations presents a significant legal, ethical, and operational challenge.

8.5.1.4 Energy consumption concerns

Public blockchains, especially those that predominantly utilize PoW consensus mechanisms, are widely recognized for their exceptionally high energy consumption [48]. While permissioned blockchains frequently employ more energy-efficient consensus algorithms (such as variants of PoA or PoS), the general perception of blockchain as an energy-intensive technology can still act as a significant deterrent for environmentally conscious businesses or those operating under strict sustainability mandates.

8.5.1.5 High initial implementation costs

The adoption of blockchain technology within existing supply chain infrastructures often demands a significant initial investment, which can deter many organizations, especially small and medium-sized enterprises. These costs generally include developing or licensing a suitable blockchain platform, integrating it with existing systems such as ERP and SCM software, and ensuring seamless compatibility with legacy IT frameworks. Additionally, companies must invest in recruiting or training personnel with the necessary blockchain expertise and conducting rigorous security audits to address compliance and legal risks. Collectively, these financial and technical requirements make the transition to blockchain a complex and costly endeavor, often restricting its adoption to larger firms with sufficient resources [49].

8.5.1.6 Management, governance, and coordination complexities

Establishing and effectively governing a consortium blockchain, particularly for a multi-party supply chain, involves significant complexities in coordination among numerous independent entities [50]. Reaching consensus and agreeing upon common technical standards, standardized data formats, clear governance rules, and robust dispute resolution mechanisms requires substantial collaborative effort and time. Issues related to collective decision-making, implementing network upgrades, and resolving conflicts among potentially competing partners can be inherently complex and time-consuming, potentially hindering the network's overall evolution and long-term effectiveness.

8.5.1.7 Data quality: the "garbage in, garbage out" problem

While blockchain meticulously ensures the integrity and immutability of data *once it is recorded on the chain*, it does not inherently guarantee the accuracy or initial quality of the data *entered* onto the chain [51]. If inaccurate, incomplete, or even fraudulent data is initially fed into the blockchain (often referred to as the "garbage in" problem), the immutable record will simply reflect that flawed data ("garbage out"). This critical limitation underscores the continued paramount importance of implementing robust data collection processes, utilizing reliable IoT sensors for automated data capture, and maintaining vigilant human oversight at the very point of data origination to ensure the veracity of information before it enters the blockchain.

8.5.1.8 User adoption and training challenges

Implementing blockchain technology necessitates a fundamental shift in mindset and the acquisition of new skill sets for employees across all tiers of the supply chain [52]. Resistance to change, a lack of comprehensive understanding of the new technology, and insufficient training programs can significantly hinder successful user adoption. Employees require thorough education on how blockchain benefits their specific roles and how to effectively interact with new systems, demanding substantial investment in comprehensive change management and dedicated training initiatives.

8.5.2 Benefits and challenges: a comprehensive overview

A balanced perspective on blockchain adoption in supply chains necessitates an acknowledgment of both its transformative benefits and its inherent challenges (Table 8.1).

8.5.3 Key vulnerabilities and attack vectors

Although blockchain networks have been designed with advanced cryptographic protections, they have not been entirely resistant to all forms of threats. Recognizing possible weaknesses and analyzing common attack vectors has been essential for developing resilient blockchain-enabled supply chain infrastructures. A clear understanding of these risks has provided a foundation for implementing more secure and reliable systems [52].

Table 8.1 Benefits and challenges of blockchain

Feature	*Benefits*	*Challenges*
Transparency	Enhanced visibility, auditability, and builds consumer trust	Privacy concerns for sensitive business data.
Immutability	Tamper-proof records, authenticity, reduced fraud/counterfeiting	Difficulty correcting erroneous entries and regulatory "right to be forgotten" issues.
Decentralization	No single point of failure, increased resilience, and reduced reliance on intermediaries	Complex governance and coordination among diverse stakeholders.
Automation (Smart contracts)	Streamlined processes, faster payments, and reduced manual errors	Smart contract bugs can lead to irreversible losses and the need for rigorous audits.
Traceability	Granular product tracking, improved recalls, and ethical sourcing verification	Requires reliable data input at the source and integration with the physical world (IoT).
Efficiency	Reduced administrative costs, faster transactions, and optimized operations	Initial implementation costs and scalability limitations for high volume.
Security	Cryptographic security, digital signatures, and integrity of data	Vulnerability to 51% attacks (public chains), private key management, and quantum computing threat.
Interoperability	Seamless data flow across partners (potential)	Fragmented blockchain ecosystem and difficulty connecting disparate networks.
Regulatory	Simplified compliance reporting (with proper design)	Evolving and inconsistent regulations and data privacy conflicts.
Sustainability	Verifiable ethical sourcing and carbon tracking	Energy consumption concerns (especially for PoW).

8.5.3.1 51% attacks

In PoW-based blockchain architectures, a **51% attack** represents a potential vulnerability that arises when a single entity, or a coordinated group of participants, gains control of more than half of the total computational power of the network, commonly referred to as the hash rate [53]. With this majority, the attacker is capable of executing several disruptive actions:

a. **Blocking Valid Transactions:** Preventing the confirmation of legitimate transactions, thereby disrupting the flow of operations across the ledger.
b. **Double Spending:** Reversing transactions that were previously confirmed, allowing the same digital asset to be fraudulently spent multiple times.
c. **Monopolizing Block Creation:** Hindering honest participants from successfully mining blocks, consolidating the attacker's dominance in block generation.

While the probability of such an attack has been extremely low for large-scale public blockchains like Bitcoin, given the enormous computational resources required, it has remained a realistic risk for smaller, less decentralized networks [54]. Additionally, consortium or private blockchains have been particularly vulnerable if governance mechanisms have been weak, or if dominant entities within the system act maliciously [55].

8.5.3.2 Smart contract bugs and exploits

Smart contracts, essentially self-executing lines of code, are not immune to errors or vulnerabilities [32]. Once deployed on an immutable blockchain, any bugs or logical flaws become extremely difficult to correct, potentially resulting in unintended outcomes or exploitation by malicious actors. High-profile incidents, such as the decentralized autonomous organization hack, have highlighted how a single flaw in smart contract code can lead to massive financial losses. Common vulnerabilities include re-entrancy attacks, where a malicious contract repeatedly invokes a vulnerable contract before the initial transaction completes; integer overflow or underflow, where arithmetic operations exceed the data type limits and yield incorrect results; and access control weaknesses, which allow unauthorized users to perform actions that should be restricted. These risks emphasize the critical need for thorough code audits, rigorous testing, and robust security practices before deploying smart contracts in real-world supply chain applications.

8.5.3.3 Fake nodes (Sybil attacks)

A Sybil attack occurs when a malicious actor creates and controls numerous false identities, or "nodes," within a peer-to-peer network in order to gain disproportionate influence or manipulate voting power [54]. Within blockchain systems, such an attacker attempts to overwhelm the network by deploying a large number of fraudulent nodes, thereby disrupting the consensus mechanism or selectively blocking legitimate transactions. Although robust public blockchains that utilize economic deterrents such as PoW or PoS have been relatively resilient to such threats, Sybil attacks have posed greater risks to smaller or permissioned blockchain environments [56]. In these contexts, if node authentication and identity verification have not been adequately enforced, adversaries could exploit the system's governance structure to undermine security, distort consensus, or degrade network performance [57].

8.5.3.4 Data feed attacks (Oracles)

Blockchains are inherently deterministic systems and cannot directly access real-world information that exists off their chain. To incorporate external, off-chain data (such as IoT sensor readings, real-time market prices, or actual shipment statuses), they rely on "oracles," trusted third-party services specifically designed to securely feed this external data onto the blockchain [55]. A critical vulnerability arises if an oracle is compromised, malfunctions, or deliberately provides incorrect or malicious data. Such an event can lead to faulty smart contract execution based on erroneous information or the recording of inaccurate records on the blockchain. This introduces a potential attack vector at the crucial interface between the physical world and the digital blockchain ledger.

8.5.4 Strategic mitigation approaches

Addressing the diverse challenges and inherent vulnerabilities of blockchain technology within supply chain contexts requires a comprehensive and multifaceted approach. This strategy must seamlessly combine robust technological safeguards with strong governance frameworks and diligently implemented operational best practices.

8.5.4.1 Rigorous code checks and audits

To effectively mitigate the risks posed by smart contract bugs and vulnerabilities, extensive and independent code audits are paramount [32]. This crucial process typically involves a combination of techniques:

a. **Static Analysis Tools:** Automated software tools that systematically check the contract code for common vulnerabilities and adherence to best practices without actually executing it.
b. **Dynamic Analysis:** Thoroughly testing smart contracts in simulated environments to observe their behavior under various conditions and identify potential flaws during execution.

c. **Manual Code Review:** Highly skilled and independent security experts meticulously examining the code line by line for subtle logical flaws, complex security weaknesses, and unintended behaviors.
d. **Formal Verification:** A more advanced, mathematically rigorous approach to prove the correctness of smart contract logic, though often computationally intensive and complex. Regular, independent audits conducted by reputable third-party security firms are absolutely essential throughout the entire smart contract development and deployment lifecycle to ensure their integrity and security.

8.5.4.2 Secure data feed mechanisms

To effectively combat data feed (oracle) attacks, it is crucial to implement highly secure and inherently reliable oracle solutions [55]. Several strategies can be employed:

a. **Decentralized Oracles:** Utilizing multiple independent oracles to source the same data and then aggregating their responses. This approach significantly reduces reliance on any single point of failure and increases data veracity.
b. **Reputation Systems:** Establishing robust reputation scores or trust mechanisms for oracles, thereby penalizing those that consistently provide incorrect or unreliable data.
c. **Cryptographic Proofs:** Employing advanced cryptographic proofs to verify the authenticity and integrity of off-chain data provided by oracles before it is accepted onto the blockchain.
d. **Hardware Security Modules:** Protecting oracle private keys and ensuring the integrity of their data processing environments through the use of tamper-resistant physical devices.

8.5.4.3 Access limitations and permissioned network design

For most enterprise-level supply chain applications, permissioned blockchains often provide a superior balance of security, performance, and privacy when compared to fully public blockchains [17]. By their very design, permissioned networks incorporate several inherent advantages:

a. **Control Participants:** Only authorized and known entities are permitted to join the network, effectively mitigating the threat of Sybil attacks by preventing the creation of numerous fake identities.
b. **Role-Based Access Control:** Data visibility and transaction permissions can be granularly controlled and assigned based on a participant's specific role within the supply chain, thereby ensuring necessary confidentiality while maintaining transparency for relevant parties.

c. **Stronger Governance:** It is inherently easier to establish, enforce, and adapt governance rules, including clear mechanisms for dispute resolution and network upgrades. This also helps mitigate potential 51% attack scenarios, especially those originating from rogue internal actors within a consortium.

8.5.4.4 Enhanced cryptographic security (quantum resistance)

To strategically prepare for the long-term, theoretical threat posed by the advent of quantum computing, which could potentially break current cryptographic standards, ongoing research and development in quantum-resistant cryptography (PQC) is vitally important [36]. While not an immediate concern for today's conventional computing capabilities, integrating PQC algorithms into future blockchain protocols will become absolutely essential to ensure the enduring security and integrity of sensitive supply chain data against potential future quantum attacks. Organizations should proactively monitor PQC advancements and meticulously plan for necessary future cryptographic upgrades.

8.5.5 Navigating regulatory impacts with solutions

The dynamic and continuously evolving regulatory landscape presents ongoing challenges for widespread blockchain adoption, particularly concerning aspects of data privacy, data sovereignty, and legal liability. Proactive and innovative solutions are therefore necessary to address these complexities.

8.5.5.1 Editable Blockchains for data compliance

The inherent immutable nature of blockchain can, at times, directly conflict with regulations such as the GDPR's "right to be forgotten" provision. To navigate this, several nuanced solutions are being actively explored:

a. **Off-Chain Storage with On-Chain Hashes:** This approach involves storing sensitive, personally identifiable information off-chain in traditional, centralized databases that allow for data modification or deletion. Only cryptographic hashes of this data are then recorded on the immutable blockchain [47]. If the off-chain data needs to be deleted, it can be, and the hash on the chain simply becomes a reference to non-existent data without altering the blockchain itself.
b. **Data Layering:** Architectures are being developed that separate sensitive data onto distinct, private layers within the blockchain ecosystem. These private layers can be modified or even removed, while core,

non-sensitive transaction data remains permanently on the immutable public or shared ledger.

c. **"Right to be Forgotten" Smart Contracts:** For specific use cases where data deletion is legally mandated, smart contracts can be designed to effectively "nullify" or cryptographically obscure certain data points, even if the raw data technically remains on the chain, rendering it unusable or unlinkable.

8.5.5.2 Advanced privacy-enhancing tools

Beyond the inherent capabilities of permissioned networks, advanced cryptographic tools are absolutely key to achieving the delicate balance between transparency and confidentiality. These include:

a. **Zero-Knowledge Proofs (ZKPs):** As previously discussed, ZKPs allow one party to prove the validity of a statement without revealing any sensitive underlying data [33]. This is invaluable for compliance without exposure (e.g., proving a product meets a specific standard without revealing proprietary ingredients).
b. **Homomorphic Encryption (HE):** This advanced technique, while still largely in early research and development for practical blockchain applications, enables computations to be performed directly on encrypted data, thereby preserving privacy during collaborative data analysis [34]. These sophisticated tools empower businesses to leverage blockchain's benefits for verification and transparency while rigorously adhering to strict privacy requirements and protecting sensitive information.

8.5.5.3 Importance of regular compliance audits

Given the dynamic and often inconsistent nature of global regulations, conducting regular and comprehensive compliance audits of blockchain implementations is absolutely crucial [47]. These audits should rigorously:

a. Assess the solution's strict adherence to relevant data privacy laws (such as GDPR and CCPA).
b. Verify the inherent integrity and robust security posture of the blockchain network itself.
c. Ensure that smart contracts function precisely as intended and remain fully compliant with applicable legal frameworks.
d. Review the established governance structures and data access controls within the network. Proactive and ongoing compliance efforts are thus essential not only to avoid significant legal penalties but also to consistently maintain trust among all stakeholders.

8.5.6 Regulatory sandboxes and pilot programs

Governments and regulatory authorities across the globe have increasingly established regulatory sandboxes and initiated pilot programs dedicated to blockchain applications [56]. These initiatives have created controlled environments in which businesses have been able to test blockchain-based solutions under temporarily relaxed regulatory conditions. Such frameworks have enabled experimentation while mitigating immediate legal or financial risks associated with non-compliance. By observing these pilot implementations, regulators have gained valuable insights into the operational, economic, and legal implications of blockchain adoption [58]. At the same time, businesses have benefited by refining their solutions to align with emerging compliance requirements. This collaborative approach has accelerated the development of regulatory clarity, fostered innovation, and laid the groundwork for more effective and supportive policies guiding blockchain integration into supply chain ecosystems [59].

8.6 BLOCKCHAIN CAPABILITIES AND MARKET DYNAMICS

Blockchain's integration into SCM is rapidly moving from a theoretical concept to a practical reality, actively enabling a new generation of sophisticated capabilities and fostering a burgeoning, dynamic market.

8.6.1 Core capabilities enabled by blockchain in supply chains

The fundamental attributes of blockchain coalesce to deliver a powerful suite of core capabilities that fundamentally transform traditional supply chain operations.

8.6.1.1 Advanced traceability and transparency

This is arguably blockchain's most immediate and profoundly impactful capability within SCM. By providing an immutable, distributed ledger, blockchain enables granular, end-to-end traceability of products, their individual components, and even raw materials from their absolute origin [18]. Every single step, from initial sourcing to manufacturing, packaging, shipment, and final delivery, can be meticulously recorded on this ledger. This unprecedented level of transparency allows all authorized participants to access and see the complete, verifiable history of a product, leading to several key benefits:

a. **Authenticity Verification:** Consumers can, for instance, scan a QR code on a product to independently verify its origin and journey, directly combating the proliferation of counterfeit goods.
b. **Recall Efficiency:** In the unfortunate event of product defects or contamination, affected batches can be precisely identified and recalled with far greater efficiency, minimizing waste and mitigating risks.
c. **Ethical Sourcing Proof:** Companies can transparently demonstrate their adherence to fair labor practices, sustainable sourcing methodologies, and compliance with environmental standards through verifiable records.

8.6.1.2 Automation and efficiency gains

The inherent programmability of blockchain, primarily through the use of smart contracts, introduces unprecedented levels of automation into various supply chain processes [16]. This automation, in turn, leads to significant and measurable efficiency gains by:

a. **Eliminating Intermediaries:** Reducing or removing the need for third-party brokers, escrow agents, or other intermediaries for contract execution, thereby streamlining transactions.
b. **Automating Payments:** Facilitating the automatic release of funds upon the verified delivery of goods or the successful completion of predetermined milestones.
c. **Streamlining Documentation:** Enabling the digitization and real-time validation of critical documents, such as bills of lading and customs forms, reducing paperwork and processing delays.
d. **Accelerating Dispute Resolution:** Leveraging automated triggers and verifiable data to significantly reduce the time and associated costs involved in resolving disagreements among supply chain partners.

This pervasive automation minimizes manual errors, substantially accelerates transaction cycles, and frees up valuable human resources, allowing them to focus on more strategic and value-added tasks.

8.6.1.3 Data privacy and security

While transparency is a foundational principle, blockchain also offers robust mechanisms for ensuring data privacy and enhancing security, particularly within permissioned network environments [29].

Cryptographic Security: The use of digital signatures and advanced hashing functions ensures the undeniable integrity of data and provides non-repudiation for transactions.

a. **Immutable Records:** Once data is recorded on the blockchain, it becomes virtually impossible to alter or delete, effectively preventing fraud and unauthorized modifications.
   i. **Permissioned Access:** In private or consortium blockchains, sophisticated access controls can be implemented to ensure that sensitive or proprietary data is only visible to specifically authorized parties, striking a crucial balance between transparency and confidentiality requirements.

b. **Privacy-Enhancing Technologies (ZKPs, HE):** Advanced techniques like ZKPs and HE enable verification or computation without revealing the underlying sensitive information.

8.6.1.4 Enhanced connectivity and interoperability

Blockchain fundamentally fosters the creation of a truly connected supply chain ecosystem by providing a common, trusted, and synchronized data layer [37].

a. **Shared Ledger:** All participating entities operate from the same, consistent, and real-time data, effectively eliminating information silos and resolving data discrepancies across the network.
b. **Standardized Data:** The adoption of blockchain naturally encourages the development and adherence to common data formats and communication protocols among network participants.
c. **Cross-Chain Potential:** As interoperability solutions continue to mature, the seamless exchange of data and assets between disparate blockchain networks will become increasingly prevalent, ultimately leading to the formation of a truly interconnected and globally integrated supply chain. This enhanced connectivity leads to superior collaboration, improved coordination, and a more comprehensive, holistic view of the entire supply chain.

8.6.1.5 Digital asset management and tokenization

Blockchain technology enables the powerful concept of **tokenization**, where physical assets can be represented as their unique digital counterparts on the blockchain ledger [25].

a. **Fungible Tokens:** These can represent quantities of interchangeable items, such as raw materials, commodities, or even standardized units of energy.
b. **Non-Fungible Tokens:** These are used to represent unique products with distinct serial numbers, high-value individual assets, or digital certificates of authenticity for specific items. Tokenization facilitates

precise tracking of ownership, enables streamlined and secure transfer of assets, and opens avenues for the creation of innovative financial instruments directly within the supply chain (e.g., fractional ownership of goods in transit or tokenized invoices for accelerated financing).

8.6.2 Components of the blockchain supply chain market

The emerging blockchain in the supply chain market comprises several key segments that collectively drive its growth, development, and widespread adoption.

8.6.2.1 Software platforms and protocols

This critical segment encompasses the foundational blockchain technologies and frameworks utilized to build diverse supply chain solutions.

a. **Enterprise Blockchain Platforms:** This includes robust platforms like Hyperledger Fabric, R3 Corda, and Quorum, which are specifically designed for consortium-based and private network deployments due to their scalability, privacy, and control features.
b. **Public Blockchain Protocols:** While less common for direct enterprise supply chain operations, protocols like Ethereum are increasingly used for specific use cases, often leveraging private sidechains or Layer 2 scaling solutions for performance and privacy.
c. **Blockchain-as-a-Service:** This offers cloud-based solutions from major providers (e.g., IBM Blockchain Platform, Microsoft Azure Blockchain Service, and Amazon Managed Blockchain), significantly simplifying the deployment, management, and scaling of blockchain networks for businesses.
d. **Middleware and APIs:** These are essential solutions that facilitate seamless integration between blockchain networks and existing traditional enterprise systems, such as ERP and SCM platforms.

8.6.2.2 Hardware (IoT and secure devices)

This segment represents the crucial physical layer that connects real-world events and physical assets to the digital blockchain ledger.

a. **IoT Sensors:** A wide array of devices that collect vital data (e.g., temperature, location, humidity, light, shock, and vibration) from physical goods and assets, and then securely feed this information onto the blockchain for immutable recording.

b. **RFID and NFC Tags:** These technologies are extensively used for unique product identification, enabling a reliable link between physical items and their corresponding digital records on the blockchain.
c. **Hardware Security Modules:** Specialized physical devices that securely store cryptographic private keys and provide trusted execution environments, crucial for maintaining the integrity and security of blockchain operations.
d. **Smart Packaging:** Innovative packaging solutions integrated with embedded sensors or unique digital identifiers that can interact with the blockchain.

8.6.2.3 Professional services

Professional services constitute a vital component for the successful planning, implementation, and ongoing management of blockchain solutions in supply chains.

a. **Consulting and Advisory Services:** Expertise offered to businesses to strategize, design, and meticulously plan their blockchain implementations, aligning technology with business objectives.
b. **System Integration Services:** Crucial services focused on seamlessly integrating new blockchain solutions with existing ERP, SCM, and other legacy IT systems.
c. **Development and Implementation Services:** Tailored services for building custom blockchain applications, developing industry-specific smart contracts, and deploying the network infrastructure.
d. **Support and Maintenance:** Ongoing technical support, system updates, and troubleshooting services essential for the continuous operation and evolution of blockchain networks.
e. **Auditing and Security Services:** Specialized services to ensure the integrity, security, and compliance of blockchain code, smart contracts, and the underlying infrastructure.

8.6.2.4 Key Industry verticals

Blockchain technology is rapidly gaining significant traction and finding diverse applications across a wide range of industry verticals due to its inherent versatility and problem-solving capabilities.

a. **Food and Agriculture:** Enhancing the traceability of produce, meat, and seafood to improve food safety and quality.
b. **Pharmaceuticals and Healthcare:** Crucial for drug anti-counterfeiting efforts, ensuring medical supply chain integrity, and verifying product authenticity.

c. **Retail and Consumer Goods:** Strengthening brand protection, enabling verifiable ethical sourcing, and providing authenticity verification for high-value goods.
d. **Logistics and Transportation:** Optimizing freight tracking, streamlining customs clearance processes, and improving port operations.
e. **Manufacturing:** Ensuring precise component traceability throughout the production process, simplifying warranty management, and enhancing quality control.
f. **Luxury Goods:** Providing irrefutable authenticity and provenance verification for high-end items.
g. **Automotive:** Managing parts traceability for assembly and aftermarket, and streamlining recall management processes.

8.6.3 Market size and growth trajectory (2025 projections)

The global market for blockchain in SCM is currently experiencing robust and accelerated growth. While exact figures may vary between different market research firms, the overarching consensus indicates a rapidly expanding sector. By 2025, the global blockchain in supply chain market is projected to reach several billion US dollars, primarily driven by increasing awareness of its extensive benefits, escalating demand for transparency and traceability, and substantial investments in digital transformation across industries [57,58]. Furthermore, the market is widely anticipated to exhibit a high compound annual growth rate well into the double digits.

8.6.3.1 Long-term market growth forecast (toward 2030)

Looking further into the future, toward 2030, the market for blockchain in supply chains is anticipated to continue its strong upward trajectory, with projections suggesting it could reach tens of billions of US dollars [59]. This substantial long-term growth will be fueled by a confluence of factors:

a. **Maturation of Technology:** Ongoing improvements in the scalability, interoperability, and overall user-friendliness of blockchain platforms will drive broader adoption.
b. **Standardization Efforts:** The continued development of industry-wide technical standards and best practices for blockchain applications within supply chains will foster greater clarity and trust.
c. **Increased Regulatory Clarity:** The emergence of more favorable and consistent regulatory frameworks globally will reduce uncertainty for businesses.

d. **Growing Enterprise Adoption:** Larger corporations are expected to move beyond initial pilot projects toward full-scale, widespread implementations of blockchain solutions.
e. **Convergence with Other Technologies:** Deeper and more seamless integration with complementary technologies such as AI, IoT, and advanced analytics will unlock new capabilities.

8.6.4 Market segmentation: by platform, application, region, and type

Understanding the intricate dynamics of the blockchain in supply chain market requires segmenting it based on various crucial criteria.

8.6.4.1 Platform segmentation

a. **Private/Consortium Blockchains:** These platforms currently hold the dominant market share for enterprise supply chain solutions. Their prevalence is attributed to their superior performance, enhanced privacy features, and greater control mechanisms [17]. Examples include Hyperledger Fabric, R3 Corda, and Quorum.
b. **Public Blockchains (with Scaling Solutions/Sidechains):** While less dominant for core, high-volume supply chain operations, public blockchains are seeing growing adoption for specific use cases. These include the tokenization of assets or public verification of credentials, often facilitated by Layer 2 scaling solutions or private sidechains (e.g., Ethereum-based solutions).

8.6.4.2 Application segmentation

a. **Traceability and Visibility:** This remains the largest and most significant segment of the market, driven by critical demands for anti-counterfeiting measures, improved food safety, and verifiable ethical sourcing.
b. **Smart Contracts and Payments:** This segment focuses on automating transactions, streamlining inter-company payments, and revolutionizing aspects of trade finance.
c. **Supply Chain Financing:** Facilitating faster, more secure, and more accessible financing options for various participants across the supply chain.
d. **Logistics and Transportation:** Aiming to optimize freight tracking, inventory management, and overall transportation efficiencies.
e. **Risk Management and Compliance:** Enhancing overall supply chain resilience and ensuring adherence to increasingly complex regulatory requirements.

8.6.4.3 *Regional market dynamics*

a. **North America:** Currently leads the market, primarily due to early technological adoption, substantial investments in research and development, and a strong presence of key technology providers and solution integrators [57].
b. **Europe:** This region is experiencing rapid growth, largely driven by strict regulatory requirements (e.g., GDPR, stringent food safety standards) and a pronounced focus on sustainability initiatives.
c. **Asia Pacific:** Anticipated to witness the highest growth rate, fueled by its vast manufacturing bases, accelerating digitalization efforts across industries, and supportive government initiatives that actively promote blockchain adoption [58].
d. **Rest of the World:** Emerging markets in other regions are progressively showing increased interest and initiating various pilot projects to explore blockchain's potential in their supply chains.

8.6.4.4 *Blockchain-type segmentation (private, public, hybrid)*

a. **Private Blockchains:** These currently command the largest market share within enterprise supply chains due to their inherent advantages in control, transaction speed, and customizable privacy features.
b. **Hybrid Blockchains:** Expected to gain significant traction, these offer a compelling balance between the transparency characteristic of public chains and the control and privacy offered by private ones. They strategically allow for selective data exposure to public networks while maintaining sensitive data in private, permissioned environments.
c. **Public Blockchains:** While less prevalent for direct, core supply chain operational processes, they are utilized for specific functions such as asset tokenization and the public verification of credentials or certifications, where full transparency is desired.

8.6.5 Key drivers of market growth

Several powerful factors are converging to collectively propel the rapid expansion and widespread adoption of blockchain within the supply chain market.

8.6.5.1 *Increasing cyber-attack incidents*

The rising sophistication and frequency of cyber-attacks targeting centralized databases have significantly increased the urgency for more secure data management systems [60]. Blockchain, with its cryptographic security mechanisms and distributed architecture, has offered resilience against malicious data alterations and unauthorized access. These features have made it a preferred solution for organizations seeking to safeguard critical supply chain information against evolving cyber threats [61].

8.6.5.2 Growing demand for transparency and traceability

Consumers, regulatory bodies, and businesses across all sectors are increasingly demanding greater transparency and end-to-end traceability across supply chains [20]. This rising demand is driven by heightened concerns over product authenticity, critical food safety issues, ethical sourcing practices, and the overall environmental impact of products. Blockchain provides the verifiable, immutable record precisely needed to meet these demands, allowing companies to substantiate their claims with undeniable proof and thereby build deeper trust with their stakeholders.

8.6.5.3 Evolving regulatory landscape

Although regulatory requirements have introduced challenges, they have simultaneously acted as a catalyst for blockchain adoption. Governments and industry regulators have progressively mandated stricter standards for traceability, accountability, and data integrity such as pharmaceutical serialization rules and conflict mineral reporting requirements [47]. Blockchain has provided a structured framework for compliance, enabling organizations to meet increasingly complex regulations with greater efficiency [62].

8.6.5.4 Proliferation of IoT devices

The exponential growth and widespread proliferation of IoT devices provide a vast and continuous stream of real-time data from the physical world [19]. This invaluable data, when securely fed onto a blockchain, profoundly enhances the accuracy, granularity, and timeliness of supply chain records, effectively moving operations toward truly intelligent and automated processes. The powerful synergy between IoT and blockchain creates a compelling combination for achieving precise, granular tracking and generating verifiable, actionable insights throughout the supply chain.

8.6.5.5 Strong consumer demand for verified products

Modern consumers are increasingly conscious and discerning regarding a product's entire journey, its ethical implications, and its environmental footprint. They are often willing to pay a premium for products that offer verifiable proof of high quality, genuine authenticity, and responsible sourcing [20]. This strong and growing consumer demand directly incentivizes and pushes brands to adopt transformative technologies like blockchain to provide the very transparency and verification that builds lasting trust and foster deep brand loyalty.

8.6.5.6 Investor interest and funding trends

Significant venture capital and substantial corporate investment are actively flowing into blockchain startups and initiatives specifically focused on supply chain applications [63]. This substantial influx of funding vigorously accelerates research and development efforts, fosters groundbreaking innovation, and enables new, sophisticated solutions to rapidly come to market. This investment trend further fuels the overall adoption rate of blockchain within SCM.

8.6.5.7 Government initiatives and standards development

Governments worldwide have recognized the strategic importance of blockchain and have launched pilot projects, funded research programs, and supported the creation of technical standards [56]. These initiatives have provided legitimacy to blockchain adoption, encouraged public–private collaboration, and reduced uncertainty by clarifying regulatory frameworks. As a result, governmental support has played a critical role in enabling wider industry adoption and establishing blockchain as a transformative force in SCM [64].

8.6.6 Proposed solution: the integrated blockchain-powered supply chain

The deployment of a blockchain-powered integrated supply chain model has been shown to deliver transformative benefits across multiple dimensions. By ensuring unmatched accuracy, security, and accessibility of data, it has redefined interactions among stakeholders. Supply chain relationships that were previously fragmented and adversarial have been reshaped into collaborative, trust-driven partnerships [65]. This transition has facilitated improved transparency, reduced disputes, accelerated decision-making, and strengthened resilience against disruptions, fundamentally shifting how supply chain ecosystems operate.

8.6.6.1 Impact of the proposed solution

The implementation of an integrated blockchain-powered supply chain model has the potential for a transformative impact across numerous dimensions, directly addressing traditional challenges and amplifying inherent strengths. This approach provides an unparalleled level of data veracity and accessibility, fostering a new era of collaborative trust. It fundamentally redefines how supply chain partners interact, moving from siloed, often adversarial relationships to a more transparent, collaborative, and efficient ecosystem.

The implementation of a comprehensive blockchain-based supply chain model has yielded measurable benefits across operational, financial, and ethical dimensions. Operational efficiency has improved through automation and real-time monitoring, while financial performance has benefited from reduced transaction costs and minimized fraud [66]. Ethical considerations, including verifiable adherence to sustainability standards and responsible sourcing, have also been reinforced, meeting both regulatory requirements and consumer expectations. Collectively, these outcomes demonstrate a strong return on investment and contribute to a more secure, transparent, and sustainable global trade environment [67].

8.7 CONCLUSION AND FUTURE OUTLOOK

Blockchain has emerged as a transformative technology capable of redefining SCM by addressing persistent inefficiencies and introducing enhanced transparency, reliability, and accountability. Although challenges such as scalability, interoperability, and regulatory uncertainty remain, the benefits of blockchain adoption outweigh the risks, making it a compelling tool for future-ready supply chains.

Adopting blockchain in supply chains offers significant opportunities. It improves traceability and visibility by enabling end-to-end tracking of products, allowing organizations to monitor goods at every stage. Increased transparency establishes a single source of truth, fostering trust and accountability among partners. Immutable records reduce counterfeiting and fraud, while smart contracts automate processes, enhancing operational efficiency. Blockchain also strengthens data security through cryptographic protections, supports proactive risk management for greater supply chain resilience, enhances consumer trust and brand loyalty with verifiable claims, accelerates cash flows through streamlined payments, and facilitates verified ethical sourcing and sustainability.

At the same time, organizations face notable challenges. Scalability limitations can affect performance under high transaction volumes, while interoperability is difficult due to fragmented blockchain platforms. Initial development, integration, and training costs can be substantial. Regulatory and legal ambiguity, particularly regarding data privacy, may complicate implementation. Energy consumption is a concern for PoW systems, and the principle of "garbage in, garbage out" highlights the importance of high-quality data, as inaccurate inputs can undermine blockchain benefits.

Managing consortium governance, addressing user resistance, and auditing smart contracts are additional operational considerations.

Successful blockchain adoption relies on several critical factors. Clearly defining a use case ensures that blockchain addresses specific supply chain challenges rather than being implemented arbitrarily. Collaborative ecosystems, with active participation from all key supply chain partners, enhance the technology's effectiveness, with consortium-based networks often being the most efficient model. A robust governance framework with transparent rules, decision-making processes, and dispute resolution mechanisms is essential. Seamless integration with existing enterprise systems such as ERP and SCM, supported by secure APIs, is necessary to maintain operational continuity while leveraging blockchain capabilities. Ensuring high-quality data entry is crucial, as blockchain preserves immutability but cannot correct errors. Regulatory compliance, particularly with evolving data privacy laws like GDPR and CCPA, must be maintained. Security measures, including smart contract auditing, network design, and private key management, should be prioritized. Scalability planning ensures platforms can handle current and future transaction volumes, and structured change management and employee training facilitate adoption. Starting with smaller Proof-of-Value projects allows organizations to demonstrate tangible benefits and refine solutions before broader deployment.

Future research is expected to focus on improving scalability through approaches such as Layer 2 solutions, sharding, new consensus mechanisms, and inter-blockchain communication protocols. Quantum-resistant cryptography will be essential for long-term data security, while deeper integration with AI will support predictive maintenance, demand forecasting, autonomous decision-making, and enhanced anomaly detection. Privacy-enhancing technologies, including ZKPs, HE, and confidential computing, will continue to evolve, allowing sensitive information to be protected without compromising transparency.

Regulatory frameworks are likely to become clearer and more harmonized, addressing data sovereignty, digital identity management, cross-border data flows, and compliance with "right to be forgotten" requirements. Over time, blockchain may enable fully decentralized and self-optimizing supply chains through innovations such as decentralized autonomous organizations, tokenized physical assets, and self-sovereign identities for products and participants.

In conclusion, blockchain represents a fundamental shift in SCM. While its adoption involves technical, operational, and regulatory challenges, the benefits and ongoing innovations suggest that blockchain-powered supply chains are poised to become the standard. This evolution promises more resilient, efficient, transparent, and sustainable global supply chains capable of managing the growing complexity of modern trade.

REFERENCES

1. M. Christopher, *Logistics & Supply Chain Management*. Pearson Education, 2016.
2. R. M. Monczka, R. J. Trent, and R. B. Handfield, *Purchasing and Supply Chain Management*. Cengage Learning, 2021.
3. D. Simchi-Levi, P. Kaminsky, and E. Simchi-Levi, *Designing and Managing the Supply Chain: Concepts, Strategies, and Case Studies*. McGraw-Hill Education, 2013.
4. K. W. Lai and S. W. M. Yuen, "A Survey of Supply Chain Management and the Challenges of Traditional Systems," *Journal of Enterprise Information* Management, vol. 18, no. 5, pp. 586–602, 2005.
5. OECD & EUIPO, *Trade in Counterfeit and Pirated Goods: Mapping the Economic Impact*. Illicit Trade, OECD Publishing, 2019.
6. C. Minervini, "The Impact of Transparency on Consumer Trust and Brand Loyalty," *Journal of Retailing and Consumer Services,* vol. 55, p. 102094, 2020.
7. A. Agrawal, S. K. Singh, and M. Singh, "Sustainable Supply Chain Management: A Review and Research Opportunities," *Journal of Cleaner Production,* vol. 187, pp. 570–588, 2018.
8. D. J. Bowersox, D. J. Closs, and M. B. Cooper, *Supply Chain Logistics Management*. McGraw-Hill Education, 2019.
9. S. Nakamoto, *Bitcoin: A Peer-to-Peer Electronic Cash System*, 2008. https://bitcoin.org/bitcoin.pdf.
10. M. Swan, *Blockchain: Blueprint for a New Economy*. O'Reilly Media, 2015.
11. M. Crosby, P. Pattanayak, M. Graham, and R. Pazandak, "Blockchain Technology: Beyond Bitcoin," *Applied Innovation,* vol. 2, pp. 6–10, 2016.
12. J. R. Narayanan and D. S. Kim, "Blockchain-Based Secure Data Storage in Cloud," *Journal of Computer Science and Technology,* vol. 35, no. 2, pp. 411–427, 2020.
13. K. Christidis and M. Devetsikiotis, "Blockchains and Smart Contracts for the Internet of Things," *IEEE Access,* vol. 4, pp. 5850–5857, 2016.
14. A. Tapscott and D. Tapscott, *Blockchain Revolution: How the Technology Behind Bitcoin Is Changing Money, Business, and the World*. Penguin, 2016.
15. Z. Zheng, S. Xie, H. Dai, X. Chen, and H. Wang, "Blockchain Challenges and Opportunities: A Survey," *International Journal of Web and Grid Services,* vol. 14, no. 4, pp. 352–375, 2018.
16. N. Szabo, "Smart Contracts: Building Blocks for Digital Markets," *Extropy: The Journal of Transhumanist Thought,* no. 16, 1996. https://www.fon.hum.uva.nl/rob/Courses/InformationInSpeech/CDROM/Literature/CfP/html/szabo/smart_contracts_2.html
17. A. Walport, *Distributed Ledger Technology: Beyond Blockchain*. Government Office for Science, 2016.
18. S. Kshetri and N. Voas, "Blockchain-enabled Applications in Supply Chain," *IEEE Internet Computing,* vol. 21, no. 6, pp. 78–83, 2017.
19. R. W. T. Lee, "Internet of Things with Blockchain for Supply Chain Management," In: *Proceedings of 2018 Internatinal Conference on Smart City and Industrial Application (ICSCIA),* Bandung, Indonesia, pp. 104–108, 2018.

20. A. Queiroz, W. S. Wang, and G. Zhou, "Blockchain for Sustainable Supply Chains: A Systematic Review and Future Research Agenda," *Journal of Cleaner Production,* vol. 269, p. 121307, 2020.
21. G. Xu, Z. Zheng, and B. Cao, "Blockchain Technology in Supply Chain Management: A Comprehensive Review," *IEEE Access,* vol. 9, pp. 110595–110617, 2021.
22. D. K. Kouhizadeh and A. P. Sarkis, "Blockchain for the Sustainable Supply Chain: A Critical Review," *Journal of Cleaner Production,* vol. 201, pp. 218–232, 2018.
23. Y. Li, G. Jiang, Z. Cao, and S. Li, "A Blockchain-Based Traceability System for Supply Chain Management," In: *Proceedings of 2018 IEEE International Conference on Systems, Man, and Cybernetics (SMC),* Seagaia Convention Center, Miyazaki, Japan, pp. 1446–1451, 2018.
24. D. Chen, M. Preston, and M. Al-Ansari, "Blockchain and Supply Chain Management: A Review of the Current Landscape," *Computers & Industrial Engineering,* vol. 160, p. 107590, 2021.
25. A. Hawlitschek, B. Notheisen, and H. Teubner, "The Limits of Blockchain for Supply Chain Management," *Logistics,* vol. 3, no. 2, p. 13, 2019.
26. A. Min and Z. H. Zheng, "Blockchain-Based Traceability System for Agri-Food Supply Chain: A Survey," *IEEE Access,* vol. 8, pp. 83888–83907, 2020.
27. Y. Yang, H. Zhang, and X. Yuan, "A Blockchain-Based Traceability System for Pharmaceutical Supply Chain," *Journal of Industrial Information Integration,* vol. 20, p. 100175, 2020.
28. B. Schneier, *Applied Cryptography: Protocols, Algorithms, and Source Code in C.* John Wiley & Sons, 1996.
29. W. A. Stallings, *Cryptography and Network Security: Principles and Practice.* Pearson, 2017.
30. N. Koblitz, "Elliptic Curve Cryptosystems," *Mathematics of Computation,* vol. 48, no. 177, pp. 203–209, 1987.
31. A. Narayanan, J. Bonneau, E. Felten, A. Miller, and S. Goldfeder, *Bitcoin and Cryptocurrency Technologies.* Princeton University Press, 2016.
32. L. Luu, D. Chu, H. Olickel, P. Saxena, and A. Hobor, "Making Smart Contracts Smarter," In: *Proceedings of 2016 ACM SIGSAC Conference on Computer and Communications Security,* Hofburg Palace, Vienna, Austria, pp. 254–269, 2016.
33. S. Goldwasser, S. Micali, and C. Rackoff, "The Knowledge Complexity of Interactive Proof Systems," *SIAM Journal on Computing,* vol. 18, no. 1, pp. 186–208, 1989.
34. C. Gentry, "Fully Homomorphic Encryption with Additively Homomorphic Encryption and Approximate Eigenvalues," *Journal of Cryptology,* vol. 23, no. 2, pp. 238–251, 2010.
35. F. Tian, "A Blockchain-Based Food Supply Chain Traceability System for Food Safety," *Computers and Electronics in Agriculture,* vol. 168, p. 105151, 2020.
36. D. J. Bernstein, J. Buchmann, and E. Dahmen, *Post-Quantum Cryptography.* Springer, 2009.

37. S. Li, J. Zhang, Y. Zhao, and G. Li, "Interoperability in Blockchain Systems: A Survey," *IEEE Access,* vol. 8, pp. 19717–19731, 2020.
38. M. Kshetri and N. Voas, "Blockchain-enabled Supply Chain: A Strategic Perspective," *Journal of Enterprise Information Management,* vol. 32, no. 6, pp. 936–953, 2019.
39. K. V. S. R. Krishna, "Blockchain and Its Applications in Supply Chain Management," *International Journal of Computer Science and Engineering,* vol. 7, no. 5, pp. 183–189, 2019.
40. A. K. Singh and V. Kumar, "Integration of Blockchain with ERP Systems for Enhanced Supply Chain Transparency and Efficiency," *Procedia Computer Science,* vol. 170, pp. 690–696, 2020.
41. Y. Min, Q. Zhang, and J. Liu, "A Blockchain-Based Traceability System for Agricultural Products Using QR Code and RFID," In: *Proceedings of 2019 IEEE 4th International Conference on Cloud Computing and Big Data Analytics (ICCCBDA),* Chengdu, Sichuan, China, pp. 317–321, 2019.
42. T. P. G. Choi, H. S. L. Lam, and R. Ning, "Blockchain in Supply Chain Management: A Review and Research Agenda," *Computers & Industrial Engineering,* vol. 150, p. 106842, 2020.
43. B. P. A. Tan, S. L. L. Pan, J. I. T. Zhang, and H. J. T. Lu, "Blockchain and Trade Finance: A Review," *Journal of Financial Economics,* vol. 14, no. 1, pp. 1–20, 2020.
44. AgriDigital, *Digitising the Global Grain Supply Chain.* 2025. https://www.agridigital.io/
45. Intel, *Intel Explores Blockchain for Food Safety and Supply Chain Transparency.* 2025. https://www.intel.com/
46. X. He, C. Liu, and Q. Zhang, "Research on Blockchain Scalability Issues," In: *Proceedings of 2019 IEEE International Conference on Blockchain and Cryptocurrency (ICBC),* Seoul, South Korea, pp. 165–170, 2019.
47. L. T. Xu, K. H. Lu, and E. Lu, "Blockchain-Enabled Data Privacy and Security in Supply Chain Management: A Review," *Journal of Cleaner Production,* vol. 272, p. 122709, 2020.
48. A. de Vries, "Bitcoin's Energy Consumption is Underestimated: A Market-Wide Survey," *Joule,* vol. 3, no. 7, pp. 1599–1605, 2019.
49. M. M. Queiroz, R. Telles, and S. H. Bonilla, "The Economic Impact of Blockchain Technology on Supply Chains," *Supply Chain Management: An International Journal,* vol. 25, no. 5, pp. 627–640, 2020.
50. A. M. Horsch, S. M. Muth, and P. Spiess, "Blockchain Governance: A Multi-Dimensional Framework," *Journal of Business Research,* vol. 116, pp. 586–599, 2020.
51. J. L. Zhou, "Data Quality and Security in Blockchain-Based Supply Chain Management: A Survey," *Journal of Industrial Information Integration,* vol. 20, p. 100185, 2020.
52. K. M. Ozkeskin and H. V. Sari, "Challenges and Drivers of Blockchain Adoption in Supply Chain Management: An Empirical Study," *International Journal of Production Economics,* vol. 237, p. 108119, 2021.

53. C. Decker and R. Wattenhofer, "A Farewell to the Wild West? Exploring the Impact of 51% Attacks on Bitcoin," In: *Proceedings of 2013 IEEE International Conference on Peer-to-Peer Computing (P2P),* Trento, Italy, pp. 1–10, 2013.
54. J. R. Douceur, "The Sybil Attack," In: *Proceedings of 1st Interntional Workshop on Peer-to-Peer Systems (IPTPS '02),* Cambridge, Massachusetts, USA, pp. 251–260, 2002.
55. C. Fan, S. Ghaemi, H. Khazaei, and P. Musilek, "Blockchain Oracles: A Review," *IEEE Access,* vol. 8, pp. 19732–19748, 2020.
56. Deloitte, *Blockchain Regulatory Sandboxes: Promoting Innovation While Managing Risks.* Deloitte Global Services Limited, 2020.
57. MarketsandMarkets, *Blockchain in Supply Chain Market by Provider, Application, Organization Size, and Region–Global Forecast to 2025.* MarketsandMarkets Research Pvt. Ltd, 2020.
58. Grand View Research, *Blockchain in Supply Chain Market Size, Share & Trends Analysis Report by Application, by End-use, by Region, and Segment Forecasts, 2020–2027.* Grand View Research Inc, 2020.
59. Statista, *Blockchain in Supply Chain Market Size Worldwide from 2020 to 2028.* 2023. https://www.statista.com/statistics/1230117/blockchain-in-supply-chain-marketsizeworldwide/.
60. IBM, *Cost of a Data Breach Report.* 2023. https://www.ibm.com/security/data-breach/.
61. S. Rezaeinejad, K. Rayman, B. Sauser, et al. ,"How can Blockchain Contribute to Cyber Resilience? Supply Chain Analytics and Digitalization Benefits", *Inf Syst Front,* 2025, https://doi.org/10.1007/s10796-025-10651-w
62. Md Ruhul Amin, Nahin Akhtar, Md Ekramul Hoque, Nabil, A. R. & Kazi Md Shahadat Hossain, " Blockchain-Enabled Traceability in Pharmaceutical Supply Chains: An Integrated Engineering and It Management Framework for Regulatory Compliance and Pandemic Resilience", *Journal of Computer Science and Technology Studies*, vol. 7, no.10, pp. 343–356, 2025. https://doi.org/10.32996/jcsts.2025.7.10.37
63. Accenture, *Blockchain Technology: A Foundation for Digital Trust.* Accenture plc, 2020.
64. K. E. Busch, *Blockchain: Novel Provenance Applications.* Congressional Research Service (CRS) Report R47064, U.S. Congress, April 12, 2022. https://www.congress.gov/crs_external_products/R/PDF/R47064/R47064.3.pdf
65. A. S. Salama and M. A. Alquraan, An Integrated Cloud-based Blockchain Model for Supply Chain Management. *Journal of Supply Chain and Customer Relationship Management*, vol. 14, no. 1, pp. 1–15, 2024. https://www.aasmr.org/jsms/Vol14/No.1/Vol.14%20No.1.21.pdf
66. R. M. Difrancesco, P. Meijer, and M. Dijk, How blockchain technology improves sustainable supply chain processes: a practical guide. *Logistics*, vol. 6, no. 3, p. 52, 2022. https://pmc.ncbi.nlm.nih.gov/articles/PMC9797894
67. T. S. Douglas, *When Is(n't) Blockchain Right? Global Supply Chain Institute White Paper,* University of Tennessee, Haslam College of Business, 2021. https://haslam.utk.edu/gsci/publication/2021-04-when-isnt-blockchain-right

Chapter 9

Blockchain and IoT

Successful enablers of sustainable food supply chain

Yuvaraj Mohanselvam and Richa Choudhary

9.1 INTRODUCTION

From an estimated 2.5 billion in 1950 to an anticipated 8.5 billion in 2030, the global population has expanded at a fast pace. Sustainability, digitization, food safety, and the critical need for a more efficient Food Supply Chain (FSC) are just a few of the many difficulties that the FSC is currently facing [1]. However, in 2019, 931 million tons of food were wasted, with 61% coming from households, 26% from restaurants, and 13% from grocery stores, according to a UNEP study [2]. Sustainability, resilience, and transparency in FSC have become a top priority in this rapidly evolving environment. The most important issues that FSC faces in the current world are information sharing and the manual entry and retrieval of data in the register. These issues create complicated situations for all the stakeholders in the FSC in tracking the food products [3]. When these issues are addressed, food safety is improved by reducing food loss and food waste. This is possible only when technology comes into the picture of FSC. Though there are many technologies available, blockchain technology (BCT) and Internet of Things (IoT) are more relevant and secure in the digital supply chain, especially the cold chain. BCT assures the stakeholders by documenting all their transactions, whether regarding origin, transport, storage, or compliance, that these cannot be tampered with. Meanwhile, IoT collects real-time data about physical surroundings with the help of devices like sensors and Radio-Frequency IDentification (RFID) tags. These sensors check and transmit valuable data, including temperatures, humidity, and location, to reduce spoilage, preserve quality, and utilize resources more efficiently [4].

The farmers are highly affected due to high storage costs and transportation costs. The farmers in the current supply chain are more negatively impacted than other supply chain partners because of a lack of information openness. Nowadays, even consumers are eager to know all the information about agri-food, such as when it is harvested, where it is coming from, and the temperature maintained throughout the value chain [5]. The adoption of technologies like IoT and BCT, to get real-time quantitative and qualitative information about fruits while storing and transporting in the value

DOI: 10.1201/9781003660163-9

chain, will help to minimize wastage to a greater extent. An IoT-enabled system is designed for monitoring real-time temperature, relative humidity, and gas concentration to increase Fruits & Vegetables (FV) profitability. The IoT system offers helpful data for bolstering the logistics network and assessing the benefits in the FSC [6]. In BCT, with smart contracts, compliance becomes automated, alerts are triggered, and recall functions are managed automatically. An IoT-based FSC to record real-time data that include temperature, gas concentration, and relative humidity is implemented in order to lower operating costs and boost FSC revenues. Logistics systems powered by IoT safeguard the cold chain's integrity. When it comes to implementing a BCT–IoT-based FSC, many obstacles remain, including the requirement for ownership and information management solutions, fresh market potential buyers, and security. This chapter addresses the advantages and challenges of integrating BCT–IoT in the FSC, and a framework is proposed for implementing the integrated BCT–IoT in the FSC to reduce food loss and food waste, thereby achieving the Sustainable Development Goal (SDG) of Zero Hunger by improving food security.

9.2 TECHNOLOGY OVERVIEW

9.2.1 BCT in FSC

A tamper-proof, transparent, and secure digital ledger called blockchain is used to record transactions. With blockchain, information added can never be removed or changed, in contrast to traditional databases where data can be changed or manipulated [7]. From farm fields to the table, BCT can provide better tracking of agri-foods. Food safety issues brought on by willful lying, shoddy planning, and a lack of regulations may be significantly reduced by BCT. This reduces the chances of fraud, data manipulation, and lack of accountability. Smart contracts are one of the most significant elements of BCT. They are self-executing contracts that automate processes such as compliance, activate quality checks, and provide auto payments in the financial settlements when specific requirements are fulfilled [8]. Traceability is another important element in BCT that creates permanent and real-time information about a product's origin, handling, and transportation. During the 2018 E. coli outbreak in the U.S. lettuce supply, traditional methods took nearly a week to trace the source. In contrast, BCT solutions trace the source in seconds [9].

Additionally, BCT helps prevent fraud. Its transparency and tamper-proof design stop counterfeiting, mislabeling, and unauthorized product substitutions. In compliance and quality management, regulatory documents and third-party certifications, like organic or safety audits, are securely digitized. This makes them auditable and accessible, reducing reliance on physical paperwork and lowering operational and compliance costs [10]. In Brazil, a BCT-based certification network for GM-free soybean exports allowed price premiums of up to 15% where consumers and importers trusted the authenticity and ethical sourcing of the food products [11]. The tamper-proof nature of BCT ensures product integrity. This increases market trust and reduces food fraud. Research in manufacturing and agri-food logistics showed that adopting BCT significantly cut inventory lead times, improved data synchronization, and reduced losses during product recalls [12]. These operational improvements directly support environmental and economic sustainability. BCT builds consumer trust by letting end-users and retailers verify product origins, environmental impact, and certifications independently. They can do this using user-friendly tools like QR codes or Near-Field Communication (NFC) tags. BCT enhances food safety, accuracy of recalls, traceability between farm and table, verifies origins, and compliance with regulations. All this contributes to constructing a more transparent and resilient FSC [13].

9.2.2 IoT in FSC

IoT is a network that consists of interconnected physical devices with sensors and software. This arrangement enables them to collect and share information among themselves. The IoT is a major factor in transforming FSC as it enables real-time tracking, automation, and intelligent decision-making. Cold chain monitoring is among the most valuable IoT applications. In this case, temperature and humidity are monitored continuously to ensure that perishable items, specifically seafood, dairy products, and fresh produce, never spoil because even minimal changes in temperature are detrimental. It has been demonstrated that such monitoring can lead to a significant decrease in food waste and better quality in storage and transport [14]. This capability significantly reduces the risks of spoilage, contamination, and foodborne diseases. Also, RFID tags and GPS systems assist in the accurate tracking of inventory, transportation, and logistics. This enhances transparency and minimizes food loss resulting from delays or misplaced goods. Precision farming is applied in agriculture to soil moisture sensors, weather trackers, and nutrient sensors. Information collected from IoT sensors poses a powerful means to improve the economic performance of the FSC. The real-time monitored parameters are converted into useful information via sensors such as the temperature sensor, carbon dioxide sensor, humidity sensor, and volatile organic compound sensor. Sensors such as visible, Fourier transform infrared, and near-infrared spectroscopy are used to quickly grade fruits and assess how fresh processed goods are [15].

The compact size and inexpensive cost of e-nose sensors are crucial because they may be mounted in shipping boxes, freezers, cold storage rooms, and carriages and can differentiate between the edible coatings on agri-foods, especially fruits and vegetables. Information on food safety is sent directly to a central repository via in-packaging printed sensors [16]. IoT sensors have been associated with 30% less fish mortality and 40% reduced disease outbreaks in the seafood supply chain, which have resulted in improved efficiency and healthier ecosystems [17]. IoT enhances customer trust in the FSC by tracking the conditions of food from farm to fork and converting the supply chain to a value chain [18].

9.2.3 The need for integration

The combination can be applied to the automated validation of the data collected via the IoT and verified by the BCT. It simplifies the process of end-to-end traceability and sustainability enhancement [19].

9.3 COMPARATIVE ANALYSIS: STANDALONE VS. INTEGRATED

Though each of the BCT and IoT can contribute to the improvement of FSC independently, their separation restricts the possibilities. Conversely, when these technologies are combined, shared benefits are formed, which enhance transparency, responsiveness, trust, and sustainability significantly, and the same have been discussed in Table 9.1.

Table 9.1 Standalone vs. integrated BCT–IoT

Aspect	*BCT only*	*IoT only*	*Integrated BCT+IoT*
Traceability	Static, historical records	Real-time, but not auditable	Real-time, verifiable, and tamper-proof
Data trust	Vulnerable to false inputs	Vulnerable to manipulation	Secured by sensor authentication and immutability
Automation	Limited without sensor triggers	Requires manual processing	Smart contracts triggered by sensor data
Sustainability reporting	Possible but unverifiable	Possible but fragmented	Auditable and evidence-based
Food waste reduction	Low impact	Medium impact	High impact via proactive alerts and secure sharing
Consumer trust	Limited to label info	Limited unless linked to secure data	High – QR-based transparency and verified origins

9.3.1 Limitations of using standalone BCT and IoT in FSC

9.3.1.1 Limited data trust or validity

BCT relies on manual or unverified data without the use of IoT. This also influences the quality of records, and records can still be spoiled due to human error or intentional fraud. For example, a BCT may show that the produce was transported at the right temperature, but without IoT sensors, this data cannot be automatically checked or fixed in real time [20].

9.3.1.2 Lack of data integrity and transparency

When IoT runs without BCT, sensor data often gets stored in centralized or isolated systems. This raises the risk of data manipulation, cyber-attacks, and selective reporting. These issues harm the transparency and accountability needed for food safety certifications, sustainability claims, and ethical sourcing verification [21].

9.3.1.3 Missed automation opportunities

Smart contracts need event triggers to carry out tasks automatically. Without IoT sensors providing real-time input, such as temperature breaches, smart contracts cannot work on their own. Also, if IoT is used without BCT, there is no decentralized rule enforcement or multi-party trust in automated actions.

9.3.1.4 Inefficient decision-making and recall processes

If data is in real time through IoT but not securely stored or shared without BCT, traceability and recalls become slow and unreliable. This is particularly troubling for perishable goods or contamination crises, where delays can cost lives and result in food waste [22].

9.3.2 Advantages of integrated BCT and IoT in FSC

BCT and IoT represent a closed-loop and intelligent supply system that is transparent when combined. The use of data is created automatically, stored safely, and used, and new data is created at a continuous pace.

9.3.2.1 End-to-end real-time traceability

The supply chain of food products can be monitored in real time by using the integrated BCT–IoT. The IoT sensors capture certain data like temperature and humidity that are very critical in perishable products.

The collected information from IoT sensors is stored safely in BCT. This stored information is transparent and accessible to all the members of the FSC [23].

9.3.2.2 *Enhanced food safety and quality control*

Automatic warnings and rerouting logistics can be issued on the smart contracts with BCT. In case of any deviation from the required conditions reported by IoT devices, such as a cold chain breach, a warning is issued to all the stakeholders in the FSC. These automatic warnings ensure proactive quality control measures, reducing the opportunities for spoiled products getting to the market and ensuring the health of consumers [24].

9.3.2.3 *Reduction in food waste and carbon emissions*

By using real-time IoT data, the participants in the supply chain will manage the inventory level, predict the shelf life, and divert excess inventory prior to spoilage. BCT can ensure the authenticity of such information, which gives the opportunity to organize the activities of interested parties. The integrated systems for reducing food waste were demonstrated in various pilot studies. This assists in environmental sustainability and helps in achieving SDGs [25].

9.3.2.4 *Trust, certification, and consumer confidence*

BCT provides authentic and verifiable assurance of sustainability, such as organic and fair trade. IoT gives supporting information like farm origin and surveillance of environmental conditions in the form of GPS. These enable transparency, in the form of QR codes at the point of sale, thereby enhancing customer trust. This enhances brand loyalty and has a positive effect on responsible consumption [26].

9.4 SUSTAINABILITY DIMENSIONS ENABLED BY INTEGRATED BCT–IoT

The integrated BCT–IoT is trending toward ensuring sustainability in FSCs. This integration is useful in operations and enhances the triple bottom line of sustainability: environmental protection, social responsibility, and economic viability.

9.4.1 Environmental sustainability

Integrated BCT–IoT helps in monitoring the key environmental factors in the FSC in real time. With the integration, energy usage can be measured, CO_2 emissions can be calculated, and changes in temperature during storage and transportation can be recorded. BCT can store this information indefinitely and can be used to report on the carbon footprint that can help companies meet SDG goals [27]. In addition, the integrated technology may enhance the efficiency of logistics and storage to reduce the spoilage and waste of food since these are the primary sources of greenhouse gas emissions. IoT devices will be able to provide automatic alerts, and smart contracts will be able to save resources and proactively manage inventory, reducing unnecessary energy and material use [28].

9.4.2 Social sustainability

An FSC has social sustainability that implies a moral sourcing operation, food security, employee rights, and consumer trust. The lifetime database of BCT will be able to verify the provenance of products, thus that which is labeled as organic, fair trade, or cruelty-free could actually be that. This improves the chain of production, farm to fork, and transparency and accountability to empower consumers to make informed decisions. In addition, by tracing food to its source, it minimizes food fraud, contamination, and recalls. Integrated together with IoT-based quality sensors, BCT systems will have an opportunity to offer food safety and transmit automated alerts and recall management to save consumers and brand image [29]. Moreover, BCT traceability may add to social causes, such as ensuring that labor standards are met and additional inclusiveness of the FSC to smallholder farmers, especially those in developing countries.

9.4.3 Economic sustainability

The integrated BCT–IoT enhances efficiency and decision-making that leads to sustainability in the long run. BCT minimizes the need for intermediaries because it establishes direct and verifiable trust between the parties, and IoT provides sufficient and timely information. This avoids conflict accelerates payment, and improves logistics [30]. Also, there are new value-making opportunities from these technologies that are driven by data monetization, digital certification services, and new financing models of agriculture. An illustration of this is the supply chain data using BCT, which has been verified and can be used by financial institutions to provide microloans or insurance to farmers based on their performance indicators and traces [31]. BCT and IoT systems reduce operational risks and increase the level of transparency, generating more confidence between investors and consumers, which is needed for the long-term sustainability of agri-food businesses.

9.4.4 Alignment with UN sustainability development goals

The integrated BCT–IoT helps in achieving the United Nations SDGs and, therefore, contributes to global sustainability efforts. Specifically, it helps with Zero Hunger, as it makes the distribution of food more equitable through the improvement of distribution efficacy and the tracing of food to the end. It also helps in achieving the second SDG, responsible consumption and production, since it helps to reduce food wastage and ensure food circular systems with real-time monitoring. This integration will facilitate monitoring greenhouse gas emissions and environmental compliance in the supply chain, which helps in the third SDG of climate action [32].

9.5 FRAMEWORK FOR IMPLEMENTING INTEGRATED BCT AND IOT IN SUSTAINABLE FSCs

To create a sustainable FSC, BCT and IoT are integrated into a framework. Figure 9.1 illustrates the proposed framework for combining BCT and IoT in sustainable FSCs. The process starts with understanding the sustainability goals and traceability needs specific to the food supply network. Next, we identify key inefficiencies and vulnerabilities, such as fraud, waste, and lack of transparency, that affect current operations. A thorough mapping of the supply chain stages and their data needs from production to consumption lays the groundwork for digital change. Afterward, we assess IoT readiness, focusing on sensor deployment, connectivity infrastructure, and data collection abilities throughout the FSC. At the same time, we evaluate how BCT technology applies, looking at data immutability, smart contract capabilities, and consensus methods.

A feasibility study includes all stakeholders and incorporates pilot tests on selected food products or regions to confirm the proposed integration. This stage also identifies key enablers, such as supportive policy frameworks, technological infrastructure, and workforce training, along with barriers like financial limitations, gaps in digital skills, and resistance to change. Based on the results from the pilot phase, we refine the implementation strategy to ensure it can scale and be effective.

The final steps involve planning for system-wide integration, adopting a phased deployment strategy, and providing continuous training. Ongoing monitoring of key performance indicators, such as transparency, waste reduction, and emissions control, is crucial to assess impact and encourage ongoing improvement. This will help facilitate the transition to a more resilient and sustainable FSC.

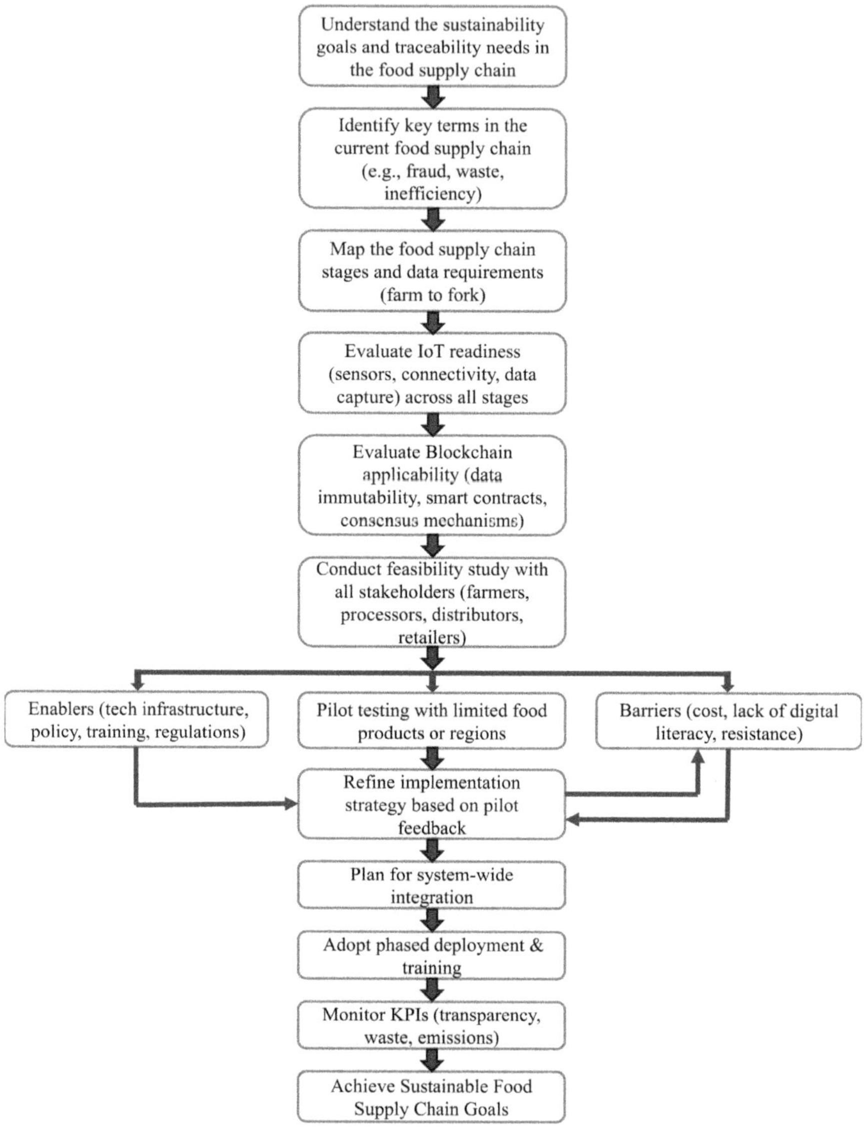

Figure 9.1 Framework for implementation.

9.5.1 Proposed technical integration of BCT–IoT in sustainable FSC

Figure 9.2 illustrates the role of IoT sensors in the FSC in collaboration with the Message Queuing Telemetry Transport (MQTT) protocol, the Ethereum BCT, and smart contracts. IoT detectors on the farm monitor the

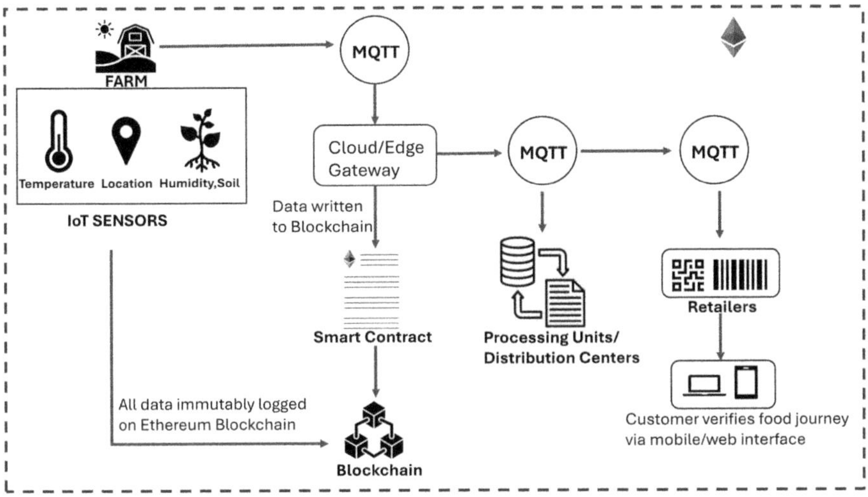

Figure 9.2 Proposed technical integration.

environmental conditions, including temperature, soil conditions, and GPS position. This streaming data is transmitted through the MQTT protocol to an edge gateway or a cloud. This gateway receives the input and transmits the data to the BCT. The cloud or edge gateway links to Ethereum BCT, where the data is stored securely using smart contracts. These contracts are automatic, and they ensure that conditions set, such as organic farming standards, are adhered to. The data is then irreversibly registered on the BCT of Ethereum, and it cannot be edited or altered in any manner. More data is gathered by IoT sensors as food items pass through processing and distribution, such as temperature and GPS position, and is transmitted via the same MQTT protocol. QR codes and RFID tags in the retail sector associate every product with the corresponding digital identity on the BCT. By reviewing this permanent BCT record, consumers can verify the path and authenticity of the food via mobile or web platforms. This system enhances traceability, food safety, reduces fraud potential, promotes sustainable operations, and contributes to consumer trust; it contributes to making the food chain more open and just.

9.6 CHALLENGES AND BARRIERS IN IMPLEMENTING INTEGRATED BCT AND IoT IN SUSTAINABLE FSC

The integration of BCT and IoT in FSCs has several challenges in practical implementation, as detailed in Table 9.2. Large-scale adoption remains limited because of technical, economic, and regulatory challenges.

Table 9.2 Major challenges and mitigation strategies in BCT–IoT integration

Category	*KPI*	*Measurement metric*
Environmental	Carbon emissions	CO_2 equivalent (kg) reduction
Economic	Waste reduction	% decrease in spoilage losses
Social	Traceability coverage	% of products fully traceable
Operational	Data latency	Average data update time (seconds)
Technological	System uptime	% availability of blockchain–IoT platform

Both technologies need good infrastructure, standardized procedures, and a high degree of digital maturity, which is lacking in rural regions. Interoperability between different systems and IoT devices is a major technical challenge. The costs incurred for implementing the integrated BCT–IoT in FSC are one of the key obstacles. Many small and medium businesses and small farm owners cannot afford the costs of the IoT hardware, BCT development, cloud services, and ongoing maintenance. The lack of a clear return on investment makes it less attractive to adopt, particularly in low-margin supply chains. Another crucial point is standardization, as the lack of global communication standards, data formats, and interoperability models results in fragmented ecosystems, making it difficult to transfer information across different platforms. The need to scale also makes adoption harder, with public blockchains potentially suffering from latency, high transaction costs, and throughput issues in dealing with the huge volumes of real-time IoT data. The performance bottlenecks limit the efficiency of the system, particularly in global supply networks. Posting sensitive information in decentralized registers not only exposes cybersecurity threats and privacy issues but also questions trustworthiness. The insecurity of IoT devices, due to the absence of standards, exposes them to cyber problems, and the lack of security poses additional security problems. Many of the participants, especially smallholders and cooperatives, do not have the skills, training, and financial aid to adopt these technologies. The majority of farmers are not ready to use these technologies unless the short-term gains are evident or the government is ready to offer incentives. To tackle these concerns, there is a need to establish a multi-stakeholder partnership to develop cost-effective models, standardized frameworks, and user-friendly and scalable solutions that may guarantee fair digital inclusion throughout the FSC.

9.7 CONCLUSION AND FUTURE SCOPE

IoT combined with the BCT will be an effective solution to the existing inefficiencies and sustainability issues of the FSCs. The integrated technology enables various sustainability goals at a higher level of traceability and real-time tracking, automation of compliance, and decisions based on data. It will entail minimization of food waste, carbon emissions, transparency, and consumer trust. The integrated technology can achieve ethical sourcing, food quality standards, and environmental compliance through the safe capture, validation, and sharing of supply chain information with different stakeholders. A framework is proposed that acts as a guide in the implementation of the integrated BCT–IoT in the FSCs. It can fit all food environments, with its general structure adaptable to small-scale networks as well as large-scale logistics. This plays a powerful tool for practitioners who aim to increase transparency, traceability, and sustainability. The framework helps in achieving the SDGs and achieving food security.

REFERENCES

1. Lezoche, Mario, Jorge E. Hernández, María del Mar Eva Alemany Díaz, Hervé Panetto, and Janusz Kacprzyk. (2020, May). Agri-food 4.0: A survey of the supply chains and technologies for the future agriculture. *Computers in Industry*, 117, 103187. https://doi.org/10.1016/j.compind.2020.103187.
2. Desai, Bharat Harshadray. (2021, December 1). United Nations Environment Programme (UNEP). *Yearbook of International Environmental Law*, 32(1), 293–298. https://doi.org/10.1093/yiel/yvac039.
3. Vasileiou, Marios, Leonidas Sotirios Kyrgiakos, Christina Kleisiari, Pantelis Z. Lappas, Christos Tsinopoulos, Georgios Kleftodimos, Amalia Ntemou, Dimitrios Kateris, Christina Moulogianni, and George Vlontzos. "Digital Transformation of Food Supply Chain Management Using Blockchain: A Systematic Literature Review towards Food Safety and Traceability." *Business & Information Systems Engineering*, June 12, 2025. doi:10.1007/s12599-025-00948-0.
4. Kolikipogu, Ramakrishna, None Shivaputra, Elangovan Muniyandy, Jnaneshwar Pai Maroor, Gundala Venkata Rama Lakshmi, Bhagavan Konduri, and R. Naveenkumar. "Improving Food Safety by IoT-Based Climate Monitoring and Control Systems for Food Processing Plants." *Remote Sensing in Earth Systems Sciences*, January 13, 2025. doi:10.1007/s41976-024-00190-4.
5. Barrett, Christopher B., Thomas Reardon, Johan Swinnen, and David Zilberman. "Agri-Food Value Chain Revolutions in Low- and Middle-Income Countries." *Journal of Economic Literature* 60, no. 4 (December 1, 2022): 1316–77. doi:10.1257/jel.20201539.
6. Spitalleri, Anita, Isaak Kavasidis, Vincenzo Cartelli, Rossella Mineo, Francesco Rundo, Simone Palazzo, Concetto Spampinato, and Daniela Giordano. "BioTrak: A Blockchain-Based Platform for Food Chain Logistics Traceability." *arXiv (Cornell University)*, January 1, 2023. doi:10.48550/arxiv.2304.09601.
7. Rajput, Devraj V., Pavankumar R. More, Preeti A. Adhikari, and Shalini S. Arya. "Blockchain Technology in the Food Supply Chain: A Way towards Circular Economy and Sustainability." *Sustainable Food Technology*, January 1, 2025. doi:10.1039/d5fb00065c.

8. Duong, Cong Doanh, Thanh Tung Dao, Trong Nghia Vu, Thi Viet Nga Ngo, and Minh Hoa Nguyen. "Blockchain-Enabled Food Traceability System and Consumers' Organic Food Consumption: A Moderated Mediation Model of Blockchain Knowledge and Trust in the Organic Food Chain." *Sustainable Futures,* September 1, 2024, 100316. doi:10.1016/j.sftr.2024.100316.
9. Irvin, Kari, Stelios Viazis, Angela Fields, Sharon Seelman, Karen Blickenstaff, Ellen Gee, Matthew E. Wise, Katherine E. Marshall, Laura Gieraltowski, and Stic Harris. "An Overview of Traceback Investigations and Three Case Studies of Recent Outbreaks of Escherichia Coli O157:H7 Infections Linked to Romaine Lettuce." *Journal of Food Protection* 84, no. 8 (April 9, 2021): 1340–56. doi:10.4315/jfp-21-112.
10. Ellahi, Rizwan Matloob, Lincoln C. Wood, and Alaa Ei-Din A. Bekhit. "Blockchain-Driven Food Supply Chains: A Systematic Review for Unexplored Opportunities." *Applied Sciences* 14, no. 19 (October 4, 2024): 8944. doi:10.3390/app14198944.
11. Lucena, Percival, Alécio P. D. Binotto, Fernanda Da Silva Momo, and Henry M. Kim. "A Case Study for Grain Quality Assurance Tracking Based on a Blockchain Business Network." *arXiv (Cornell University)*, January 1, 2018. doi:10.48550/arxiv.1803.07877.
12. Gupta, Anchal, Rajesh Kumar Singh, and Muhammad Mustafa Kamal. "Blockchain Technology Adoption for Secured and Carbon Neutral Logistics Operations: Barrier Intensity Index Framework." *Annals of Operations Research,* February 26, 2024. doi:10.1007/s10479-024-05824-w.
13. Hastig, Gabriella M., and Man Mohan S. Sodhi. "Blockchain for Supply Chain Traceability: Business Requirements and Critical Success Factors." *Production and Operations Management* 29, no. 4 (December 16, 2019): 935–54. doi:10.1111/poms.13147.
14. Da Costa, Tamíris Pacheco, James Gillespie, Xavier Cama-Moncunill, Shane Ward, Joan Condell, Ramakrishnan Ramanathan, and Fionnuala Murphy. "A Systematic Review of Real-Time Monitoring Technologies and Its Potential Application to Reduce Food Loss and Waste: Key Elements of Food Supply Chains and IoT Technologies." *Sustainability* 15, no. 1 (December 29, 2022): 614. doi:10.3390/su15010614.
15. Rejeb, Abderahman, John G. Keogh, Suhaiza Zailani, Horst Treiblmaier, and Karim Rejeb. "Blockchain Technology in the Food Industry: A Review of Potentials, Challenges and Future Research Directions." *Logistics* 4, no. 4 (October 26, 2020): 27. doi:10.3390/logistics4040027.
16. Sharma, Shrawan Kumar. "Smart Agriculture Revolution." In *Smart Agriculture Revolution*, 253–88, 2025. doi:10.1002/9781394287062.ch10.
17. Kumar, Jeetendra, Rashmi Gupta, Suvarna Sharma, Tulika Chakrabarti, Prasun Chakrabarti, and Martin Margala. "IOT-Enabled Advanced Water Quality Monitoring System for Pond Management and Environmental Conservation." *IEEE Access* 12 (January 1, 2024): 58156–67. doi:10.1109/access.2024.3391807.
18. Feng, Huanhuan, Xiang Wang, Yanqing Duan, Jian Zhang, and Xiaoshuan Zhang. "Applying Blockchain Technology to Improve Agri-Food Traceability: A Review of Development Methods, Benefits and Challenges." *Journal of Cleaner Production* 260 (March 12, 2020): 121031. doi:10.1016/j.jclepro.2020.121031.

19. Ahamed, N. Nasurudeen, Ramakrishnan Vignesh, and Tanweer Alam. "Tracking and Tracing the Halal Food Supply Chain Management Using Blockchain, RFID, and QR Code." *Multimedia Tools and Applications* 83, no. 16 (November 3, 2023): 48987–12. doi:10.1007/s11042-023-17474-4.
20. Bhat, Showkat Ahmad, Nen-Fu Huang, Ishfaq Bashir Sofi, and Muhammad Sultan. "Agriculture-Food Supply Chain Management Based on Blockchain and IoT: A Narrative on Enterprise Blockchain Interoperability." *Agriculture* 12, no. 1 (December 30, 2021): 40. doi:10.3390/agriculture12010040.
21. Kaur, Amanpreet, Gurpreet Singh, Vinay Kukreja, Sparsh Sharma, Saurabh Singh, and Byungun Yoon. "Adaptation of IoT with Blockchain in Food Supply Chain Management: An Analysis-Based Review in Development, Benefits and Potential Applications." *Sensors* 22, no. 21 (October 25, 2022): 8174. doi:10.3390/s22218174.
22. Sharma, Mahak, Rose Antony, Suniti Vadalkar, and Alessio Ishizaka. "Role of Industry 4.0 Technologies and Human-Machine Interaction for de-Carbonization of Food Supply Chains." *Journal of Cleaner Production* 468 (June 15, 2024): 142922. doi:10.1016/j.jclepro.2024.142922.
23. Wang, Gaofeng, Shuai Li, Yang Yi, Yingying Wang, and Changhoon Shin. "Digital Technology Increases the Sustainability of Cross-Border Agro-Food Supply Chains: A Review." *Agriculture* 14, no. 6 (June 6, 2024): 900. doi:10.3390/agriculture14060900.
24. Tsang, Yung Po, King Lun Choy, Chun Ho Wu, George to Sum Ho, and Hoi Yan Lam. "Blockchain-Driven IoT for Food Traceability with an Integrated Consensus Mechanism." *IEEE Access* 7 (January 1, 2019): 129000–17. doi:10.1109/access.2019.2940227.
25. Rejeb, Abderahman, and Suhaiza Zailani. "Blockchain Technology and the Circular Economy: A Systematic Literature Review." *Journal of Sustainable Development of Energy Water and Environment Systems* 11, no. 2 (September 1, 2022): 1–25. doi:10.13044/j.sdewes.d10.0436.
26. Duan, Keru, Helen Onyeaka, and Gu Pang. "Leveraging Blockchain to Tackle Food Fraud: Innovations and Obstacles." *Journal of Agriculture and Food Research,* September 1, 2024, 101429. doi:10.1016/j.jafr.2024.101429.
27. Camel, Afaf, Amine Belhadi, Sachin Kamble, Sunil Tiwari, and Fatima Ezahra Touriki. "Integrating Smart Green Product Platforming for Carbon Footprint Reduction: The Role of Blockchain Technology and Stakeholders Influence within the Agri-Food Supply Chain." *International Journal of Production Economics* 272 (April 16, 2024): 109251. doi:10.1016/j.ijpe.2024.109251.
28. Liu, Zhaolong, Xinlei Yu, Nan Liu, Cuiling Liu, Ao Jiang, and Lanzhen Chen. "Integrating AI with Detection Methods, IoT, and Blockchain to Achieve Food Authenticity and Traceability from Farm-to-Table." *Trends in Food Science & Technology,* February 1, 2025, 104925. doi:10.1016/j.tifs.2025.104925.
29. Wang, Yanze, Shanshan Li, Huikun Liu, He Zhang, and Bo Pan. "A Reference Architecture for Blockchain-Based Traceability Systems Using Domain-Driven Design and Microservices." *IEEE Xplore*, December 1, 2022, 269–78. doi:10.1109/apsec57359.2022.00039.
30. Adewusi, None Adebunmi Okechukwu, None Njideka Rita Chiekezie, and None Nsisong Louis Eyo-Udo. "Blockchain Technology in Agriculture: Enhancing Supply Chain Transparency and Traceability." *Finance & Accounting Research Journal* 5, no. 12 (December 30, 2023): 479–501. doi:10.51594/farj.v5i12.1514.

31. Karthikeyyan, Periyasami, Sarveshwaran Velliangiri, and Iwin Thanakumar Joseph Swamidason. *Review of Blockchain Based IoT Application and Its Security Issues*. In: *2019 2nd International Conference on Intelligent Computing, Instrumentation and Control Technologies (ICICICT)*, Vimal Jyothi Engineering College, Kannur, Kerala, India, 2019. doi:10.1109/icicict46008.2019.8993124.
32. Arabi, Hojatallah Sharifpour, Hamidreza Fallah Lajimi, Maghsoud Amiri, Mohammad Hashemi-Tabatabaei, and Seyed Khadije Hosseini. "Enhancing Food Supply Chain Sustainability: Exploring the Role of IoT Enablers and Sustainable Development Goals." *Sustainable Futures*, July 1, 2025, 100979. doi:10.1016/j.sftr.2025.100979.

Chapter 10

Deep learning-based barcode and quick response code recognition for real-time supply chain management applications

Gajalakshmi Kumar and Saravanan Somasundaram

10.1 INTRODUCTION

In the competing nature of the market prevailing today, proficient supply chain management (SCM) is inevitable. To necessitate effective operation, barcode and quick response (QR) codes, the two principal methods, are frequently used in the supply chain due to their competitive advantages such as effective monitoring and better control. The salient features, inherent advantages, and best practices of barcodes and QR codes can be understood easily and are ready for fitting into the supply chain to improve its visibility.

In the early 1950s, one-dimensional (1D) barcodes were developed and found application in the retail business due to the rapid growth of packaged goods and their inventory management. Since then, the barcodes continue to dominate the retail market. As the years progress, companies are looking for alternative ways for continuous contact with customers, though at different levels and through different means. However, the companies and retailers involved in handling consumer packaged goods depend on conventional 1D barcode (black and white) to achieve efficient inventory management. Unfortunately, 1D barcodes fail to cope with the modern world's demands; better accountability and transparency is quite impossible due to the complex nature of SCM processes. The above demands force the end user to attempt to use QR codes and two-dimensional (2D) barcodes, which revolutionize the supply chains all over the world, as they are more successful than 1D codes. The company's packaging, delivery, and product offerings meet the needs of clients is one way to deliver the best achievable client experience. The 2D barcodes are very useful in helping increase the efficiency of the supply chain while providing conscious customers with data about the product. Compared to conventional barcodes, which have a space of 20–25 characters, 2D barcodes can store an amazing amount of information (7,089 characters) in a small package. Businesses can now manage operations more effectively and build stronger relationships with

DOI: 10.1201/9781003660163-10

customers using these barcodes. Because there is very little tracking and monitoring possible with traditional barcodes, mistakes and uncertainties, such as delivering spoiled goods and upsetting a customer's expectations of the product they are receiving, are possible. Consider the same situation with 2D barcodes. Upon implementing 2D barcodes on all products to enhance tracking capabilities, logistics teams can obtain more detailed information about a product's batch codes or expiration date by scanning the barcodes.

Every company now has better accountability and traceability of the goods they ship. Customers will receive a safer product with more detailed data that keeps the supplier and customer informed. Suppliers can now feel more secure and transparent while guaranteeing the integrity of the delivered goods throughout their journeys, thanks to 2D barcodes. Additionally, by strengthening their relationships with customers, they can offer more individualized experiences and improved product authentication. As a result, companies can benefit from having better data and visibility in their order-to-delivery process, thanks to 2D barcodes. That involves accurate confirmation of product descriptions, warehouse locations, product health, and more. The most popular devices for reading barcodes are handheld barcode scanners [1]. Integrated decoders that can read multiple barcode types are a feature of every barcode scanner that IDAutomation recommends and sells.

The barcode pen scanner is a straightforward device that organizes, stores, and reads barcodes. Currently, a far greater variety of applications uses QR codes. These include ticketing products, loyalty marketing, and commercial tracking. Numerous such apps are targeted at mobile phone users. Using QR codes, users can email or text, send money, open a webpage, store contact details on their device, or receive text messages [2]. Additionally, Reed–Solomon error-correcting features allow for unbeaten decoding even in cases where the code is up to 30% distorted or broken. The built-in cameras of mobile phones have the ability to scan QR codes, making it simple to connect interactive digital experiences with physical marketing materials or items. QR codes can be classified as dynamic or static.

Poor print quality, uneven ink or toner levels, or inadequate contrast between the code's components (e.g., non-black bars on non-white backgrounds) affects the scanner. On mobile devices or automated platforms, moving objects in the camera regularly face two main issues: out-of-focus photos and motion blur from phenomena that significantly lower the detection accuracy of camera-based recognition systems. Images with complicated or cluttered backgrounds, different resolutions, and ambient noise are frequently difficult to handle using traditional image processing techniques.

Inconsistent detection rates may result from these factors interfering with the precise localization and segmentation of the codes. The issue of concurrently detecting and categorizing several symbologies (both 1D barcodes and 2D QR codes) inside a single image, particularly when mounted on a product, adds a distinct layer of complexity beyond the difficulties associated with recognizing individual codes.

A strong recognition system must be able to distinguish between the essentially different visual features of 2D QR codes, which are distinguished by their square grid structure and noticeable finder and alignment patterns, and 1D barcodes, which are mainly composed of parallel black-and-white lines. To do this, the model must acquire abstract representations that can differentiate between these various patterns. The codes can appear at various scales, distances, and orientations within an image. While QR codes are designed for omni-directional readability, 1D codes can only be read in a specific orientation for decoding information. It may overlap several codes if codes nearly appear on a single product. To avoid false predictions and missed actual detections, the spatial density makes individual code recognition, precise bounding box regression, and subsequent classification necessary. Convolutional neural network (CNN) frameworks have become a game-changing idea for barcode and QR code recognition in recent years.

Deep learning gives greater accuracy, robustness, and inference speed, which surpass many of the drawbacks of traditional methods. Deep learning-based approaches can work on sensitive environmental parameter modifications because they automatically extract robust and relevant features from large datasets.

MobileNet marks a substantial idea shift in the availability and application of artificial intelligence (AI), making it more than just another CNN. It can detect codes both offline and in real time on smart devices or scanners. This pervasive on-device intelligence is crucial for scalable and effective applications in vital industries such as manufacturing, consumer services, retail, and logistics, where prompt and accurate identification is crucial.

The novelty of this research pioneers a deep learning-centered methodology for barcode and QR code recognition that is specifically optimized for real-time operations in SCM, addressing limitations of existing optical scanning systems. Detecting and decoding codes from various angles, damaged surfaces, and low-light conditions are common challenges in dynamic warehouse, logistics, and retail environments. This framework improves throughput, scalability, and operational flexibility in SCM by eliminating hardware dependencies, reducing scanning time, and ensuring robust code reading even under non-ideal conditions such as motion blur or occlusion.

10.2 RELATED WORK

10.2.1 QR code/barcode

Based on cameras and laser scanners, there are two main categories of common technologies for barcode recognition. The latter is primarily used to identify 1D barcodes, with low recognition rates and a small range of recognition devices. As a result, this task only considers barcode identification using the camera. Furthermore, camera-based detection methodologies are classified as deep learning and traditional digital image processing. First, researchers used regular methods to process digital images to identify barcodes, facilitating scientific breakthroughs. To simplify the image information, Ohbuchi et al. [3] adopted a gray threshold to the input image (barcode) and a spiral scanning approach to accurately determine important black pillars.

They chose the barcode area and determined the orientation perpendicular to these black bars. Wachenfeld et al. [4] created a unique recognition algorithm for barcodes optimized for chambered mobile phones, tested its effectiveness on public datasets, reaching an accuracy of over 90% and exhibiting strong resistance to typical image distortion. In a new approach, Katona [5] introduced a barcode identification method that uses morphological manipulation in conjunction with filters in the lower part. They compared possible zones of barcodes and used unparalleled Evlidov calculations to show significant improvements over previous approaches.

In recent years, researchers have increasingly considered employing deep learning-based detection algorithms for barcode detection applications, as deep learning has demonstrated significant accuracy and inference speed across various fields, including pedestrian detection, road defects, and drone operation [5][6][7][8][9]. Hansen et al. [10] developed a hybrid method that integrates the YOLO algorithm with the regression network to ensure effective barcode identification and precise forecasting of the orientation angles. The outcome confirms that the combined method outperforms multiple previous conventional methods.

Tian et al. [11] reported higher computational efficiency, along with accurate barcode detection, while attempting a two-step neural network integrating the barcode region proposal network and VGG16. Meanwhile, Jia et al. [12] improved the accuracy of predictions in deformed codes by enhancing the region proposal network of Faster-RCNN. Zhang et al. [13] concentrated on enhancing the layers in Faster-RCNN by connecting the bounding box regression to the coordinates of quadrilateral vertices, resulting in improved performance and compliance of barcode identification outcomes.

Xu et al. [14] employed the enhanced YOLOv4 model alongside OCR to capture barcode details and three-segment codes, achieving a recognition rate of 98.5% (package information). However, the size of the sample during experiments was limited, and the dataset was not publicly as well available. Cai et al. [15] highlight how QR code-based mobile payments are transforming traditional retail by enhancing transactional convenience and customer engagement. Prior literature primarily focused on NFC and card-based systems, with limited emphasis on QR technology's retail impact.

Yan et al. [16] and De Luna et al. [17] separately explore how different mobile payment technologies (e.g., NFC, QR code, and SMS) influence user adoption behaviors. They reveal that perceived ease of use and trust vary significantly depending on the technology implemented. Elaskari et al. demonstrate the effectiveness of barcode [18] systems in automating attendance tracking and asset management in academic settings. It emphasizes improved operational efficiency, data accuracy, and real-time monitoring through barcode integration.

Tan et al. propose [19] a QR code-based XOR-ed visual secret sharing scheme to securely embed and transmit confidential data. The approach enhances both security and robustness while preserving the visual meaning of shadow images. Nuhi et al. introduce [20] a smart attendance system that impacts QR codes to streamline and automate student attendance processes. The proposed approach boosts efficiency, reduces manual errors, and supports real-time data access for academic institutions. Küng et al. [21] investigated the contribution of barcode technology in enhancing the safety of medication preparation in healthcare settings. Findings show a noteworthy decline in medication errors, highlighting the role of barcode systems in improving patient safety.

In contrast to the established digital image processing techniques, systems that utilize deep learning networks do not require the setting of limiting parametric values vulnerable to ecological aspects. This notably reduces the influence of ecological aspects on the real-time application of the identification techniques. However, owing to issues such as the complexity of the algorithms and the necessity to increase detection speed, current deep learning-based detection techniques have limited use in the real world.

The authoritative parallel processing abilities of GPUs are capable of considerably enhancing the training ability of the model along with inference speeds, accommodating larger datasets and new intricate model architectures owing to advances in computing technology. Consequently, it is essential to conduct studies on deep learning-centered barcode identification algorithms. This study introduces a rapid barcode recognition method that employs deep learning-based object detection models. Because bi-staged models make candidate object bounding boxes in the initial stage and subsequently sort and refine those bounding boxes in the following stage, their detection speed is slower than that of single-stage models. Consequently, the one-stage model MobileNet served as the reference for this study. They endeavor to undertake barcode classification challenges in multi-faceted and assorted real-world circumstances by enhancing the network architecture, reducing the dimension of the model, and boosting the inference speed.

10.2.2 Image classification

Image classification is the primary move in image processing used for classifying images into defined classes and labels. The main aim of these machines is to spot and categorize specific regions, locations, and patterns inside the images. Classification-centered data analysis involves grouping data points into correct groups or classes in manually selected features and rules. Typically various traditional methods were widely attempted by earlier peer researchers for the classification, such as support vector machines, decision trees, naïve Bayes, and k-nearest neighbors. Researchers or experts decide on feature selection, meticulous design, and opt for features that summarize applicable information from the data. Most of the time, these features are picked according to subject knowledge and understanding, with the goal of capturing unique traits that help separate classes apart. Manual feature engineering is time-consuming because if one feature set is not working, then it is necessary to select another set of features by testing one by one. Depending on the patterns and relationships of datasets, features gathering changes. These selected features are considered as inputs for classification methodologies that use well-defined classes to allocate data points to explicit classes.

Ismael et al. [22] introduced a sophisticated method that exploits the resources of residual networks (ResNets) for segmenting brain tumors (medical images). Their outcomes engaged a wide-ranging assessment of a yardstick dataset containing 3,064 MRI images of three discrete brain tumor varieties. An extremely accurate 99% result of this model was achieved, surpassing earlier work in the same domain. In this context, Xu et al. [23] shifted the focus of deep learning toward remote sensing-based image classification. They combined two methods, viz., recurrent neural networks and random forests, to attain cross-validation. In their study, they used the UC Merced land dataset and reported 87% accuracy.

Aggarwal and Kumar [24] stressed the importance of texture analysis and its classification. Their research employed two deep learning-based models that revolve around CNNs. While comparing, the first model achieved 92.2% accuracy, while the second model surpassed the first model, resulting in 96.36%. In an attempt to classify skin cancer disease based on two different datasets, Abdar et al. [25] employed three-way decision theory with the aid of a Bayesian deep learning model, uncertainty quantification (UQ) methods, and deep neural networks, and reported better accuracy and F1-score.

Ibrahim et al. [26] explored a deep learning method for COVID-19, pneumonia, and healthy CXR picture classification that depends on a well-taught AlexNet model. The scenario of medical diagnostics moved another stride upward as this model demonstrated prominent performance in three- and four-way categorization, accomplishing better accuracy, sensitivity, and specificity. In the image classification arena under resource restrictions, an emerging Deep CNN categorization technique involving an information transfer model was introduced by Ma et al. [27]. This method reported greater correlation than the established histogram-based methods, attaining a notable 93.4% classification accuracy.

Similarly, an accuracy of 96% in automatic identification and categorization of three cast iron grades with respect to the microstructure by two machine learning techniques (GLCM and LBP) is reported by Gajalakshmi et al. [28]. In a similar approach, Gajalakshmi and Saravanan [29] extended the above techniques to classify ductile and brittle fractures in the weld joints. They obtained 96% accuracy. In another study, Gajalakshmi et al. [30] successfully identified good and defective weld joints employing YOLO V8, a deep learning model, with an impressive 95% accuracy.

Magsi et al. [31] extended a cross CNN–RNN methodology for quality appraisal and fruit sorting. The novel model demonstrated exceptional efficiency in quality assessment and sorting of fruits and was recommended for various agricultural applications. In a similar fruit classification attempt, Abu-Jamie et al. [32] applied VGG16A, another novel deep learning approach, and reported a remarkable 100% accuracy, emphasizing the potential of deep learning-based applications in the day-to-day demands of small- and large-scale consumers.

In a different attempt, Sharma et al. [33] extended the CNN and transfer learning to diagnose breast cancer, highlighting the potential for further exploration in medical image analysis. The researchers reported an optimistic 98.4% accuracy, promoting further research in medical sciences. Hussain et al. [34] accomplished a notable 97.75% accuracy in identifying Alzheimer's disease with the aid of a 12-layer CNN model on the OASIS dataset, exceeding the competing CNN models and displaying a successful methodology for Alzheimer's detection. Khullar et al. [35] lead the way in employing resting-state functional MRI (re-fMRI) data in classifying ADHD and distinguishing it from wellbeing parameters. They built a hybrid 2D CNN–LSTM model to improve the accuracy, specificity, sensitivity, F1-score, and AUC compared to the competing models. This advancement disseminates the possibility of combining deep learning and re-fMRI to craft an influential means for detecting neurological issues. In a similar attempt, Skouta et al. [36] reported a superior 95.5% accuracy in classifying normal and proliferative diabetic retinas using CNNs. Their research leveraged a wider dataset to successfully train the model and warranted precise classification by identifying complex features of the images.

In a different application, Yang et al. [37] employed multiple CNN models to recognize urban wetland and identified the DenseNet121 algorithm as superior among the attempted CNN models pertaining to land images. They obtained a higher Kappa (a statistical measure to assess the closeness among two or more classifiers) and overall accuracy, accentuating the possibility of employing deep learning in land cover classification.

Gao et al. [38] extended the CNN model, having multi-convolution and max-pooling layers, to detect defects in fabric manufactured in a textile factory and achieved an overall accuracy of 96.52%, emphasizing the ability of CNN in industrial environments.

The review of literature sheds light on the existing opportunity of employing deep learning in different areas such as medical science, texture examination, image identification, and the detection of neurological states. Though the conventional techniques offer definite rewards, they largely depend on expert knowledge for identifying significant features and improving the algorithms. These methodologies frequently face critical issues while handling complex, multi-dimensional datasets [39], where the determination of relevant features itself is challenging. In addition, the manual feature extraction requires significant effort and is likely not flexible enough while acclimatizing to new varieties of data.

In contrast, deep learning algorithms automatically get trained over the hierarchical feature representations obtained from the raw input, thereby suppressing the dependence on manual feature selection. This capability has modified the classification regime by facilitating newer algorithms to extract complex patterns and textures that evade conventional methodologies. To be specific, CNN models have demonstrated robust performance in various image classification tasks, while architectures such as MobileNet have established their effectiveness in handling sequential data. In a nutshell, deep learning methods constantly exceed established techniques across a wide range of fields.

10.2.3 Prevailing issues in barcode/QR code categorization

Though scanning either barcodes or QR codes is inevitable during testing of connected devices, achieving this automation has few limitations. Hence, manual testing is often required even is it requires a long time and varies depending on the personnel involved. In addition, the testing phase also requires examining a considerable quantum of codes. Hence, automating this phase is cumbersome, specifically when operating with mobile devices. This can be attributed mainly to these factors: imitating human motion to maintain stability during the scanning process. Automating the scanning procedure necessitates mimicking human-like physical actions to correctly align the phone position above the code for scanning. A novel approach to deal with this problem involves incorporating a robotic arm that is operated using Python. This support accurately positions the phone's camera with the monitor, allowing seamless code scanning. An alternative method for mechanized barcode and QR code scanning is employing a physical (manual) device to align the camera with the screen. This mechanism is easier to establish rather than affixing a robotic arm and is economically feasible as well. Improving the automation coverage facilitates exploring a higher number of codes. However, managing and conserving the codes during automation is a demanding task. Automation of the testing process through the internet (web automation) demands the generation of dynamic code. For this purpose, tools like Selenium, which enable the creation of automation scripts for specific applications, are employed. On the other hand, mobile automation is a reliable alternative capable of seamless scanning based on the dynamic codes. These web and mobile automation combinations simplify the process of maintaining the coding process and also enable complete testing on all devices. For accomplishing rapid and competent automation, a suitable alliance of discrete mechanisms is inevitable. This integration is achieved by combining web and mobile automation along with device control over multiple movements.

Aligning different mechanisms is crucial for faster and more effective automation. This can be accomplished by combining web and mobile automation with control of device movements. The approach is to create a cohesive test automation framework. This framework serves as the foundation, orchestrating the integration of mobile, web, and physical movement automation elements. This method can ensure consistency and coherence, leading to a smooth end-to-end automation experience.

Two technologies capable of transforming QR code/barcode scanning automation: The sector holds a highly positive perspective regarding the future of automated QR code and barcode scanning on mobile devices. Two main developments that transform the codes are AI integration and cloud-based automation.

AI: The precision and efficiency of QR code/barcode automatic scanning are enhanced. Hence, broken and damaged QR codes and barcodes are processed and recognized by AI integration. It also reduces manual intervention, saving time and costs.

Cloud-based Automation: It is possible to handle the storage of data in a central cloud when the size of the business is enlarged. Cloud storage is more flexible and fits immediately into the ever growing demands. In addition, the maintenance and troubleshooting processes are efficient. The payback of automating QR and barcode scanning in a mobile phone or tablet is listed below.

Trim Down Human Intervention and Errors: Automatic scanning removes the necessity of physical scanning, reduces errors, and improves accuracy.

Testing Workflows Streamlined: Streamline testing workflows by automated scanning. It can save time and resources.

Efficiency: This scanning process reduces the time and effort required to scan QR/barcodes. Automated scanning can enhance collective efficiency.

Reliable and Accurate: Following the pre-defined instructions meticulously results in consistent and precise results. This is especially beneficial when scanning larger QR codes and barcodes.

High-Speed Execution Time: Real-time scanning aids in providing a flawless consumer experience and ensures rapid implementation by reducing the time taken to wait in a queue to finish the scan.

New Possibilities for Mobile Automation: Enabling barcode/QR code integration promotes new ventures for mobile automation.

10.3 PROBLEM ASSERTION

The classification of barcodes, QR codes, and both on the product pack using MobileNetV3 in product applications involves deep learning models to improve the efficacy and correctness of categorization. MobileNetV3, recognized for its lightweight architecture, is particularly appropriate for mobile applications where computational resources are minimal. This approach is beneficial in various consumer and retail environments where rapid and accurate code classification is essential for product identification and information retrieval.

10.4 METHODOLOGY

The novel method to classify the barcodes and QR codes using the MobileNetV3 architecture is presented in Figure 10.1. The study attempts to generate and assess a lightweight deep learning model for effective and precise categorization.

10.4.1 Data curation

Data was collected across multiple product categories during packaging, transit, or warehouse stages of electronics (phones, laptops, chargers), food and beverages (canned goods, bottled drinks), pharmaceuticals (medicine boxes, blister packs), household items (cleaning supplies, tools), apparel and footwear, e-commerce shipments (multi-label products from Amazon, Flipkart, etc.), and also Kaggle datasets. Each product was photographed under natural and artificial light in varying conditions (blur, rotation, low-light, partial occlusion).

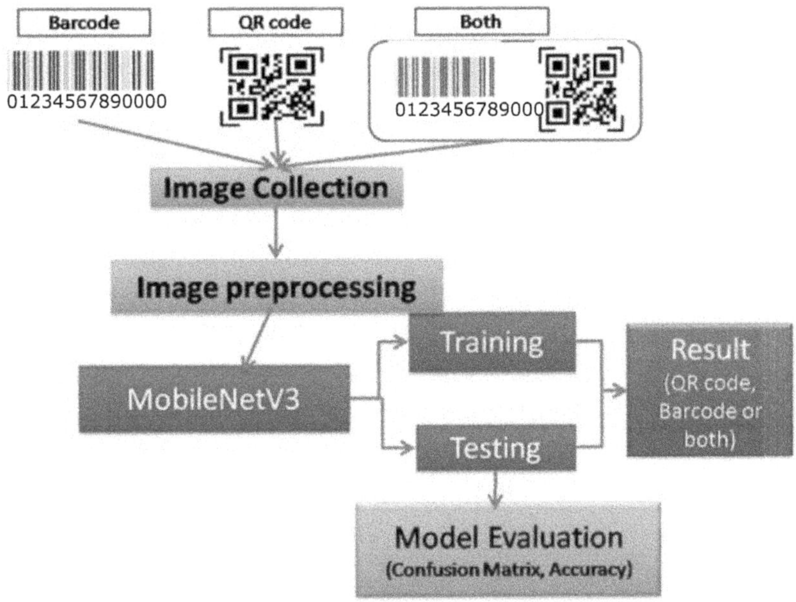

Figure 10.1 Workflow of the proposed work.

The image capture of a QR code on a mobile device indicates the possibility of capturing continuous images in short form or in a sequence. In addition, multiple images organized in a sequence are recognized judiciously during QR code recognition. The scanned images are subjected to "super-resolution," a novel technique that transforms multiple input low-resolution images into a high-resolution image. The augmentation in image quality is due to complementing pixel information caused by the displacement of pixels achieved by camera movements. With the aid of the super-resolution technique, even low-resolution QR codes (Figure 10.2) are recognized.

In this study, a database comprising 900 images of barcodes and QR codes, comprising various distortions, viz. non-uniform lighting, reflections, and image blurring resulting from the unexpected movement of the camera. The database is available without restraint for other researchers, allowing for advanced exploration and validation of the barcode recognition algorithm (Figure 10.3).

10.4.2 Data preprocessing

The graphical presence of code types determines the labeling of each image: class labeling in a classification task involves assigning meaningful and identifiable names to each class the model predicts. Class 0, Class 1, and

Figure 10.2 Sample QR codes and barcodes utilized during the model development.

Class 2 are stand-ins for the dataset's actual classifications. These labels need to be interpreted or replaced based on the specific application domain.

10.4.2.1 Class labels

Class labels play a critical role in making the model's predictions interpretable to users, visualizing performance meaningfully, and evaluating specific areas where the model might need improvement. Thus, assigning accurate, domain-relevant class labels enhances the clarity of analysis and ensures the results are aligned with the real-world context of the data.

Class 0: Barcode only, Class 1: QR Code only, Class 2: Both. Mobile NetV3 models can be trained using this straightforward three-class categorization to automatically distinguish between different codes types for usage in last-mile delivery or warehouse decision-making.

Figure 10.3 Sample input images for the model.

Several preprocessing steps are used to prepare the data for efficient modeling when classifying barcode, QR code, and “both”. Particularly, scaling 300X300, rotating, and normalization are required for images taken in different sizes and lighting conditions.

10.4.3 Modeling using MobileNet

MobileNet is one of the forms of resource-efficient neural architecture for mobile and embedded applications, developed with CNN as the foundation and explicitly intended for mobile and resource-constrained environments. Of its different versions or improvements, MobileNetV3 is superior due to its lightweight nature, capable of achieving results faster. In addition, the proposed architecture significantly reduces computational cost and memory usage without compromising accuracy, making it highly suitable for edge computing tasks.

MobileNetV3 is built based on the successful earlier CNN models such as AlexNet, VGGNet, GoogLeNet, ResNet, and MobileNetV1. Though these architectures contributed to improvements in accuracy and efficiency, MobileNetV3 introduces key innovations that further optimize performance for real-world, low-power devices.

10.4.3.1 Key innovations in MobileNetV3

The core component of MobileNetV3 is the application of depth-wise separable convolutions, originally introduced in MobileNetV1. This technique decomposes a regular convolution into two parts: depth- (employing a filter in each channel) and point-wise (the outputs of each input layer combined) convolutions.

The separation of both depth- and point-wise convolutions drastically reduces the influencing parameters and floating-point operations by a factor of 8–9 compared to standard convolutions, while achieving higher accuracy levels.

Linear Bottlenecks: Conventional machine learning/deep learning techniques largely use ReLU, a non-linear activation function, subsequent to reduced dimensionality. Conversely, the ReLU algorithm loses critical information due to its truncation behavior when employed in low-dimensional feature spaces. To counter this loss, the MobileNetV3 algorithm utilizes linear layers as an alternative to the ReLU algorithm during the bottleneck. This alteration preserves the diverse nature of the data, eliminates distortions, and warrants vital features are kept abreast during the compression and expansion operations.

10.4.3.2 Inverted residual blocks

The MobileNetV3 algorithm compresses the input images followed by expansion, rather than the traditional method of expansion followed by compression. The inversion process and the usage of skip connections between bottlenecks provide gradient flow and restrict information loss, resulting in superior training ability along with lesser computational cost.

10.4.3.3 Model architecture

The MobileNetV3 architecture spins in the region of bottleneck residual blocks that incorporate:

1. Depth-wise separable convolutions
2. Linear bottlenecks
3. Inverted residual connections

Each block of the algorithm leverages an input tensor (size $h \times w \times k$) and is analyzed in a sequence of operations, resulting in a compressed and transformed tensor. The design exploits the ReLU6 activation function and employs a 1×1 convolution to correct the dimensions of the channels. In this attempt, MobileNetV3 starts with a 32-filter convolutional layer followed by 19 bottleneck blocks that apply diverse expansion factors and stride values.

The proposed model supports flexible scaling through width multipliers and input resolution adjustments, enabling it to operate effectively across a wide spectrum of devices, from low-power smartphones to high-end embedded systems.

10.4.3.4 Trade-offs and hyperparameter tuning

MobileNetV3 performs fine-tuning based on specific hardware constraints. For instance:

A configuration having an input resolution of 96×96 and a width multiplier of 0.5 yields a smaller model with fewer operations, suitable for extremely constrained environments. However, by augmenting the resolution to 224×224 and doubling the width multiplier to 1 or above, accuracy improves, with a compromise on computational cost. The striking balance between efficiency and accuracy makes MobileNetV3 adaptable for multiple/diverse applications for the end user.

10.4.3.5 Application in object detection

MobileNetV3 is employed as a feature extractor in object detection tasks using Single Shot Detector (SSD) and its lightweight variant SSD Lite on the COCO dataset. While MobileNetV3 excels in extracting features, SSD Lite is optimized for predicting bounding boxes and object classes. Together, they form an efficient pipeline. Compared to models like YOLOv2, MobileNetV3+SSD Lite utilizes significantly fewer parameters (e.g., 2.1M vs 36.1M in SSD300) and requires less computational activity while maintaining competitive accuracy, which is another major benefit for real-time mobile applications such as face mask detection.

10.4.3.6 Application in semantic segmentation

In semantic segmentation, MobileNetV3 is evaluated with DeepLabv3, a widely used segmentation architecture [39][40][41]. When compared with ResNet-101 and MobileNetV1 as backbone feature extractors, MobileNetV1 achieves an mIOU of 78.56% with 4.9–5.7× fewer Multiply-Adds (MAdds) than ResNet-101. MobileNetV3 performs similarly while requiring 2.5× fewer operations than MobileNetV1 by leveraging intermediate feature maps (e.g., 320-channel maps instead of 1280).

10.4.4 Design model MobileNetV3

The MobileNetV3 architecture incorporates residual layers having a stride of 1 and downsampling layers with a stride of 2, both utilizing ReLU activations. The residual and downsampling layers are composed of three sub-layers. In this architecture, MobileNetV3 is leveraged to extract features at lower levels. The extracted feature maps are then decimated to decrease resolution while conserving critical spatial information. These 2D feature maps are subsequently flattened into 1D vectors, enabling their input into fully connected (dense) layers. A dense layer having 128 neurons along with a ReLU activation is added to help the system learn deeper relationships among the extracted features. During training, dropout regularization is applied, randomly deactivating 20% of the neurons to improve model generalization and prevent overfitting. For the final classification task, the network includes three output neurons, one for each target group, and applies a softmax activation function to convert the unprocessed output scores into class probabilities.

10.5 EXPERIMENT AND RESULTS

The experiments conducted in the present study are assisted by ImageNet, an image dataset.

The inputs for the MobileNetV3 model for image classification are sized at 300×300×3 pixels. The input dataset contains 992 images, the primary source being Kaggle, and other locations. The subsections of the total images include 423 barcode images, 511 QR code images, and 58 images containing both. The dataset is partitioned into training and testing by maintaining an 80:20 ratio, thereby allocating 794 images for training and 198 images for testing.

10.5.1 Fine-Tuned MobileNetV3

Fine-tuning, a machine learning process in which a pre-trained AI model is exposed further and trained in a new environment of dataset to adapt to the newer one for a particular task, improves the speed of training and reduces the potential overfitting issue occurring in an insufficient dataset. Though the pre-trained MobileNetV3 is competent enough to accept the input image (300×300×3 pixels), fine-tuning is an empirical substitute for conventional layers with either two or three dense layers along with the ReLU activation function. The fine-tuned model requires less duration for training and testing the ReLU function, which congregates rapidly and can be activated sparsely. Though the pre-trained models are capable of reducing overfitting, fine-tuned fully connected models result in superior performance. The fine-tuned models demonstrate lesser execution duration and memory utilization with higher classification accuracy. The sample output images are displayed in Figure 10.4. The models subjected to both pre-training and fine-tuning are capable of achieving 100% training accuracy and 97.5% accuracy during testing or validation of three different classes (Barcode,

Figure 10.4 Sample output images emanating from the attempted model.

QR code, and a mixture of both bar and QR codes) within 50 epochs. The whole process of training and testing lasts less than 15 min. The proposed approach generates a multi-label classification matrix that operates similarly to a 2D confusion matrix but is designed for multi-label classifiers, which are capable of assigning multiple labels to a single occurrence. This novel method reduces the uncertainty in traditional methods by systematically analyzing all combinations of true and predicted labels, leading to a judicial determination of true positives (TPs), false positives (FPs), true negatives (TNs), and false negatives (FNs) for improved performance evaluation using metrics such as precision, recall, and F-score.

10.5.2 Validation

The validation phase evaluates the capability of the developed model on unknown/untrained test data from real-time data to determine its potential to attain generalization. In the validation phase, the prediction performance of the developed system is determined with the aid of a confusion matrix, which has two ordinates, the first being the actual result and the other being the outcome of the prediction. The fundamental structure of

Table 10.1 General structure of the confusion matrix

		Predicted		
Actual	A_l	A_j	A_n	
	A_1	N_{11}	N_{1j}	N_{1n}
	A_j	N_{i1}	N_{ij}	Ni_n
	A_n	N_{n1}	N_{nj}	N_{nn}

a multi-category confusion matrix, with the categories A_1, A_2, and A_n, is illustrated in Table 10.1. In the confusion matrix, N_{ij} corresponds to the quantity of samples that actually belong to class A_i but are misclassified as class A_j.

The confusion matrix (Table 10.1) graphically shows the outcome of the categorization attempted in the present study with respect to each of the three individual code systems attempted: the vertical axis stands for the correct category in each validation image, and the horizontal axis denote the category predicted by the classifier. The correct classifications lie alongside the diagonal line of the confusion matrix.

Earlier peer researchers in the past [42–46] made significant efforts to classify two categories: either positive (*P*) or negative (*N*). A classification model maps cases to their predicted categories. Certain classification models generate a continuous output that allows for various thresholds to be utilized for predicting the exact nature of the image considered. Few methodologies yield a distinct class label that signifies solely the expected class of the instance. To differentiate between the true and the expected outcomes, alphabets {*Y*, *N*} are used based on the estimation of the model. With a classifier and a sample, four potential results are probable. If the instance is positive and classified as positive, it is denoted as a TP; if classified as negative, it is considered a FN. If the example is negative and categorized as negative, it is regarded as a TN; if it is categorized as positive, it is considered an FP.

In this classification task involving three classes and a total of 198 images, the model accomplished a remarkable 97.5% accuracy, correctly predicting 193 out of 198 instances. A confusion matrix was created to gain deeper insight into the performance of the model across the individual categories (Table 10.2).

The confusion matrix shows that, with only five misclassified samples, the model consistently performed well across all classes in a balanced dataset with 66 images per class (Class 0, 1, and 2). One image was incorrectly identified as Class 1, and another as Class 2. In contrast, 64 photographs were accurately classified as Class 0. In the same manner, Class 1 had 64 accurate predictions, whereas Class 0 and Class 2 each had one incorrectly classified image. With 65 photographs, Class 2 had the highest number of

Table 10.2 Results of correct class prediction in a Confusion Matrix

Actual\predicted	*Class 0*	*Class 1*	*Class 2*	*Total*
Class 0	64	1	1	66
Class 1	1	64	1	66
Class 2	0	1	65	66
Total	65	66	67	198

correctly categorized images, while only one image was incorrectly labeled as Class 1. This matrix indicates that the model maintains balanced performance, avoiding any significant bias toward one particular class. The small number of misclassifications is likely due to overlapping features or minor variations in the dataset that made certain samples more ambiguous.

10.5.3 Multi-class classification

In a multi-class categorization greater than 2 class situation, determining the F1 score per category is significant instead of calculating a single general F1 score for the whole model. Every class in the multi-class problem is considered a binary classification problem, considering any one category as positive and the remaining as negatives. Compute the F1 score. TP are the cases correctly classified as a particular class, FP are cases erroneously classified as that class, and FN are cases of that class imperfectly classified as other classes. Yin et al. [47] predicted the shear, tensile, and impact strengths of the aluminum-stainless steel welded joints using four machine learning algorithms, viz., k-nearest neighbor regression, decision tree regression, random forest regression, and extra tree regression, which displayed favorable results during correct prediction. Based on the confusion matrix, performance is determined via accuracy, precision, recall, specificity, and F-score [29,48].

TP: Predicting the category correctly in a positive case.
FP: Incorrect prediction in a positive case.
TN: Right prediction in a negative case.
FN: Incorrect prediction in a negative case.

10.5.3.1 Precision

Precision is the defined as the ratio of correct predictions (TP) to the summation of all the positive predictions (TP+FP). It symbolizes the accuracy of the positive predictions made by the model. Precision is mathematically represented by:

$$\text{precision} = \frac{\text{TP}}{\text{TP} + \text{FP}} \tag{12.1}$$

10.5.3.2 Recall

Recall is the ratio between the TP and the actual total positives, denoted by

$$\text{recall} = \frac{\text{TP}}{\text{TP} + \text{FP}} \tag{12.2}$$

10.5.3.3 F-Score

$$\text{F-score} = \frac{2}{1/\text{precision} + 1/\text{recall}} \tag{12.3}$$

10.5.3.4 Accuracy

In machine/deep learning models, accuracy quantifies the frequency of accurate predictions. It is computed by summing the number of correct predictions (TP+FN) divided by the total number of predictions (total attempted). This is an uncomplicated way to characterize the overall precision of the algorithm. The attainment of higher accuracy means the attempted model is accurate; however, this parameter becomes uncertain for uneven datasets.

$$\text{Accuracy} = \frac{\text{TP} + \text{TN}}{\text{P} + \text{N}} \tag{12.4}$$

The classification results show a total of 193 correctly classified instances: 64 each from classes 0 and 1, and 65 from class 2. However, 5 instances were misclassified, including two each in classes 0 and 1, and the remaining from class 2. These misclassifications are represented by off-diagonal entries. Furthermore, this assessment could be further enhanced by precision, recall, and F1-score metrics. The present model exhibits a robust classification capacity with little class confusion, suggesting its applicability for real-world implementation in tasks requiring high accuracy and consistency.

Table 10.3 sums up the performance of a categorization across three conditions by means of precision, recall, F1-score, and accuracy. The model shows consistently high precision values (around 0.9697–0.9701) across all classes, indicating accurate positive predictions. Recall is also strong, with Class 2 achieving the highest at 0.9848, signifying that the model is particularly effective at identifying instances of this class. The F1-scores reflect the existence of an unbiased swap between precision and recall, with Class 2 again slightly outperforming the others. The average values across all

Table 10.3 Overall performance metrics for all classes

Parameter	*Class 0*	*Class 1*	*Class 2*	*Average*
Precision	0.9697	0.9697	0.9701	0.9698
Recall	0.9697	0.9697	0.9848	0.9747
F1-Score	0.9697	0.9697	0.9774	0.9723
Accuracy		97.47%		

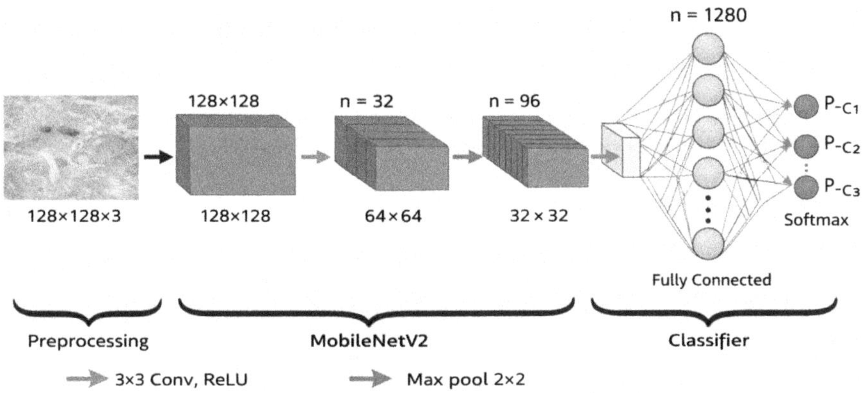

Figure 10.5 Performance metrics of all classes.

metrics remain high, and the overall accuracy of 97.47% confirms that the model performs reliably and consistently in classifying the data.

The classification performance of a model across three classes (Class 0, Class 1, and Class 2), using four evaluation metrics: precision (blue), recall (red), F1-score (green), and accuracy (purple), is exhibited in Figure 10.5. For Class 0 and Class 1, the values of all metrics are nearly identical and consistently balanced just below 0.97, indicating balanced and reliable performance. In contrast, Class 2 stands out with a significantly higher recall, nearing 0.985, and a correspondingly high F1-score, showing the capability of the model toward exact recognition of most actual cases of Class 2. The average bars reflect the overall high performance of the model, with accuracy and F1-score being the highest among the metrics. This graph visually confirms that the model performs uniformly across Class 0 and Class 1 and excels particularly in detecting Class 2.

10.5.4 Accuracy plot

The curves pertaining to training and validation accuracy show a stable progress during the initial epochs. The training accuracy begins at nearly 80% and goes beyond 97% by epoch 5 (Figure 10.6a). The trend of the

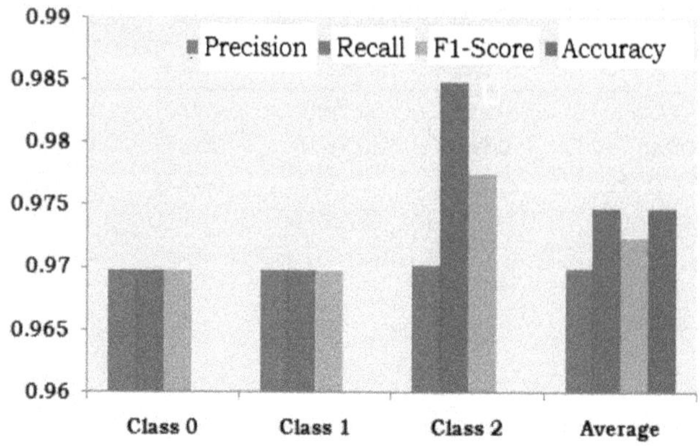

Figure 10.6 (a) Accuracy plot and (b) loss plot of the model.

curve continues to improve slightly and stabilizes just below 100%, signifying that the model is fitting well to the data used for training. Similarly, the validation accuracy follows a similar path, peaking around 97.5% (Figure 10.6b), which is consistent with the earlier confusion matrix results. The results show minor fluctuations in the later segments of the epochs but remain relatively stable, suggesting good generalization.

10.5.5 Loss plot

The training loss drops sharply from above 2.5 to near zero within the first few epochs, indicating that the model swiftly learns from the training data. It remains very low throughout, reinforcing the high training accuracy obtained. The validation loss is low in general, displaying significant fluctuations compared to the training loss. The validation loss values drop in the initial segment (below 0.5); however, it tends to amplify after 10 epochs. The increase indicates a marginal overfitting, where the model keeps on learning the training data but probably begins to drop performance on a new set of data.

10.6 CONCLUSIONS AND FUTURE PERSPECTIVES

The technical progress in mobile computing, neural networks, robotics, data storage, and cloud-based learning has pushed the usage of AI in diverse business organizations. AI technologies aid multiple businesses by eliminating many levels of manual actions such as promotions, assortments, and supply chains. The e-commerce industry applies AI for trend prediction, price fixation of warehouses and logistics, and personalizes promotions, leading to smart manufacturing. However, numerous alterations are obligatory to obtain the salient advantages of AI; more specifically, these modifications will force companies to accept innovative approaches in either modifying the plant design, reshaping the manufacturing footsteps, or working out novel supply chain models. In addition, companies require revolutionizing the manner in which they carry out business, as there will be a changeover from human operators to AI-facilitated machines and robots. The worldwide trend in industrial operation backed by AI is escalating exponentially, which suggests the preparedness of all industries for AI.

In the current competitive market scenario, companies are forced to manage their supply chains in warehouses efficiently with minimal manpower without affecting efficiency. Judicial management is implemented through proper planning and real-time management of their warehouses and logistics. This chapter attempts to establish a mechanism for effective warehouse management based on MobileNetV3 that addresses data acquisition issues (either damaged or broken) and reduces the likelihood of errors during data collection. In this study, an attempt is made to classify and identify the mode of scanning based on MoblieNet, an efficient network architecture, to build a precise model for the identification of barcodes and QR codes.

The output of this study in terms of frames per second is 34 fps, which is 31% faster than OpenCV. Initially, training the model on the customized dataset yielded 98.3% accuracy, and during testing, the model resulted in 97.47% accuracy. It is inferred that this methodology is efficient for categorizing QR codes and barcodes separately and in images involving both barcodes and QR codes, making it a simple, portable solution as scanning can easily be done by mobile devices and conserves time.

The promising attempt of the attempted model indicates noteworthy prospects for its larger application in supply chain companies. Though the present system attained accuracy above 97% with a response time of less than 30 ms, further progress is demanded to attain a versatile system. Future studies on enhancing performance in real-time confrontations such as low illumination, partially damaged printed codes, glare from digital images, and scanning circumstances can be focused on. Furthermore, the present study can be extended to handle multiple barcodes and QR codes concurrently, which is useful during the handling a large number of goods in warehouses and retail shops. The integration of the attempted system with handheld scanners or edge computing devices eliminates the reliance on network connections. Similarly, connecting this system with Internet of Things frameworks will reinforce data security and traceability, providing transparent progress of products within the supply chain. The system aids effectively in building faster, more reliable, and automated supply chain environments in the near future.

REFERENCES

1. Blanger, L., & Hirata, N. S. (2019, September). An evaluation of deep learning techniques for QR code detection. In *2019 IEEE International Conference on Image Processing (ICIP),* Taipei International Convention Center, Taipei, Taiwan (pp. 1625–1629). IEEE.
2. Tribak, H., Moughyt, S., Zaz, Y., & Schaefer, G. (2016, October). Remote QR code recognition based on HOG and SVM classifiers. In *2016 International Conference on Informatics and Computing (ICIC),* Mataram, Indonesia (pp. 137–141). IEEE.
3. Ohbuchi, E., Hanaizumi, H., & Hock, L. A. (2004, November). Barcode readers using the camera device in mobile phones. In *2004 International Conference on Cyberworlds,* Tokyo, Japan (pp. 260–265). IEEE.
4. Wachenfeld, S., Terlunen, S., & Jiang, X. (2008, December). Robust recognition of 1-d barcodes using camera phones. In *2008 19th International Conference on Pattern Recognition,* Tampa Convention Center, Tampa, Florida, USA (pp. 1–4). IEEE.
5. Katona, M., & Nyúl, L. G. (2013, May). Efficient 1D and 2D barcode detection using mathematical morphology. In *International Symposium on Mathematical Morphology and Its Applications to Signal and Image Processing,* Uppsala, Sweden (pp. 464–475). Berlin, Heidelberg: Springer Berlin Heidelberg.
6. Aceto, G., Ciuonzo, D., Montieri, A., & Pescapé, A. (2019). Mobile encrypted traffic classification using deep learning: Experimental evaluation, lessons learned, and challenges. *IEEE Transactions on Network and Service Management, 16*(2), 445–458.
7. Li, J., Sun, A., Han, J., & Li, C. (2020). A survey on deep learning for named entity recognition. *IEEE Transactions on Knowledge and Data Engineering, 34*(1), 50–70.
8. Chen, J., Wen, Y., Nanehkaran, Y. A., Zhang, D., & Zeb, A. (2023). Multiscale attention networks for pavement defect detection. *IEEE Transactions on Instrumentation and Measurement, 72*, 1–12.
9. Liu, M., Jiang, J., Zhu, C., & Yin, X. C. (2023). Vlpd: Context-aware pedestrian detection via vision-language semantic self-supervision. In *Proceedings of the IEEE/CVF Conference on Computer Vision and Pattern Recognition,* Vancouver Convention Center, Vancouver, BC, Canada (pp. 6662–6671).
10. Hansen, D. K., Nasrollahi, K., Rasmussen, C. B., & Moeslund, T. B. (2017). Real-time barcode detection and classification using deep learning. In *International Joint Conference on Computational Intelligence,* Funchal, Madeira, Portugal (pp. 321–327). SciTePress.
11. Tian, Y., Che, Z., Zhai, G., & Gao, Z. (2018, December). BAN, a barcode accurate detection network. In *2018 IEEE Visual Communications and Image Processing (VCIP),* Taichung, Taiwan (pp. 1–5). IEEE.
12. Jia, J., Zhai, G., Zhang, J., Gao, Z., Zhu, Z., Min, X., & Guo, G. (2019). EMBDN: An efficient multiclass barcode detection network for complicated environments. *IEEE Internet of Things Journal, 6*(6), 9919–9933.
13. Zhang, J., Min, X., Jia, J., Zhu, Z., Wang, J., & Zhai, G. (2021). Fine localization and distortion resistant detection of multi-class barcode in complex environments. *Multimedia Tools and Applications, 80*(11), 16153–16172.

14. Xu, X., Xue, Z., & Zhao, Y. (2022). Research on an algorithm of express parcel sorting based on deeper learning and multi-information recognition. *Sensors*, *22*(17), 6705.
15. Cai, H., Li, J., Hu, M., Gan, C., & Han, S. (2023). Efficientvit: Lightweight multi-scale attention for high-resolution dense prediction. In *Proceedings of the IEEE/CVF International Conference on Computer Vision,* Paris, France (pp. 17302–17313).
16. Yan, L. Y., Tan, G. W. H., Loh, X. M., Hew, J. J., & Ooi, K. B. (2021). QR code and mobile payment: The disruptive forces in retail. *Journal of Retailing and Consumer Services*, *58*, 102300.
17. De Luna, I. R., Liébana-Cabanillas, F., Sánchez-Fernández, J., & Muñoz-Leiva, F. (2019). Mobile payment is not all the same: The adoption of mobile payment systems depending on the technology applied. *Technological Forecasting and Social Change*, *146*, 931–944.
18. Elaskari, S., Imran, M., Elaskri, A., & Almasoudi, A. (2021). Using barcode to track student attendance and assets in higher education institutions. *Procedia Computer Science*, *184*, 226–233.
19. Tan, L., Lu, Y., Yan, X., Liu, L., & Zhou, X. (2020). XOR-ed visual secret sharing scheme with robust and meaningful shadows based on QR codes. *Multimedia Tools and Applications*, *79*(9), 5719–5741.
20. Nuhi, A., Memeti, A., Imeri, F., & Cico, B. (2020, June). Smart attendance system using qr code. In *2020 9th Mediterranean Conference on Embedded Computing (MECO),* Budva, Montenegro (pp. 1–4). IEEE.
21. Küng, K., Aeschbacher, K., Rütsche, A., Goette, J., Zürcher, S., Schmidli, J., & Schwendimann, R. (2021). Effect of barcode technology on medication preparation safety: a quasi-experimental study. *International Journal for Quality in Health Care*, *33*(1), mzab043.
22. Ismael, S. A. A., Mohammed, A., & Hefny, H. (2020). An enhanced deep learning approach for brain cancer MRI images classification using residual networks. *Artificial Intelligence in Medicine*, *102*, 101779.
23. Xu, X., Bao, S., Shao, H., & Shi, P. (2024). A multi-sensor fused incremental broad learning with DS theory for online fault diagnosis of rotating machinery. *Advanced Engineering Informatics*, *60*, 102419.
24. Aggarwal, A., & Kumar, M. (2021). Image surface texture analysis and classification using deep learning. *Multimedia Tools and Applications*, *80*(1), 1289–1309.
25. Abdar, M., Samami, M., Mahmoodabad, S. D., Doan, T., Mazoure, B., Hashemifesharaki, R., & Nahavandi, S. (2021). Uncertainty quantification in skin cancer classification using three-way decision-based Bayesian deep learning. *Computers in Biology and Medicine*, *135*, 104418.
26. Ibrahim, A. U., Ozsoz, M., Serte, S., Al-Turjman, F., & Yakoi, P. S. (2024). Pneumonia classification using deep learning from chest X-ray images during COVID-19. *Cognitive Computation*, *16*(4), 1589–1601.
27. Kong, Y., Ma, X., & Wen, C. (2022). A new method of deep convolutional neural network image classification based on knowledge transfer in small label sample environment. *Sensors*, *22*(3), 898.
28. Gajalakshmi, K., Palanivel, S., Nalini, N. J., & Saravanan, S. (2018). Automatic classification of cast iron grades using support vector machine. *Optik*, *157*, 724–732.

29. Gajalakshmi, K., & Saravanan, S. (2025). Automatic identification of brittle, elongated and equiaxed ductile fracture modes in weld joints through machine learning. *Sigma: Journal of Engineering & Natural Sciences 43*(2), 441–451.
30. Gajalakshmi, K., Krishnan, S. B., & Saravanan, S. (2025). Automated weld defect detection using YOLOv8 algorithm. *Welding International*, 39(9), 649–655.
31. Magsi, A., Mahar, J. A., Razzaq, M. A., & Gill, S. H. (2020, November). Date palm disease identification using features extraction and deep learning approach. In *2020 IEEE 23rd International Multitopic Conference (INMIC),* Bahawalpur, Pakistan (pp. 1–6). IEEE.
32. Abu-Jamie, T. N., Abu-Naser, S. S., Alkahlout, M. A., & Aish, M. A. (2022). Six fruits classification using deep learning. *International Journal of Academic Information Systems Research,* 6(1),1–8.
33. Sharma, T., Nair, R., & Gomathi, S. (2022). Breast cancer image classification using transfer learning and convolutional neural network. *International Journal of Modern Research,* 2(1), 8–16.
34. Hussain, E., Hasan, M., Hassan, S. Z., Azmi, T. H., Rahman, M. A., & Parvez, M. Z. (2020). Deep learning–based binary classification for Alzheimer's disease detection using brain MRI images. In *Proceedings of the 15th IEEE Conference on Industrial Electronics and Applications (ICIEA 2020),* Kristiansand, Norway (pp. 1115–1120).
35. Khullar, V., Salgotra, K., Singh, H. P., & Sharma, D. P. (2021). Deep learning-based binary classification of ADHD using resting state MR images. *Augmented Human Research,* 6(1), 5.
36. Skouta, A., Elmoufidi, A., Jai-Andaloussi, S., & Ochetto, O. (2020). Automated binary classification of diabetic retinopathy by convolutional neural networks. In *Advances on Smart and Soft Computing: Proceedings of ICACIn 2020,* Casablanca, Morocco (pp. 177–187). Singapore: Springer Singapore.
37. Yang, R., Luo, F., Ren, F., Huang, W., Li, Q., Du, K., & Yuan, D. (2022). Identifying urban wetlands through remote sensing scene classification using deep learning: A case study of Shenzhen, China. *ISPRS International Journal of Geo-Information,* *11*(2), 131.
38. Gao, C., Zhou, J., Wong, W. K., & Gao, T. (2018, June). Woven fabric defect detection based on convolutional neural network for binary classification. In *International Conference on Artificial Intelligence on Textile and Apparel,* Hong Kong, China (pp. 307–313). Cham: Springer International Publishing.
39. Koonce, B. (2021). *Convolutional Neural Networks with Swift for Tensorflow.* Apress.
40. Zhao, L., & Wang, L. (2022). A new lightweight network based on MobileNetV3. *KSII Transactions on Internet and Information Systems (TIIS),* *16*(1), 1–15.
41. Wang, H., Bhaskara, V., Levinshtein, A., Tsogkas, S., & Jepson, A. (2020, August). Efficient super-resolution using mobilenetv3. In: *European Conference on Computer Vision,* Glasgow, UK (pp. 87–102). Cham: Springer International Publishing.

42. Yan, L., Zhong, B., & Ma, K. K. (2019, December). Confusion-aware convolutional neural network for image classification. In *International Conference on Neural Information Processing,* Sydney, NSW, Australia (pp. 151–161). Cham: Springer International Publishing.
43. Ichim, L., & Popescu, D. (2017, July). Retinal image segmentation based on weighted local detectors and confusion matrix. In *2017 40th International Conference on Telecommunications and Signal Processing (TSP),* Prague, Czech Republic (pp. 654–658). IEEE.
44. Hay, A. M. (1988). The derivation of global estimates from a confusion matrix. *International Journal of Remote Sensing*, *9*(8), 1395–1398.
45. Kamila, I. P., Sari, C. A., Rachmawanto, E. H., & Cahyo, N. R. D. (2024). A good evaluation based on confusion matrix for lung diseases classification using convolutional neural networks. *Advance Sustainable Science, Engineering and Technology*, *6*(1), 0240102.
46. Spence, M. A., Barry, J., Bartos, T., Blackwell, R., Scott, J., & Pitois, S. (2025). Avoiding confusion: Modelling image identification surveys with classification errors. *Methods in Ecology and Evolution*, *16*(3), 625–639.
47. Yin, X., Li, W., Li, Z., & Yi, L. (2022). Recognition of grape leaf diseases using MobileNetV3 and deep transfer learning. *International Journal of Agricultural and Biological Engineering*, *15*(3), 184–194.
48. Saravanan, S., Gajalakshmi, K., Kumararaja, K., Sivaraman, B., & Raghukandan, K. (2023). Prediction of aluminium–stainless steel explosive clad strength through machine learning. *Sādhanā*, *48*(3), 140.

Chapter 11

Revolutionizing supply chain management with AI-driven digital twin networks in 6G era

Afsha Anjum and Saravanan Chinnappan

11.1 INTRODUCTION TO DIGITAL TWIN NETWORKS AND 6G TECHNOLOGIES

Modern supply chains are no longer simple, linear systems. They have grown into large, interconnected networks shaped by globalization, rising customer expectations, and the fast pace of technological change. The traditional approaches depending on buffer inventories, rigid forecasting, and step-by-step information flows struggle to cope with present-day disruptions such as pandemics, trade conflicts, and climate-related shocks. Firms now need systems that can adjust quickly and continue operating even when conditions change unexpectedly.

Digital twins (DTs) and artificial intelligence (AI) are two technologies that have begun to address this gap. A DT creates a live virtual model of a physical process or asset, while AI helps analyze the continuous flow of data and guides decisions. Together, they provide tools for monitoring, simulation, and optimization in real time. Practical uses are already visible: Dalsey, Hillblom and Lynn (DHL) has tested logistics twins to streamline last-mile delivery, while Maersk employs smart container twins to improve tracking and visibility in global shipping.

For these systems to scale across entire supply chains, a stronger communications backbone is required. Fifth-generation (5G) networks have improved connectivity but fall short when millions of devices and instant decision-making are involved. Sixth-generation (6G) networks are designed to close this gap by offering extremely low latency (below 1 ms), support for dense device ecosystems, and built-in intelligence at the network edge. These network capabilities enable DTs to act independently of cloud latency, a necessity in high-speed logistics or manufacturing.

Figure 11.1 depicts physical systems feeding data into DTs, connected via 6G edge cloud, with AI decision loops.

6G will also support immersive interfaces such as extended reality (XR) and haptic feedback, allowing managers to engage directly with digital environments. For instance, a warehouse manager could conduct a virtual walk-through of a distant facility using holographic twins, monitoring robots, adjusting schedules, and responding to alerts in real time (Table 11.1).

DOI: 10.1201/9781003660163-11

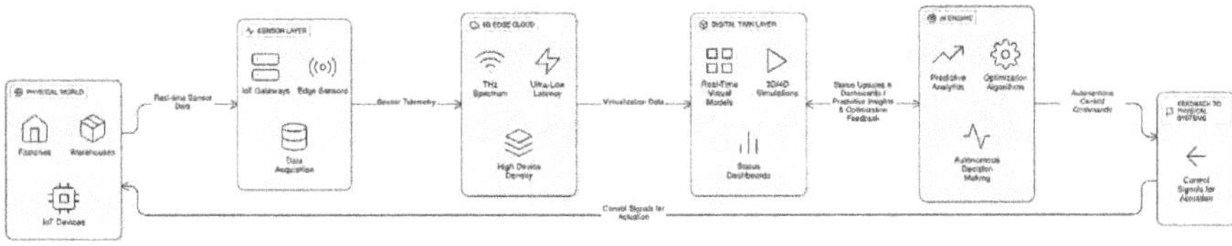

Figure 11.1 Conceptual architecture of a digital twin-6G-enabled smart supply chain.

Table 11.1 5G vs 6G in supply chain environments

Feature	*5G*	*6G (Projected)*
Latency	~10 ms	<1 ms
Max bandwidth	~10 Gbps	>1 Tbps
Device density	~1 M devices/km^2	>10M devices/km^2
Edge AI integration	Partial (offloaded to cloud)	Native, distributed at edge
Application example	Real-time tracking	Real-time decision-making

These advances open the door to what can be called the cognitive supply chain—a self-aware, self-optimizing, and self-healing system that functions with minimal human intervention.

11.2 EVOLUTION OF SUPPLY CHAIN MANAGEMENT: BEFORE AND AFTER AI

Supply chain management (SCM) was not always viewed as a core strategic function. In the early decades of the 20th century, it was treated mainly as a set of basic logistics tasks: procurement, warehousing, and transportation, each carried out separately using manual labor and basic planning tools.

After the Second World War, international trade expanded rapidly. The adoption of containerized shipping and large-scale manufacturing systems allowed goods to move more efficiently across borders. While this increased output, supply chains still tended to be rigid and struggled to adjust when disruptions occurred. By the 1980s and 1990s, new approaches such as just-in-time, developed most notably by Toyota, encouraged firms to keep lean inventories, coordinate more closely with suppliers, and cut down on waste. Around the same time, enterprise resource planning systems began linking internal functions, and the introduction of satellite-based Global Positioning Systems into logistics provided better tracking of shipments. These technologies improved efficiency but remained centralized and limited in flexibility [1].

In the 2000s, global sourcing and outsourcing became widespread. Companies often relied on multi-tier supplier networks that reduced costs but also introduced fresh risks, including longer lead times, political instability, and compliance with varying regulations. The rapid growth of e-commerce, led by firms such as Amazon, placed further demands on supply chains, as customers came to expect faster delivery and higher levels of product customization.

11.2.1 Digitalization and the first wave of SCM intelligence

The early stage of digitalization in SCM focused mainly on automating basic functions and improving visibility. Technologies such as barcode scanning, Radio-Frequency Identification (RFID) tagging, and simple dashboard systems allowed managers to keep better track of goods and transactions. These tools reduced manual errors and improved record-keeping but did not provide the ability to act on real-time changes. Forecasting continued to depend on scheduled reviews and static models that often lagged behind actual demand. By the 2010s, with the wider use of Internet of Things (IoT) devices, cloud computing, and big data platforms, firms started gathering far more detailed information across procurement, production, and distribution processes. Even so, much of this data was stored in separate systems, leaving departments isolated from one another and limiting the speed of decision-making.

11.2.2 Disruptive catalysts: COVID-19, climate change, and beyond

The COVID-19 crisis brought these shortcomings into sharp focus. Factory closures, transport delays, and worker shortages created widespread supply bottlenecks. Companies in sectors such as pharmaceuticals and automotive faced serious challenges in securing key materials and parts [2]. At the same time, external pressures around sustainability increased. Climate-related disruptions, new regulatory requirements, and changing consumer expectations pushed firms to prove that their operations were not only cost-effective but also environmentally responsible.

Together, these pressures revealed the risks of relying on supply chains that were digitized but still fragmented. Businesses began to recognize the need for systems offering continuous visibility across the chain, better predictive forecasting, and adaptive responses—capabilities that extend beyond what traditional IT solutions can deliver.

11.2.3 Emerging trends in the modern supply chain

The transition into the Industry 4.0 era has accelerated four defining shifts:

i. Supply chains are moving away from forecast-heavy approaches toward real-time sensing, enabled by IoT devices and DT environments.

Table 11.2 Traditional vs modern SCM characteristics

Dimension	*Traditional SCM*	*Modern SCM (digital twin era)*
Visibility	Limited, after-the-fact	Real-time, predictive
Decision-making	Centralized, manual	Distributed, AI-assisted
Inventory strategy	Safety-stock heavy	Just-in-case with simulation models
Risk management	Reactive	Proactive with scenario modeling
Customer customization	Minimal	Mass customization at scale

ii. Instead of rigid linear chains, firms now coordinate through flexible digital networks that connect suppliers, partners, and even end-customers.
iii. Planning is shifting from occasional batch reviews to continuous adjustment cycles, supported by AI tools that update forecasts as data flows in.
iv. Routine choices are being automated, while humans retain oversight for exceptions, strategic priorities, and high-stakes risks (Table 11.2).

These shifts illustrate that SCM is no longer about efficiency alone; it is about resilience, adaptability, and sustainability. The evolution of the supply chain, from fragmented workflows to intelligent and connected ecosystems, sets the stage for the integration of AI-driven DTs and 6G networks.

11.3 ROLE OF 6G IN REALIZING REAL-TIME DTS FOR SCM

The effectiveness of DT networks (DTNs) in SCM is fundamentally tied to the quality, timeliness, and granularity of data exchange. For a DT to faithfully represent and respond to its physical counterpart, it requires uninterrupted streams of high-fidelity sensor data, seamless edge-to-cloud coordination, and ultra-fast feedback loops. These requirements surpass the capabilities of 4G and even mainstream 5G technologies, creating a compelling case for 6G networks as the next enabling infrastructure.

6G is expected to roll out commercially between 2030 and 2035 and is envisioned as a fully AI-native, ultra-intelligent wireless architecture with dramatically improved latency, bandwidth, and device density. Unlike its predecessors, 6G is being designed to embed intelligence directly at the network edge, facilitating decentralized decision-making, continuous learning, and immersive interaction in near-zero latency environments.

11.3.1 Key capabilities of 6G relevant to DTs

a. **Ultra-Low-Latency and Real-Time Responsiveness:** While 5G offers latency in the 10ms range, 6G promises to reduce this to under 1ms. This makes real-time control of critical processes such as autonomous robotics, drone-based inventory, or cold chain logistics highly feasible. DTs can now operate as live control loops, not just passive monitors.

b. **Terahertz (THz) Spectrum and High Bandwidth:** 6G leverages the THz spectrum to support bandwidths exceeding 1 Tbps. This enables massive data transfer, including 3D/4D model updates, live video feeds from smart factories, and XR-based supply chain interactions. DTs can now integrate multimodal sensor data (temperature, visual, RFID, vibration) without bottlenecks.

c. **Edge AI and Federated Learning:** One of the most disruptive features of 6G is its native support for edge intelligence. AI models can be trained and updated directly at edge nodes (e.g., factory gateways and warehouse hubs), reducing reliance on centralized cloud servers. This empowers DTs to function autonomously with on-site learning and adapt to contextual changes in real time.

d. **Massive Device Density and Scalability:** 6G is expected to support up to 10 million devices per square kilometer, making it ideal for sensor-rich environments like ports, smart cities, and mega distribution centers. DTs can now scale across geographies and asset types with seamless interoperability.

e. **Integrated Security and Blockchain Compatibility:** Security concerns have often limited the adoption of DT ecosystems. 6G is projected to include zero-trust network principles, quantum-safe encryption, and compatibility with blockchain for data immutability. These enhancements ensure secure synchronization between physical and virtual nodes.

Preliminary simulation studies by Ericsson [3] and Huawei [4] have demonstrated 6G prototypes achieving throughput above 1 Tbps and latency around 0.8 ms in controlled trials. These early results indicate that the ultra-low-latency communication required for distributed DTNs is technically attainable.

11.3.2 Enabling a cognitive supply chain

With the above capabilities, 6G transforms DTs from monitoring tools into cognitive agents capable of learning, predicting, and controlling. In supply chains, this has profound implications:

i. **Adaptive Logistics:** Delivery routes can be optimized in real time based on traffic, weather, and warehouse load—controlled by DTs of fleets and infrastructure.
ii. **Self-Healing Warehouses:** Equipment breakdowns are predicted by AI, visualized through the DT, and corrected automatically or via operator prompts enabled by edge AI.
iii. **Supplier Risk Management:** Live simulations running on federated twins evaluate geopolitical, financial, and environmental, social, and governance (ESG) risks to reconfigure supplier networks proactively.
iv. **Immersive Collaboration:** Stakeholders from different continents can interact in shared DT environments via haptic and XR interfaces over 6G.

11.3.3 Industry impact: from reaction to anticipation

The true promise of 6G-powered DTNs is a shift from reactive to anticipatory operations. This is particularly important for high-stakes industries such as

i. **Pharmaceutical Logistics:** Strict monitoring of temperature-sensitive drugs and vaccines is essential, along with rapid adjustment to sudden shifts in global demand.
ii. **Retail and E-Commerce:** Firms can use DTs to personalize orders, manage inventory more tightly, and streamline the often-complex process of returns.
iii. **Agriculture and Food Chains:** 6G-enabled monitoring makes it possible to track freshness, humidity, and transport times, while also documenting sustainability metrics for regulatory and consumer transparency.

As these examples show, the value of 6G-powered DTNs lies in enabling operations that are proactive rather than reactive where each participant in the chain provides not only goods or services but also real-time data that strengthens decision-making.

11.4 APPLICATIONS OF AI-DRIVEN DTs IN SCM

The combination of DTNs, AI, and advanced wireless technologies like 6G is no longer a futuristic idea; it is already influencing how supply chains are designed and operated. Instead of functioning as static, siloed systems, supply chains are beginning to resemble dynamic ecosystems capable of adapting to real-time events and disruptions.

This integration is proving particularly useful in areas that have traditionally been difficult to manage, such as forecasting sudden shifts in demand, reducing transportation bottlenecks, preventing equipment breakdowns, and anticipating risks that cascade across networks. The subsections below highlight some of the practical applications where this convergence is delivering measurable improvements.

11.4.1 Intelligent inventory management

Balancing inventory has always been a challenge: carrying too much stock ties up capital, while shortages risk losing customers. DTs, when paired with AI tools and real-time data from connected sensors, help supply chains strike a more accurate balance. These systems not only track quantities but also monitor factors such as product freshness, movement speed, and seasonal demand trends.

A useful example comes from **Walmart**, which has experimented with DT technology to replicate warehouse layouts and test different stocking strategies. In practice, data from RFID and IoT sensors is streamed into DT models, creating a live picture of stock levels [5]. AI algorithms then adjust reorder points dynamically, while future high-speed networks like 6G are expected to make these updates seamless, even in very large facilities with dense sensor deployments.

11.4.2 Predictive maintenance and equipment uptime

Downtime in logistics hubs, production lines, or distribution centers can significantly disrupt supply chain continuity. DTs replicate the behavior of physical equipment, tracking wear and tear, operational anomalies, and environmental stress. Combined with AI-based predictive models, potential failures can be forecasted and addressed proactively.

For example, **DHL** has deployed AI-powered DTs of conveyor systems in their smart warehouses. These systems detect early signs of motor degradation and recommend maintenance actions, which are instantly communicated to field operators via 6G edge networks, reducing downtime by up to 30% [6].

11.4.3 Real-time logistics and route optimization

Fleet management and transportation logistics are ripe for optimization through this tri-technology framework. DTs of delivery vehicles, drones, or maritime containers allow for real-time tracking of location, fuel consumption, temperature, and load conditions. AI algorithms optimize routes dynamically based on live traffic, weather, and cost constraints.

Maersk, a global shipping leader, is piloting DT models for its container fleets. Combined with AI and edge communication systems, these twins predict optimal port entry times, reroute vessels in case of congestion, and ensure compliance with sustainability mandates by minimizing fuel consumption [7].

11.4.4 Supplier and risk network management

Multi-tier supplier networks are inherently prone to cascading risks. DT models of supplier relationships allow firms to simulate supply chain disruptions caused by natural disasters, geopolitical tensions, or raw material shortages. AI enhances this by forecasting risks based on socio-economic data and real-time news feeds, while 6G ensures fast communication of threat alerts across the network.

For example, **BASF** utilizes DT-based risk models to monitor over 6,000 suppliers globally. AI models rank suppliers based on ESG compliance, delivery consistency, and geopolitical exposure [8]. When combined with high-speed alert systems, procurement teams can reconfigure sourcing strategies in near real time.

11.4.5 Demand forecasting and fulfillment automation

Demand planning has traditionally been one of the most error-prone areas in SCM. By integrating point-of-sale systems, customer behavior analytics, and market signals into DTs of demand ecosystems, AI models can forecast with higher precision. These forecasts are fed directly into automated fulfillment systems powered by robotics and IoT devices.

Zara has implemented predictive analytics embedded within its supply chain twins to align manufacturing cycles with fashion trends. 6G communication facilitates real-time synchronization between customer-facing apps and warehouse control systems, leading to a rapid design-to-shelf cycle with minimal inventory waste [9].

11.4.6 Cold chain and perishables monitoring

In industries such as pharmaceuticals and food, maintaining temperature-controlled supply chains is critical. DTs of cold storage units and transport vehicles, combined with AI-based anomaly detection, enable continuous monitoring of conditions like humidity, vibration, and temperature thresholds.

Pfizer, in its global vaccine distribution effort, used IoT-based cold chain twins integrated with machine learning models [10]. With 6G's potential, such systems can scale to billions of doses without bottlenecks, ensuring regulatory compliance and public safety.

11.5 INDUSTRIAL CASE STUDIES

11.5.1 Case study: Maersk—DT-enabled terminal operations

Maersk, the global leader in container shipping and logistics, has emerged as a pioneer in adopting DT frameworks to transform its port terminal and logistics operations. Rather than building a monolithic twin of the entire enterprise, Maersk strategically developed a network of modular vertical twins for terminals, vessels, and warehouses. These modules are progressively being integrated into a horizontal twin ecosystem, enabling real-time collaboration, predictive analytics, and closed-loop control across supply chain layers.

11.5.1.1 Terminal-level twin workflow and data integration

At select terminals, Maersk has implemented a DT simulation platform that models terminal operations using real-time data streams from sensors, schedule inputs, and equipment telemetry. As shown in Figure 11.2, the system ingests live data such as

- Crane movement and operational status
- Container position, condition, and stacking plans
- Environmental parameters (wind, humidity, temperature)
- Vessel arrival/departure schedules
- Workforce availability and shift planning

This data is processed by the DT core to simulate resource utilization, container handling times, and potential congestion points. Predictive

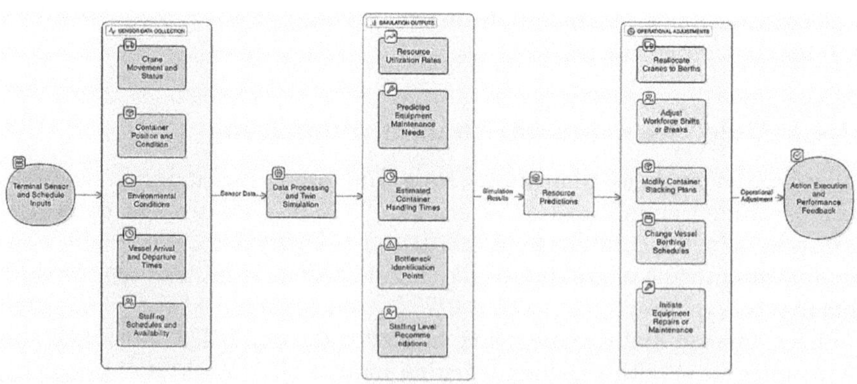

Figure 11.2 Maersk's digital twin workflow for terminal operations.

models embedded in the simulation estimate maintenance requirements, flag bottlenecks, and generate staffing recommendations based on upcoming workloads.

11.5.1.2 Actionable outputs and feedback control

The outputs of the twin simulation feed into a resource prediction engine, which guides operational decision-making. Through a control loop, the system recommends

- Reallocation of cranes and berth positions
- Adjustments to staffing shifts and break times
- Modifications in container stacking logistics
- Dynamic changes in vessel berthing schedules
- Initiation of predictive maintenance or equipment repairs

These recommendations are then enacted via Maersk's terminal control system, with performance feedback continuously fed back into the simulation loop, enabling a self-correcting, adaptive terminal management system.

11.5.1.3 Autonomous warehouse twins with Dexory

Beyond terminals, Maersk is piloting warehouse-level DTs to enhance visibility and automation. In collaboration with Dexory, a logistics robotics startup, the company built a Proof-of-Concept twin that maps real-time inventory movement and robotic stock-taking operations inside warehouses [11]. The twin enables

- Real-time process simulation
- Exploration of "what-if" scenarios for layout or staffing changes
- Faster root–cause analysis during fulfillment delays
- Scalable rollout across multiple facilities as maturity grows

This project represents a move toward autonomous warehouse ecosystems powered by AI and DT logic.

11.5.1.4 Smart container networks for global visibility

Maersk is also deploying over 300,000 smart containers equipped with Global Positioning Systems, humidity, and temperature sensors [12]. These smart units feed data into DT models of the end-to-end shipping journey from port of loading to customer delivery. The integration of 6G-capable networks (under research) with these twins is expected to support ultra-low-latency, real-time visibility and alerting for perishable goods, sensitive cargo, and high-value shipments.

11.5.1.5 Impact and key outcomes

Maersk's adoption of DT technology, supported by AI and real-time connectivity, has shown tangible benefits:

- Optimized crane allocation and berth scheduling, leading to reduced vessel turnaround times
- Smarter workforce scheduling, improving labor efficiency and reducing idle time
- End-to-end container visibility, reducing spoilage and delays
- Improved fuel efficiency, contributing to Maersk's decarbonization goals
- Closed-loop feedback control, enhancing adaptability in high-variability operations

Figure 11.2 depicts Maersk's DT workflow for terminal operations. The system integrates sensor inputs, DT simulations, and AI-based predictions to optimize resource allocation and operational performance.

11.5.2 Case study: Pfizer—cold chain twin for vaccine distribution

Pfizer's global vaccine distribution during the COVID-19 pandemic revealed the critical need for ultra-precise cold chain logistics. To overcome risks related to temperature fluctuations, spoilage, and delivery delays, Pfizer deployed a DT-based cold chain network integrated with real-time IoT monitoring and AI prediction systems.

Each vaccine shipment was embedded with smart sensors capturing temperature, humidity, location, and vibration. These data streams were fed into a DT model that simulated each container's condition and path from origin to destination. Machine learning algorithms detected anomalies and provided early warnings for threshold breaches allowing rerouting or restocking before spoilage occurred.

This closed-loop system, when combined with predictive analytics, enabled Pfizer to maintain vaccine viability across thousands of shipments globally. In the future, 6G integration is expected to support greater data density, faster alerts, and broader global connectivity, particularly in remote regions. The twin framework not only optimized logistics performance but also helped Pfizer meet strict regulatory compliance across multiple national jurisdictions.

11.5.3 Case study: Unilever—sustainability DTs

Unilever, a consumer goods giant, is leveraging DTs to track and reduce carbon emissions across its packaging, manufacturing, and distribution operations. Their DTs simulate end-to-end product lifecycles, from material sourcing to consumer delivery, assessing environmental impact at each stage.

By modeling different packaging materials, shipping routes, and energy sources, AI-enhanced DTs enable Unilever to select the lowest-emission alternatives without compromising efficiency. These simulations have supported their shift to recyclable packaging, electric delivery fleets, and renewable energy adoption in plants.

Unilever's twin platform also generates real-time dashboards for sustainability reporting, aligned with ESG goals. The firm aims to become carbon neutral by 2039, and its DT-driven operational transparency plays a vital role in meeting that target [13,14]. As Unilever experiments with federated AI and edge computing, 6G technology is seen as a future enabler of cross-continental emissions tracking.

11.5.4 Case study: Procter & Gamble—resilient global operations

During the COVID-19 pandemic, Procter & Gamble (P&G) utilized a cloud-based DT ecosystem to maintain continuity across its global operations. With over 130 manufacturing sites and thousands of suppliers, P&G modeled its entire supply chain network to evaluate potential risks in real time [15,16].

Their system simulated supplier shutdowns, transport delays, and demand surges using real-time data inputs and predictive AI. This allowed P&G to dynamically reallocate production, identify fallback suppliers, and rebalance inventory in affected regions. The twin also monitored supplier health, material availability, and regulatory constraints across different geographies.

P&G's use of DTs significantly reduced disruption and prevented revenue losses. The company has since integrated these tools permanently into its enterprise planning systems. Looking forward, 6G's ability to offer global, seamless, low-latency communication is expected to enhance the reach and resilience of P&G's DT framework, particularly in high-risk or emerging markets.

11.6 CHALLENGES, RISKS, AND ETHICAL CONSIDERATIONS

Despite the transformative potential of AI-driven DTNs in SCM, several challenges, risks, and ethical dilemmas must be carefully addressed to ensure responsible and sustainable adoption.

11.6.1 Technical challenges in implementation

- **Data Heterogeneity:** DTs require integration of diverse data sources—IoT sensors, historical databases, real-time streams—which often exist in silos or incompatible formats.
- **Edge-to-Cloud Latency:** Even with anticipated 6G advances, real-time control decisions depend on ultra-low latency, which is difficult to ensure across all geographies and network conditions.
- **Computational Demands:** Running complex AI models and twin simulations in real time requires massive computing power at the edge or in distributed clouds, leading to infrastructure and energy constraints.
- **Standardization Gaps:** Lack of global standards for DT architectures, application programming interfaces, and interoperability protocols leads to fragmented implementations and vendor lock-in.

11.6.2 Organizational and operational limitations

On the organizational side, several factors can slow down or complicate the move toward AI-driven DT systems:

- **Adoption Barriers:** Many companies struggle with internal pushback when replacing long-established enterprise resource planning or Manufacturing Execution System (MES) platforms. Employees may be hesitant to trust unfamiliar tools or change established workflows, which can delay adoption.
- **Skill Shortages:** Building and managing DT environments requires expertise in areas such as AI, data analytics, and cyber-physical systems. Not all firms have access to these skills internally, and recruitment or training often lags behind technological needs.
- **Upfront Investment:** Deploying sensors, building AI models, and upgrading digital infrastructure involve substantial costs. For small and mid-sized enterprises, this financial burden can make large-scale implementation difficult.
- **Data Protection Risks:** Running DTs depends on sharing continuous streams of operational data with partners and cloud systems. This introduces vulnerabilities, including exposure to cyberattacks, industrial espionage, or regulatory breaches in cross-border trade.

11.6.3 Ethical and societal considerations

As AI and DT technologies become deeply embedded in decision-making, several ethical concerns emerge:

- **Autonomous Decision Risks:** Over-reliance on AI predictions could lead to decisions without human oversight, risking unforeseen outcomes or ethical blind spots.
- **Bias in AI Models:** If training data is biased or incomplete, predictions from AI engines can reinforce existing supply chain inequalities favoring certain vendors, routes, or markets over others.
- **Workforce Displacement:** The automation enabled by DTs and AI may result in job displacement, especially in low-skill logistics, warehousing, and transport roles.
- **Surveillance and Transparency:** Ubiquitous tracking and visibility may cross ethical lines if employee behavior, personal devices, or sensitive locations are monitored without consent.

11.6.4 Regulatory and policy gaps

Governments and international bodies are still catching up with the rapid evolution of DT and AI technologies:

- **Lack of Compliance Frameworks:** Most supply chains operate across multiple jurisdictions. The absence of unified AI governance laws makes it hard to ensure compliance in areas like data sharing, automated decision-making, or sustainability claims.
- **Cross-Border Data Restrictions:** Some countries restrict the real-time export of operational data, which can break the continuity of global DTNs.
- **Accountability and Liability:** In AI-driven systems, it is often unclear who is legally accountable for automated errors—the AI developer, the data owner, or the operator?

11.6.5 Environmental and sustainability trade-offs

While DTs are often promoted as enablers of green logistics, their own environmental footprint must be acknowledged:

- **Energy-Intensive AI Models:** Large-scale simulations and deep learning models are computationally expensive, leading to increased carbon emissions from data centers.
- **E-Waste from IoT Hardware:** The proliferation of sensors, edge devices, and communication modules contributes to electronic waste, especially in short-lifecycle deployments.
- **Over-Digitalization Risks:** Pushing for digitization without assessing return on investment can result in technology for technology's sake, creating redundant systems and sunk costs.

11.7 FUTURE PROSPECTS AND CONCLUSION

The fusion of AI, DTNs, and 6G communication technologies is still in its early stages, but its trajectory signals a major transformation in supply chain design, execution, and resilience. The next decade will likely witness accelerated innovation in the following key areas:

a. **Hyper-Personalized Supply Chains:** By integrating AI-driven DTs with real-time consumer data, companies will be able to dynamically adjust production and logistics based on hyper-local demand signals. This will enable mass customization, reduce waste, and improve customer satisfaction particularly in retail, food, and healthcare sectors.
b. **Federated DT Ecosystems:** Rather than operating in silos, future DTs will collaborate across organizations and nations using federated models. Secure multi-party learning and blockchain integration will ensure data integrity and traceability across shared logistics networks from suppliers to manufacturers to retailers.
c. **AI-Driven Autonomous Decision Engines:** Advances in explainable AI and reinforcement learning will drive the development of DTs that can self-optimize, make autonomous decisions in volatile conditions, and learn from global supply chain disruptions in real time.
d. **6G-Powered Multi-Sensory Twins:** With projected support for terahertz communication, holographic interfaces, and sub-millisecond latency, 6G will enable DTs that aren't just visual dashboards but immersive, interactive virtual environments where planners and operators can walk through simulations, run scenario testing, and train autonomous systems with precision.
e. **Net-Zero Logistics through AI + Twins:** As sustainability becomes a global imperative, DTs will help supply chains achieve carbon transparency by tracking emissions at every node. Combined with AI, they'll suggest eco-optimal routing, packaging, and energy usage, pushing logistics closer to carbon neutrality goals.

11.7.1 Recommendations for practitioners and researchers

To ensure responsible and scalable adoption of these emerging systems, the following recommendations are proposed:

- **For Industry Leaders:** Start with pilot deployments focused on measurable return on investment (e.g., maintenance savings and delivery speed) before scaling across the network.
- **For Tech Developers:** Prioritize interoperable platforms with modular architectures, so organizations can plug into DT ecosystems without vendor lock-in.

- **For Researchers:** Explore topics like AI fairness in logistics, 6G protocol optimization, and cross-border data governance to fill knowledge gaps.
- **For Policymakers:** Develop unified regulatory standards for AI-driven decision-making in global logistics, particularly around liability, data rights, and environmental claims.

11.8 CONCLUSION

We have explored the transformative potential of AI-driven DTNs in the 6G era, specifically within the context of SCM. From intelligent inventory to predictive maintenance, from smart logistics to resilient networks, DTs are reshaping how goods move, how decisions are made, and how systems adapt.

While case studies from Maersk, Pfizer, Unilever, and P&G demonstrate real-world momentum, significant challenges remain spanning technical complexity, ethical risks, and regulatory uncertainty. However, with responsible innovation, collaborative ecosystems, and inclusive strategies, these challenges can be navigated.

As we move toward an era of hyper-connected, autonomous, and sustainable supply chains, DTNs powered by the intelligence of AI and the speed of 6G will serve as the nervous system of next-generation global commerce.

This convergence is not merely a technological upgrade; it represents a paradigm shift in how value is created, delivered, and sustained across the planet.

REFERENCES

1. S.Wamba-Taguimdje, L. Wamba, J. Kala Kamdjoug and J. Tchatchouang Wanko, "Influence of Artificial Intelligence (AI) on Firm Performance: The Business Value of AI-Based Transformation Projects," *Business Process Management Journal*, vol. 26, no. 7, pp. 1893–1924, 2020.
2. R.Vincent, "Pfizer's COVID-19 Vaccine Distribution: Cold Chain Technology and Innovation," *Pharmaceutical Technology*, 2022. https://www.pharmaceutical-technology.com/features/pfizer-covid-vaccine-cold-chain/
3. Ericsson, "Exploring the Industry Potential of 6G," *Ericsson*, 2022. https://www.ericsson.com/en/reports-and-papers/white-papers/6g-industry-potential
4. Huawei, "6G: The Next Horizon, Huawei", 2021. https://www.huawei.com/en/huaweitech/future-technologies/6g-the-next-horizon
5. D. Priest, "How Walmart is Using AI and IoT to Transform Inventory," *Forbes,* Feb. 2023.
6. DHL, "Logistics Trend Radar," *DHL Innovation Center*, 2022. https://www.dhl.com/global-en/home/insights-and-innovation/thought-leadership/trend-reports/logistics-trend-radar.html

7. A. V. Krishnamoorthy, "Maersk's Digital Twin for Logistics Optimization," *Port Technology International*, Oct. 2023. https://www.porttechnology.org/news/maersk-advances-digital-twin-integration/
8. BASF SE, "Supply Chain Transparency with Digital Twins," *Sustainability Report*, 2023. https://report.basf.com
9. Vaibhav, "AI Case Study: Zara — How AI Predicts Fashion Trends & Reinvents Retail", November, 2025. https://vaidikai.com/ai-case-study-zara-how-ai-predicts-fashion-trends-reinvents-retail/
10. LogiPharma, "Here's How Pfizer is Using IoT to Support Vaccine Distribution", 2021, https://logipharmaeu.wbresearch.com/blog/how-pfizer-is-using-iot-to-support-vaccine-distribution
11. Dexory, "How AI Is Transforming Digital Twins in Warehousing", *Dexory*, 2025. https://www.dexory.com/insights/whats-next-for-digital-twins
12. Maersk, "Maersk Embraces Smart Containers and Visibility Technology," *A.P. Moller–Maersk*, 2022. https://www.maersk.com/news/articles/2022/11/22/smart-container-visibility
13. Kitty Wheeler, "How Unilever Uses AI & Digital Twins For Sustainability", 2025 https://technologymagazine.com/articles/how-unilever-uses-ai-digital-twins- for-sustainability
14. A. Phillips, "How Unilever is Using Digital Twins to Meet Sustainability Targets," *GreenBiz*, June 2023.
15. Procter & Gamble, "S&P 500 Giant's Supply Chain Resilience and Full Visibility: Procter & Gamble", Procter & Gamble, 2026. https://truthcartographer.com/sp-500-giants-supply-chain-resilience-procter-gamble/
16. S. Turner, "Digital Twin Strategies to Build Global Resilience: Lessons from P&G," *Supply Chain Digital*, July 2022. https://supplychaindigital.com/articles/pg-digital-twin-resilience

Index

For Product Safety Concerns and Information please contact our EU
representative GPSR@taylorandfrancis.com
Taylor & Francis Verlag GmbH, Kaufingerstraße 24, 80331 München, Germany

www.ingramcontent.com/pod-product-compliance
Ingram Content Group UK Ltd.
Pitfield, Milton Keynes, MK11 3LW, UK
UKHW022217190926
475702UK00003B/412

* 9 7 8 1 0 4 1 1 1 4 7 7 2 *